T0177038

PRINCIPLES OF MULTISCALE MODELING

Physical phenomena can be modeled at varying degrees of complexity and at different scales. Multiscale modeling provides a framework, based on fundamental principles, for constructing mathematical and computational models of such phenomena, by examining the connection between models at different scales.

This book, by one of the leading contributors to the field, is the first to provide a unified treatment of the subject, covering, in a systematic way, the general principles of multiscale models, algorithms, and analysis. The book begins with a discussion of the analytical techniques in multiscale analysis, including matched asymptotics, averaging, homogenization, renormalization group methods, and the Mori–Zwanzig formalism. A summary of the classical numerical techniques that use multiscale ideas is also provided. This is followed by a discussion of the physical principles and physical laws at different scales. The author then focuses on the two most typical applications of multiscale modeling: capturing macroscale behavior and resolving local events. The treatment is complemented by chapters that deal with more specific problems, ranging from differential equations with multiscale coefficients to time scale problems and rare events. Each chapter ends with an extensive list of references to which the reader can refer for further details.

Throughout, the author strikes a balance between precision and accessibility, providing sufficient detail to enable the reader to understand the underlying principles without allowing technicalities to get in the way. Whenever possible, simple examples are used to illustrate the underlying ideas.

WEINAN E's research is concerned with developing and exploring the mathematical framework and computational algorithms needed to address problems that arise in the study of various scientific and engineering disciplines, ranging from mechanics to materials science to chemistry. He has held positions at New York University, the Institute for Advanced Study in Princeton, Peking University, and Princeton University, where he currently is Professor in the Department of Mathematics and in the Program in Applied and Computational Mathematics. His research has been recognized by numerous awards, including the 2003 Collatz Prize of ICIAM, and the 2009 Ralph E. Kleinman Prize, from SIAM.

PRINCIPLES OF
MULTISCALE MODELING

WEINAN E
Princeton University

CAMBRIDGE
UNIVERSITY PRESS

University Printing House, Cambridge CB2 8BS, United Kingdom

One Liberty Plaza, 20th Floor, New York, NY 10006, USA

477 Williamstown Road, Port Melbourne, VIC 3207, Australia

314–321, 3rd Floor, Plot 3, Splendor Forum, Jasola District Centre, New Delhi – 110025, India

79 Anson Road, #06–04/06, Singapore 079906

Cambridge University Press is part of the University of Cambridge.

It furthers the University's mission by disseminating knowledge in the pursuit of education, learning and research at the highest international levels of excellence.

www.cambridge.org
Information on this title: www.cambridge.org/9781107096547

© Weinan E 2011

This publication is in copyright. Subject to statutory exception and to the provisions of relevant collective licensing agreements, no reproduction of any part may take place without the written permission of Cambridge University Press.

First published 2011
Reprinted 2021

Printed in the United Kingdom by TJ Books Limited, Padstow Cornwall

A catalogue record for this publication is available from the British Library

ISBN 978-1-107-09654-7 Hardback

Cambridge University Press has no responsibility for the persistence or accuracy of URLs for external or third-party internet websites referred to in this publication, and does not guarantee that any content on such websites is, or will remain, accurate or appropriate.

Contents

Preface

Traditional approaches to modeling focus on one scale. If our interest is the macroscale behavior of a system in an engineering application, we model the effect of the smaller scales by some constitutive relations. If our interest is in the detailed microscopic mechanism of a process, we assume that there is nothing interesting happening at the larger scales, for example, that the process is homogeneous at larger scales.

Take the example of solids. Engineers have long been interested in the macroscale behavior of solids. They use continuum models and represent atomistic effects by constitutive relations. Solid state physicists, however, are more interested in the behavior of solids at the atomic or electronic level, often working under the assumption that the relevant processes are homogeneous at the macroscopic scale. As a result, engineers are able to design structures and bridges without acquiring much understanding about the origins of the cohesion between the atoms in the material. Solid state physicists can provide such an understanding at a fundamental level. But they are often quite helpless when faced with a real engineering problem.

The relevant constitutive relations, which play a key role in modeling, are often obtained empirically, on the basis of very simple ideas such as linearization, Taylor expansion and symmetry. It is remarkable that such a simple-minded approach has had so much success: most of what we know in the applied sciences and virtually all of what we know in engineering is obtained using such an approach. Indeed the hallmark of deep physical insight has been the ability to describe complex phenomena using simple ideas. When successful, we hail such a work as "a stroke of genius," as we often describe Landau's work, say.

A very good example of a constitutive relation is that for simple or Newtonian fluids, which is obtained using only linearization and symmetry and gives rise to the well-known Navier–Stokes equations. It is quite amazing that such a linear constitutive relation can describe almost all the phenomena of simple fluids, which are often very nonlinear, with remarkable accuracy.

However, extending these simple empirical approaches to more complex systems has proven to be a difficult task. A good example is complex fluids or non-Newtonian fluids – that is, fluids whose molecular structure has a non-trivial effect on its macroscopic behavior. After many years of effort, the results of trying to obtain the constitutive relations for such fluids by guesswork or by fitting a small set of experimental data are quite mixed. In many cases, either the functional form becomes too complicated or there are too many parameters to fit. Overall, empirical approaches have had limited success for complex systems or for small-scale systems in which the discrete or finite-size effects are important.

The other extreme is to start from first principles. As was recognized by Dirac immediately after the birth of quantum mechanics, almost all the physical processes that arise in applied sciences and engineering can be modeled accurately using the principles of quantum mechanics. Dirac also recognized the difficulty of such an approach, namely, the mathematical complexity of quantum mechanics principles is so great that it is quite impossible to use them directly to study realistic chemistry or, more generally, engineering problems. This is true not just for the true first principle, the quantum many-body problem, but also for other microscopic models such as those in molecular dynamics.

This is where multiscale modeling comes in. By considering simultaneously models at different scales we hope to develop an approach that shares the efficiency of the macroscopic models as well as the accuracy of the microscopic models. This idea is far from new. After all, there have been considerable efforts to try to understand the relations between microscopic and macroscopic models; for example, computing from molecular dynamics models the transport coefficients needed in continuum models. There have also been several classical success stories of combining physical models at different levels of detail for the efficient and accurate modeling of complex processes of interest. Two of the best known examples are the quantum-mechanics–molecular-mechanics (QM–MM) approach in chemistry and the Car–Parrinello molecular dynamics. The former is a procedure for modeling chemical reactions involving large molecules, by combining quantum mechanics models in the reaction region and classical models elsewhere. The latter is a way of performing molecular dynamics simulations using forces that are calculated from electronic structure models "on-the-fly," instead of using empirical interatomic potentials. What has prompted the increase in interest in recent years in multiscale modeling is the recognition that such a philosophy is useful for all areas of science and engineering, not just chemistry and material science. Indeed, compared with the traditional approach of focusing on one scale, looking at a problem simultaneously from several different scales and different levels of detail could be said to be a more mature way of constructing models. It represents a fundamental change in the way we view modeling.

The multiscale multi-physics viewpoint opens up unprecedented opportunities for modeling. It provides the opportunity to put engineering models on a solid footing. It allows us to connect engineering applications with basic science. It offers a more unified view of modeling, by focusing more on the different levels of application of physical laws and the relations between them, with specific applications as examples. In this way we will find that our approaches to solids and fluids are very much parallel to each other, as we explain later in the book.

Despite the exciting opportunities offered by multiscale modeling, one thing we have learned during the past decade is that we should not expect quick results. Many fundamental issues have to be addressed before its expected impact becomes a reality. These issues include:

(1) a detailed understanding of the relation between the different levels of physical models;
(2) boundary conditions for atomistic models such as molecular dynamics;
(3) systematic and accurate coarse-graining procedures.

Without properly addressing these and other fundamental issues, we run the risk of simply replacing one set of ad hoc models by another, or by ad hoc numerical algorithms. This would hardly be a step forward.

This volume is intended to present a systematic discussion of the basic ideas in multiscale modeling. The emphasis is on the fundamental principles and common issues, not on specific applications. Selecting the materials to be covered proved to be a very difficult task since the subject is so vast and, at the same time, quickly evolving. When deciding what should be included in this volume, I asked the following question: what are the basics that one has to know in order to get a global picture about multiscale modeling? Since multiscale modeling touches upon almost every aspect of modeling in almost every scientific discipline, it is inevitable that some discussions are very brief and some important aspects are neglected entirely. Nevertheless, we have tried to make sure that the fundamental issues are indeed covered.

The book can be divided into two parts: background materials and general topics. The background materials include:

(1) an introduction to the fundamental physical models, ranging from continuum mechanics to quantum mechanics (Chapter 4);
(2) basic analytical techniques for multiscale problems, such as averaging methods, homogenization methods, renormalization group methods, and matched asymptotics (Chapter 2);
(3) classical numerical techniques that use multiscale ideas (Chapter 3).

For the second part, we chose the following topics.

(1) Examples of multi-physics models. These are analytical models that use multi-physics coupling explicitly (Chapter 5). *It is important to realize that multiscale modeling is not just about developing algorithms, it is also about developing better physical models.*
(2) Numerical methods for capturing the macroscopic behavior of complex systems with the help of microscopic models, in cases when empirical macroscopic models are inadequate (Chapter 6).
(3) Numerical methods for coupling macroscopic and microscopic models locally in order to better resolve localized singularities, defects or other events (Chapter 7).

We have also included three more specific examples: elliptic partial differential equations with multiscale coefficients, problems with both slow and fast dynamics, and rare events. The first was selected to illustrate problems that involve spatial scales. The second and third are selected to illustrate problems that involve time scales.

Some of the background materials mentioned above have been discussed in numerous textbooks. However, each such textbook contains materials that are enough for a one-semester course, and most students and researchers will not be able to afford the time to go through them all. Therefore, we decided to discuss these topics in a simplified fashion, to extract the longwinded ideas necessary for a basic understanding of the relevant issues.

The book is intended for scientists and engineers who are interested in modeling and doing it right. I have tried to strike a balance between precision, which requires a considerable amount of mathematics and detail, and accessibility, which requires glossing over some details and making compromises on precision. For example, we will discuss asymptotic analysis quite systematically, but we will not discuss the related theorems. We will always try to use the simplest examples to illustrate the underlying ideas, rather than dwelling on particular applications.

Many people helped to shape my views on multiscale modeling. Björn Engquist introduced me to the subject while I was a graduate student, at a time when multiscale modeling was not a fashionable area to work on. I have also benefitted from the other mentors I have had at different stages of my career; they include Alexandre Chorin, Bob Kohn, Andy Majda, Stan Osher, George Papanicolaou, and Tom Spencer. I am very fortunate to have Eric Vanden-Eijnden as a close collaborator. Eric and I discuss and argue about the issues discussed here so often that I am sure his views and ideas are reflected in many parts of this volume. I would like to thank my other collaborators: Assyr Abdulle, Carlos Garcia-Cervera, Shanqin Chen, Shiyi Chen, Li-Tien Cheng, Nick Choly, Tom Hou, Zhongyi Huang, Tiejun Li, Xiantao Li, Chun Liu, Di Liu, Gang Lu, Jianfeng Lu, Paul Maragakis, Pingbing Ming, Cyrill Muratov, Weiqing Ren, Mark Robbins, Tim Schulze, Denis Serre, Chi-Wang Shu, Qi Wang, Yang Xiang, Huanan Yang, Jerry Yang, Xingye Yue, Pingwen Zhang, and more. I have learned a lot from them, and it is my good fortune to have the opportunity to work with them. I am grateful to the many people that I have consulted at one point or another on the issues presented here. They include Achi Brandt, Roberto Car, Emily Carter, Shi Jin, Tim Kaxiras, Yannis Kervrekidis, Mitch Luskin, Felix Otto, Olof Runborg, Zuowei Shen, Andrew Stuart, Richard Tsai, and Chongyu Wang.

My work has been consistently supported by the Office of Naval Research. I am very grateful to Wen Masters and Reza Malek-Madani for their confidence in my work, particularly at the time when multiscale, multi-physics modeling was not quite as popular as it is now.

Part of this volume is based on lecture notes taken by my current or former students Dong Li, Lin Lin, Jianfeng Lu, Hao Shen, and Xiang Zhou. Their help is very much appreciated.

Finally, I would like to thank Rebecca Louie and Christina Lipsky for their help in typing the manuscript.

1
Introduction

1.1 Examples of multiscale problems

Whether we explicitly recognize it or not, multiscale phenomena are part of our daily lives. We organize our time in days, months and years, as a result of the multiscale dynamics of the solar system. Our society is organized in a hierarchical structure, from towns to states, countries and continents. Such a structure has its historical and political origins but it is also a reflection of the multiscale geographical structure of the earth. Moving into the realm of modeling, an important method for studying functions, signals or geometrical shapes is to decompose them according to their components at different scales, as in Fourier or wavelet expansion. From the viewpoint of physics all materials at the microscale are made up of nuclei and electrons, whose structure and dynamics are responsible for the macroscale behavior of the material, such as its transport and wave propagation properties and its deformation and failure.

In fact, it is not an easy task to think of a situation that does not involve multiscale characteristics. Therefore, broadly speaking, it is not incorrect to say that multiscale modeling encompasses almost every aspect of modeling [29]. However, adopting such a position would make it impossible to carry out serious discussion in any kind of depth. Therefore we will take a narrower view and focus on a number of issues for which the multiscale character is the dominating issue and is exploited in the modeling process. This includes analytical and numerical techniques that exploit the disparity of scales, as well as multi-physics problems. Here the term "multi-physics problems" is perhaps a misnomer; what we have in mind are problems that involve physical laws at different levels of detail,

1

Figure 1.1. An image with large-scale edges and small-scale textures (from 3D Nature's Visual Nature Studio, used with permission).

such as quantum mechanics and continuum models. We will start with some simple examples.

1.1.1 Multiscale data and their representation

A basic multiscale phenomenon is that signals (functions, curves, images) often contain components at disparate scales. One such example is shown in Figure 1.1, which displays an image that contains large-scale edges as well as textures with small scale features. Such an observation motivates the decomposition of signals into different components according to their scales. Classical examples of this include Fourier and wavelet decomposition [21].

1.1.2 Differential equations with multiscale data

Propagation of wave packets Consider the wave equation

$$\partial_t^2 u = \Delta u. \tag{1.1.1}$$

This is a rather innocent-looking differential equation that describes wave propagation. Consider now the propagation of a wave packet that is a solution of (1.1.1) with initial condition

$$u(\mathbf{x}, 0) = A(\mathbf{x})e^{iS(\mathbf{x})/\varepsilon} \tag{1.1.2}$$

where A, S are smooth functions (see Figure 1.2). As always, we will assume that the scale parameter $\varepsilon \ll 1$. There are clearly two scales in this

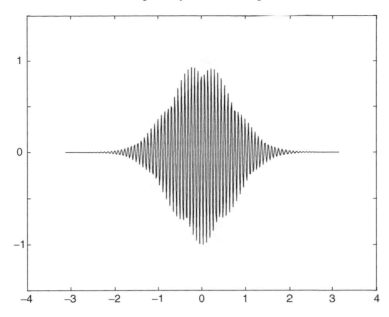

Figure 1.2. An example of a wave packet.

problem: the short wavelength ε and the scale of the envelope $A(\mathbf{x})$ of the wave packet, which is $\mathcal{O}(1)$. One can exploit this disparity between the two scales to find a simplified treatment of the problem, as is done in geometric optics. For a review of the different numerical algorithms for treating this kind of problem, we refer to [40].

Mechanics of composite materials The mechanical deformation of a material is described by the equations of elasticity theory:

$$\nabla \cdot \boldsymbol{\tau} = 0,$$
$$\boldsymbol{\tau} = \lambda(\nabla \cdot \mathbf{u})\mathbf{I} + \mu\big(\nabla\mathbf{u} + (\nabla\mathbf{u})^{\mathrm{T}}\big),$$

where $\mathbf{u}$ is the displacement field and $\boldsymbol{\tau}$ is the stress tensor. The first equation describes the force balance. The second equation is the constitutive relation, in which λ and μ are the Lamé constants that characterize the material.

To model composite materials we may simply take

$$\lambda(\mathbf{x}) = \lambda^{\varepsilon}(\mathbf{x}), \quad \mu(\mathbf{x}) = \mu^{\varepsilon}(\mathbf{x}),$$

where ε is again a small parameter that measures the scale of the heterogeneity of the material. For example, for a two-phase composite

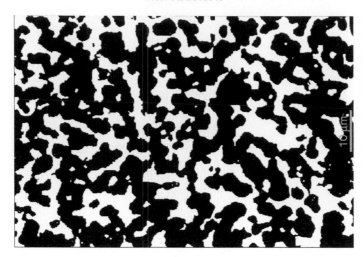

Figure 1.3. An example of a two-phase composite material (courtesy of Sal Torquato).

(see Figure 1.3), λ and μ take one set of values in one phase and another set of values in the other phase. If the microstructure happens to be periodic, then we have

$$\lambda^{\varepsilon}(\mathbf{x}) = A\left(\frac{\mathbf{x}}{\varepsilon}\right), \quad \mu^{\varepsilon}(\mathbf{x}) = B\left(\frac{\mathbf{x}}{\varepsilon}\right),$$

where A and B are periodic functions. However, in most cases in practice, the microstructure tends to be random rather than periodic. A detailed account of such issues can be found in [45].

1.1.3 Differential equations with small parameters

Consider the Navier–Stokes equation for incompressible flows at large Reynolds numbers:

$$\rho_0(\partial_t \mathbf{u} + (\mathbf{u} \cdot \nabla)\mathbf{u}) + \nabla p = \frac{1}{\text{Re}}\Delta \mathbf{u},$$

$$\nabla \cdot \mathbf{u} = 0.$$

Here $\mathbf{u}$ is the velocity field, p is the pressure field, ρ_0 is the (constant) density of the fluid and $\text{Re} \gg 1$ is the Reynolds number. In this example the highest-order derivative in the differential equation has a small coefficient. This has some rather important consequences. These include:

(1) the occurrence of boundary layers;

(2) the occurrence of turbulent flows with vortices over a large range of scales.

Partial differential equations of this type, in which the highest-order derivatives have small coefficients, are examples of singular perturbation problems. In most cases the solutions to such problems contain features at disparate scales.

1.2 Multi-physics problems

1.2.1 Examples of scale-dependent phenomena

We will now discuss some well-known examples in which the system response exhibits a transition as a function of the scales involved.

Black-body radiation Our first example is black-body radiation. Let $e_T(\lambda)$ be the energy density of the radiation of a black body at temperature T and wavelength λ. Classical statistical mechanics considerations lead to Rayleigh's formula

$$e_T(\lambda) = \frac{8\pi}{\lambda^4} k_{\mathrm{B}} T,$$

where k_{B} is the Boltzmann constant. This result fits very well with experimental results for long wavelengths but fails drastically at short wavelengths. The reason, as was discovered by Planck, is that at short wavelengths quantum effects become important. Planck postulated that the energy of a photon has to be an integer multiple of $h\nu$, where h is now known as the Planck constant and ν is the frequency of the photon. If quantum effects are taken into consideration, one arrives at Planck's formula

$$e_T(\lambda) = \frac{8\pi hc}{\lambda^5} \frac{1}{e^{hc/(k_{\mathrm{B}} T\lambda)} - 1},$$

where c is the speed of light. This result agrees very well with experimental results at all wavelengths [35]. Note that Planck's formula reduces to Rayleigh's formula at long wavelengths.

Knudsen-number dependence of the heat flux in channel flows Consider a gas flowing between two parallel plates, one of which is stationary and the other moving with a constant speed. The two plates are held at uniform temperatures T_0 and T_1 respectively. We are interested in the heat flux across the channel as a function of the Knudsen number, which is the ratio between the mean free path of the gas particles (the average distance that a gas particle travels before colliding with another gas particle) and the channel width. What is interesting is that the dependence of the heat flux as a function of the Knudsen number is non-monotonic: it is an increasing function for small values of the Knudsen number but a decreasing function at large values [43]. This phenomenon can be understood as follows. When the Knudsen number is very small, there is little heat conduction since in this case the gas is very close to local equilibrium. In this regime the dynamics of the gas is very well described by Euler's equations of gas dynamics; heat conduction is a higher-order effect (see Section 4.3). If the Knudsen number is very large then effectively the gas particles undergo free streaming. There is little momentum exchange between the particles since collisions are rare. Hence there is not much heat flux either. In this case the dynamics of the gas can be found by solving Boltzmann's equation with the collision term neglected (see Section 4.3). We see that the origins of the reduced heat conduction in the two regimes are very different.

The reverse Hall–Petch effect In the early 1950s, Hall and Petch independently found that, as the size of the grains that make up a material decreases, the yield strength of the material increases according to the equation

$$\sigma = \sigma_0 + \frac{k}{\sqrt{d}},$$

where k is a strengthening coefficient characteristic of the material and d is the characteristic grain size [31]. Roughly speaking, this is due to the fact that grain boundaries impede dislocation motion.

Recent experiments on many nanocrystalline materials demonstrate that if the grains reach a critical size, which is typically less than 100 nm, the yield stress either remains constant or decreases with decreasing grain size (see Figure 1.4) [20, 42]. This is called the *reverse Hall–Petch effect*. The exact mechanism may depend on the specific material. For example, in copper it is thought that the mechanical response changes from dislocation-mediated plasticity to grain boundary sliding as the grain size is decreased to about 10 to 15 nanometers [42].

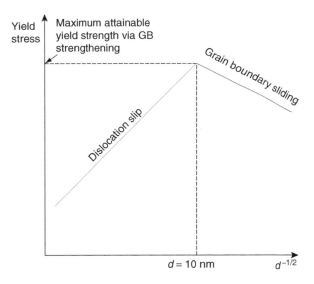

Figure 1.4. Yield stress vs. $d^{-1/2}$, where d is the grain size, illustrating the Hall–Petch and reverse Hall–Petch effects (adapted from Wikipedia).

All three examples discussed above exhibit a change in behavior as the scales involved change. The change in behavior is the result of the change in the dominating physical effects in the system.

1.2.2 Deficiencies of the traditional approaches to modeling

Central to any kind of coarse-grained model is a constitutive relation, which represents the effect of the microscopic processes at the macroscopic level. In engineering applications the constitutive relations are often empirically based on very simple considerations such as:

(1) the second law of thermodynamics;
(2) symmetry and invariance properties;
(3) linearization or Taylor expansion.

Such empirical models have been very successful in applications. However, in many cases they are inadequate, either because of the complexity of the system or because they lack crucial information about how the microstructure influences the macroscale behavior of the system.

Take incompressible fluid flow as an example. Conservation of mass and momentum gives

$$\rho_0(\partial_t \mathbf{u} + (\mathbf{u} \cdot \nabla)\mathbf{u}) = \nabla \cdot \boldsymbol{\tau}, \qquad (1.2.1)$$

$$\nabla \cdot \mathbf{u} = 0, \qquad (1.2.2)$$

where $\mathbf{u}$ is the velocity field. Here, in order to write down the momentum conservation equation we have introduced $\boldsymbol{\tau}$, the stress tensor. This is an object introduced at the continuum level to represent the short-range interactions between the molecules. Imagine a small piece of surface inside the continuous medium: the stress $\boldsymbol{\tau}$ represents the force due to the interaction between the molecules on each side of the surface. Ideally the value of $\boldsymbol{\tau}$ should be obtained from information about the dynamics of the molecules that make up the fluid. This would be a rather difficult task. Therefore in practice $\boldsymbol{\tau}$ is often modeled empirically, through intelligent guesses and calibration with experimental data. For a simple isotropic fluid, say one made up of spherical particles, isotropy and Galilean invariance implies that τ should not depend on $\mathbf{u}$. Hence a simple guess is that $\boldsymbol{\tau}$ is a function of $\nabla \mathbf{u}$. In the absence of further information about the behavior of the system, it is natural to start with the very simplest choice, namely that $\boldsymbol{\tau}$ is a linear function of $\nabla \mathbf{u}$. In any case, every function is approximately linear in appropriate regimes. Using isotropy, we arrive at the *constitutive relation* for *Newtonian fluids*:

$$\boldsymbol{\tau} = -p\mathbf{I} + \mu(\nabla \mathbf{u} + (\nabla \mathbf{u})^{\mathrm{T}}) \qquad (1.2.3)$$

where p is the pressure. Substituting this constitutive relation into the conservation laws leads to the well-known *Navier–Stokes equation*.

It is remarkable that such simple-minded considerations yield a model, namely the Navier–Stokes equation, which has proven to be very accurate in describing the dynamics of simple fluids under a very wide range of conditions. Partly for this reason, considerable effort has gone into extending such an approach to the modeling of complex fluids such as polymeric fluids. However, the results there are rather mixed [12]. The accuracy of the constitutive laws is often questionable; in many cases, they become quite complicated since many parameters need to be fitted and their physical meaning becomes obscure. Consequently the original appeal of their simplicity and universality is lost. In addition, there is no systematic way of developing such empirical constitutive relations. So a constitutive relation obtained by calibrating against one set of experimental data may not be useful in a different experimental setting. More importantly, such an approach does not contain any explicit information about the interaction between fluid flow and the conformation of the molecules, which might be exactly the kind of information in which we are most interested.

The situation described above is to some extent generic. Indeed, empirical modeling by constitutive relations is a very popular tool and is used in many areas, including:

(1) elasticity and plasticity in the theory of solids;
(2) empirical potentials in molecular dynamics;
(3) empirical models of reaction kinetics;
(4) hopping rates in kinetic Monte Carlo models;
(5) collision cross sections in the kinetic theory of gases.

These constitutive relations are typically quite adequate for simple systems, such as small deformations of solids, but fail for complex systems. When empirical models become inadequate, we have to replace them by more accurate models that rely more on a detailed description of the microscopic processes.

The other extreme is *quantum many-body theory*, which could be said to be the true first principle. Assume that our system of interest has M nuclei and N electrons. At the level of quantum mechanics, this system (neglecting spin) is described by a wavefunction of the form

$$\Psi = \Psi(\mathbf{R}_1, \ldots, \mathbf{R}_M, \mathbf{r}_1, \ldots, \mathbf{r}_N)$$

The Hamiltonian for this system is given by

$$\mathcal{H} = -\sum_{J=1}^{M} \frac{\hbar^2}{2M_J} \nabla_{\mathbf{R}_J}^2 - \sum_{k=1}^{N} \frac{\hbar^2}{2m_e} \nabla_{\mathbf{r}_k}^2$$
$$+ \sum_{J<K} \frac{Z_J Z_K}{|\mathbf{R}_J - \mathbf{R}_K|} - \sum_{J,k} \frac{Z_J}{|\mathbf{R}_J - \mathbf{r}_k|} + \sum_{i<j} \frac{1}{|\mathbf{r}_i - \mathbf{r}_j|} \quad (1.2.4)$$

where Z_J is the nuclear charge of the Jth nucleus. Here all we need in order to analyze a given system is to input the atomic numbers Z_J for the atoms in the system and then solve the Schrödinger equation. There are no empirical parameters to fit! The difficulty, however, lies in the complexity of its mathematics. This situation is very well summarized by the following remark of Paul Dirac, made shortly after the discovery of quantum mechanics [22]:

> *The underlying physical laws necessary for the mathematical theory of*
> *a large part of physics and the whole of chemistry are thus completely*
> *known, and the difficulty is only that the exact application of these laws*
> *leads to equations much too complicated to be soluble.*

The mathematical complexity noted by Dirac is indeed quite daunting: wave functions have as many (or, rather, three times as many) independent variables as the number of electrons and nuclei in the system. If the system has 10^{30} electrons, then the wave function has 3×10^{30} independent variables. Obtaining accurate information for such a function is a quite impossible task. In addition, such a description often gives far too much information, most of which is not really of any interest. Getting the relevant information from this vast amount of data would also be quite a challenge.

This dichotomy is the kind of situation that we encounter for many problems. On the one hand, empirically obtained macroscale models are very efficient but are often not accurate enough or lack crucial microstructural information in which we are interested. Microscopic models, on the other hand, may offer better accuracy but they are often too expensive to be used to model systems of real interest. It would be nice to have a strategy that combines the efficiency of macroscale models and the accuracy of microscale models. This is a tall order but is precisely the motivation for multiscale modeling.

1.2.3 The multi-physics modeling hierarchy

Between macroscale models and the quantum many-body problem lie a hierarchy of other models, which are better suited at the appropriate scales. This is illustrated in Figure 1.5. A more detailed description of this modeling hierarchy is shown in Table 1.1.

In traditional approaches to modeling we tend to focus on one particular scale: the effects of smaller scales are modeled through the constitutive relation; the effects of larger scales are neglected by assuming that the system is homogeneous at these scales. The philosophy of multiscale, multi-physics modeling is the opposite. It is based on two ideas.

(1) Any system of interest can always be described by a *hierarchy of models of different complexity*. This allows us to think about more detailed models when a coarse-grained model is no longer adequate. It also gives us a basis for understanding coarse-grained

Table 1.1. *The multi-physics hierarchy*

Gases, plasmas	Liquids	Solids
gas dynamics MHD	hydrodynamics (Navier–Stokes)	elasticity models plasticity models dislocation dynamics
kinetic theory	kinetic theory Brownian dynamics	kinetic Monte Carlo
particle models	molecular dynamics	molecular dynamics
quantum mechanics	quantum mechanics	quantum mechanics

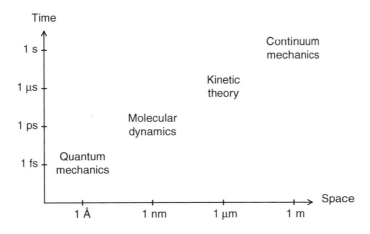

Figure 1.5. Commonly used models of physics at different scales. Strictly speaking, this figure is somewhat misleading: quantum mechanics models are not just valid at the scales indicated but also at larger scales. However, much simplified models are often sufficient at larger scales.

models from more detailed ones. In particular, when empirical coarse-grained models are inadequate, one might still be able to capture the macroscale behavior of the system with the help of microscale models.

(2) In many situations the system of interest can be described adequately by a coarse-grained model except in *some small regions, where more detailed models are needed.* These small regions may contain singularities, defects, chemical reactions or some other interesting events. In such cases, by coupling models of different

complexity in different regions we may be able to develop modeling strategies that have an efficiency comparable with coarse-graining as well as an accuracy comparable with that of the more detailed models.

Such strategies have been explored for some years. Two classical examples are the QM–MM (quantum mechanics–molecular mechanics) modeling of chemical reactions involving macromolecules, initiated in the 1970s by Warshel and Levitt [49], and first-principle-based molecular dynamics, initiated in the 1980s by Car and Parrinello [16]. Both methods have been very successful and have become standard tools in computational science. However, what has triggered the widespread interest in multiscale modeling in recent years is the realization that such strategies are useful not just for a few isolated problems but for a very wide spectrum of problems in all areas of science and engineering. This is indeed a conceptual breakthrough and arguably a more mature viewpoint on modeling.

To make this really useful we need to understand the relation between different models and how to formulate coupled models. Much of the present volume is devoted to these issues.

1.3 Analytical methods

We will begin with a discussion of the most important analytical techniques for multiscale problems. We will focus on the tools of asymptotic analysis since they tend to be the most effective. It should be emphasized that these asymptotic techniques are not blank checks. To use them effectively one often has to have some intuition about how the solutions behave. The asymptotic techniques can then be used as a systematic procedure for refining that intuition and giving quantitatively accurate predictions.

Matched asymptotics This is useful for problems whose solutions undergo rapid variations in localized regions. These localized regions are called inner regions. Typical examples are problems with boundary layers or internal layers. Matched asymptotics is a systematic way of finding approximate solutions to such problems, by introducing stretched variables in the inner region and matching the approximate solutions in the inner and outer regions. As a result, we often find that the profile of the solution in the inner region is rather rigid and is parametrized by a few parameters or functions which can be obtained from the solutions in the outer region.

Averaging and homogenization methods This is a systematic procedure based on multiscale expansion [9, 33]. In the case of ordinary differential equations (ODEs) the objective is to find effective dynamical equations for the slow variables by averaging out the fast variables. There are natural extensions of such techniques, to partial differential equations (PDEs), for example, discussed in [10].

Scaling and renormalization group methods Another important technique is scaling. In its simplest form, it is dimensional analysis. A more systematic approach is renormalization group analysis, which has become a very powerful tool in analyzing complex systems.

1.4 Numerical methods

1.4.1 Linear scaling algorithms

We will distinguish two different classes of multiscale algorithms. The first contains algorithms that aim at resolving efficiently the details of solutions, including the detailed small-scale behavior. The multigrid method (MG), the fast multipole method (FMM) and adaptive mesh refinement (AMR) are examples of this type. They all make heavy use of the multiscale structure of the problem. We call them *classical multiscale algorithms*. They are typically linear scaling algorithms in the sense that their computational complexity scales linearly with the number of degrees of freedom necessary to represent the detailed microscale solution. We will also include the domain decomposition method (DDM). Even though it relies less on the multiscale structure of the problem it does provide a platform on which multiscale methods can be constructed, particularly for the kind of problem discussed in Chapter 7 for which the solutions behave very differently on different domains.

1.4.2 Sublinear scaling algorithms

Recent efforts on multiscale modeling have focused on sublinear scaling algorithms. These are algorithms whose computational complexity scales sublinearly with the number of degrees of freedom necessary to represent the detailed microscale solution:

$$\frac{\text{computational complexity}}{\text{number of degrees of freedom for the microscale model}} \ll 1.$$

This may seem too good to be true. Indeed it is only possible to construct such algorithms

(1) if our aim is to resolve certain features of the microscale solution, such as some averaged quantities, rather than its detailed behavior; and

(2) if the microscale model has some particular features of which we can take advantage. A typical example of such a particular feature is scale separation.

Even though multiscale problems are typically quite complicated, a main objective of multiscale modeling is to identify special classes of such problems for which it is possible to develop simplified descriptions or algorithms. The efficiency of these algorithms relies on the special features of the problem. Therefore they are less general than the linear scaling algorithms discussed earlier. A good example for illustrating this difference is the Fourier transform. The well-known fast Fourier transform (FFT) scales almost linearly and is completely general. However, if the Fourier coefficients are sparse then one can develop sublinear scaling algorithms, as was done in [28]. Of course it is understood that the improved efficiency of such sublinear scaling algorithms can only be expected for special situations, i.e. when the Fourier coefficients are sparse.

In practice there may not be a sharp division between linear and sublinear scaling algorithms.

1.4.3 Type A and type B multiscale problems

It is also helpful to distinguish two classes of multiscale problems. The first are problems with local defects or singularities, such as dislocations, shocks and boundary layers, for which a macroscale model is sufficient for most of the physical domain; the microscale model is only needed locally around the singularities or heterogeneities. The second class of problems contains those for which a microscale model is needed everywhere as a supplement to the macroscale model. This occurs, for example, when the microscale model is needed to supply the missing constutitive relation in the macroscale model. We will call the former type A problems and the latter type B problems [24]. The central issues and the main difficulties can be quite different for the two classes of problems. The standard approach for type A problems is to use the microscale model near defects or heterogeneities and the macroscale model elsewhere. The main question is how to couple these two models at their interface. The standard approach for type B problems is to couple the macro and micro models everywhere in the computational domain, as is done in the heterogeneous multiscale method (HMM) [24]. Note that for type A problems the macro–micro coupling is localized. For type B problems, coupling is done over the whole computational domain.

1.4.4 Concurrent versus sequential coupling

Historically, multiscale (multi-physics) modeling was first used in the form of *sequential (or serial) coupling*. The macroscopic model is determined first except for some parameters, or functions, which are computed or tabulated using a microscopic model. Function values that are not found from the table can be obtained using interpolation. At the end of this procedure one has a macroscopic model, which can then be used to analyze the macroscopic behavior of the system under different conditions. For obvious reasons, this coupling procedure is also called *precomputing* or *microscopically informed modeling*. A typical example is found in gas dynamics, where one often precomputes the equation of state using kinetic theory and stores it as a look-up table. This information is then used in Euler's equations of gas dynamics to simulate gas flow under different conditions. When modeling porous-medium flows, a common form of upscaling is to precompute the effective permeability tensor $K(\mathbf{x})$ using the *representative averaging volume (RAV)* technique [8]: one considers a domain of suitable size (i.e. the representative averaging volume) around $\mathbf{x}$ and solves the microscale model in that domain. The result is then suitably averaged to give $K(\mathbf{x})$ [23]. Finally, this effective permeability tensor is used in Darcy's law,

$$\mathbf{v}(\mathbf{x}) = -K(\mathbf{x})\nabla p(\mathbf{x}), \quad \nabla \cdot \mathbf{v}(\mathbf{x}) = 0,$$

to compute the pressure field on a larger scale. More recently, such a sequential multiscale modeling strategy has been used to study macroscopic properties of fluids and solids using parameters that are obtained successively starting from models of quantum mechanics [19, 47].

If the missing information is a function of many variables then precomputing such a function might be too costly. For this reason, sequential coupling has been limited mainly to situations when the needed information comprises just a small number of parameter values or functions of very few variables. Indeed, for this reason, sequential coupling is often referred to as *parameter passing*.

An alternative approach is to obtain the missing information "on the fly" as the computation proceeds. This is commonly referred to as the *concurrent coupling* approach [1]. It is to be preferred when the missing information is a function of many variables. Assume that we are solving a macroscopic model in d dimensions, with mesh size Δx, and we need to obtain some data from a microscopic model at each grid point. The number of data evaluations in a concurrent approach will be roughly $\mathcal{O}((\Delta x)^{-d})$. In sequential coupling, assuming that the needed data is a function of m variables, the number of forced evaluations in the precomputing stage will be $\mathcal{O}(h^{-m})$ if a uniform grid of size h is used in the computation. Assuming that $h \sim \Delta x$, the concurrent approach is more efficient if $m > d$. This is certainly the case in Car–Parrinello molecular dynamics (see Chapter 6), where the interatomic forces are functions of all the atomic coordinates. Hence m can easily be on the order of thousands.

However, as was pointed out in [27], the efficiency of precomputing can be improved with better look-up tables and interpolation techniques. For example, if sparse grids are used and if the grid size is h then the number of grid points needed is $\mathcal{O}(|\log(h)|^{m-1}/h)$. This is significantly less than the cost on a uniform grid, making precomputing quite feasible if m is below 8 or 10. Examples of the application of sparse grids to sequential coupling can be found in [27].

The most appropriate approach for a particular problem depends to a large extent on how much we know about the macroscale process. Take again the example of incompressible fluids. We know that the macroscale model should be in the form of (1.2.1). The only question is the form or expression of $\boldsymbol{\tau}$. We can distinguish three cases.

(1) A linear constitutive relation is sufficiently accurate. We only need to know the value of the viscosity coefficient μ in (1.2.3).

(2) It is enough to assume that $\boldsymbol{\tau}$ depends on (the symmetric part of) $\nabla \mathbf{u}$. We then need to know the function $\boldsymbol{\tau} = \boldsymbol{\tau}(\nabla \mathbf{u})$.

(3) The stress tensor $\boldsymbol{\tau}$ depends on many more variables than $\nabla\mathbf{u}$, e.g. $\boldsymbol{\tau}$ may depend on some additional order parameters.

In the first case we may simply precompute the value of μ using the microscale model. In the second case, $\boldsymbol{\tau}$ is effectively a function of five variables. Precomputing such a function is much harder but might still be feasible using some advanced techniques such as sparse grids [51, 15]. In the third case, it is most likely that precomputing will be practically impossible, and we have to obtain $\boldsymbol{\tau}$ "on the fly" as the computation proceeds.

Most current work on multiscale modeling focuses on such concurrent coupling, as will the present volume. We should note, though, that in many cases the technicalities in sequential and concurrent coupling are quite similar. For example, one main problem in sequential coupling is how to set up the microscale model in order to compute the needed constitutive relation. This is also a major problem in concurrent coupling.

1.5 What are the main challenges?

Even though there is little doubt that multiscale modeling will play a major role in scientific modeling, the challenges of putting multiscale modeling on a solid foundation are quite daunting. Here is a brief summary of some of these challenges.

1. Understanding the physical models at different scales. Before we start coupling different levels of physical models, we need to understand the basic formulation and properties of each one. Although our understanding of macroscopic models (say from continuum mechanics) has reached some degree of maturity we cannot draw the same conclusion about our understanding of microscopic models, such as electronic structure or molecular dynamics models. One example is the matter of boundary conditions. Typically, it is only understood how to apply periodic or vacuum boundary conditions for these models. Since we are limited to rather small systems when conducting numerical calculations using these models, it is quite unsatisfactory that we have to waste a large amount of computational resources just to make the system artificially periodic. Just imagine what it would be like if we only knew how to deal with periodic conditions for continuum models! An example of what might happen when other boundary conditions are used is displayed in Figure 1.6. It shows

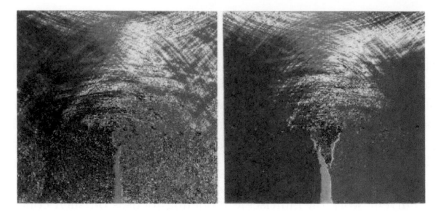

Figure 1.6. Phonon reflection at a boundary in a molecular dynamics simulation of crack propagation (from [32], copyright (1995) by the American Physical Society).

the results of a molecular dynamics simulation of crack propagation using Dirichlet-type boundary conditions, i.e. the positions of the atoms at the boundary are prescribed. As the simulation proceeds, phonons (lattice waves) are generated; they quickly reach the boundary of the simulation domain and are then reflected. The reflected phonons interfere with the dynamics of the crack. In extreme situations, phonon reflection may cause the crystal to melt.

2. Understanding how models of different complexity are related to each other. Ideally, we should be able to derive coarse-grained models from more detailed ones in the appropriate limit. This would provide a link between the macroscopic and microscopic models.

3. Understanding how models of different complexity can be coupled together smoothly, without creating artifacts. When different models are coupled together, large errors often occur at the interface of the two models and sometimes these errors can pollute a much larger part of the physical domain and even cause numerical instabilities.

4. Understanding how to formulate models at intermediate levels of complexity, or "mesoscale models." One example is the Ginzburg–Landau model of phase transitions. It is at an intermediate level of detail between the atomistic and hydrodynamic models and is very convenient for many purposes since it is a continuum model. This is why phase-field models have become quite popular even though most of them are quite empirical. Such mesoscopic models are needed particularly for heterogeneous systems such as macromolecules.

1.6 Notes

It is very hard to do justice to the massive amount of work that has been done on multiscale modeling. On the mathematical side, Fourier expansion is clearly a form of multiscale representation. On the physical side, it is not stretching the truth to say that multiscale, multi-physics ideas have played a role in many major developments in theoretical physics. As we explained earlier, Planck's theory of black body radiation is a good example of multi-physics modeling. Statistical physics has provided a microscopic foundation for thermodynamics. Cauchy, for example, explored the connection between elasticity theory and atomistic models of solids and derived the celebrated Cauchy relation between the elastic moduli of solids for a certain class of atomistic models (see Chapter 4 for more details). Cauchy's work was continued by Born and others and has led to a microscopic foundation for the continuum theory of solids [13]. Multiscale ideas were certainly quite central to the kinetic theory of gases developed by Boltzmann, Maxwell and others, which has led to a microscopic foundation for gas dynamics [38]. In the 1950s Tsien pushed for a program, called "physical mechanics," aimed at establishing a solid microscopic foundation for continuum mechanics [46]. This tradition has been continued, notably in the mathematical physics community through its work on understanding the thermodynamic, kinetic and hydrodynamic limits [39, 44].

In another direction, mesoscopic models such as phase-field models and lattice Boltzmann methods, which were initially developed as approximations to existing macroscopic models motivated by microscopic considerations, now have a life of their own and are sometimes used as fundamental physical models [2, 17], with or without justification.

The development of analytical techniques for tackling multiscale problems also has a long history. Boundary layer analysis, initiated by Prandtl more than a century ago, is an example. Averaging methods for systematically removing secular terms in ODEs with multiple time scales are also early examples. Another well-known classical example is the WKB method. These asymptotic analysis techniques have now become standard tools used in many different areas. They have also formed a subject of their own [9, 33]. Of particular importance for the modern development of multiscale modeling is the homogenization method, which can be considered as an extension of the averaging technique for PDE problems [10, 37].

A very powerful tool in harmonic analysis, a subject of mathematics, is to (dyadically) decompose functions or objects into components with different scales. Starting with the work of Haar and Stromberg, this effort eventually led to the wavelet representation, which is now widely used [21]. Multiscale analysis has also been an important tool in other branches of mathematics, such as mathematical physics, dynamical systems and partial differential equations.

Another important tool for analyzing scales is scaling. It has its origin in the dimensional analysis techniques commonly used in engineering and science as a quick way to grasp the most important features of a problem. The most spectacular application of scaling, to this day, is still the work of Kolmogorov on the small-scale structure of fully developed turbulent flows. A significant step was made by Wilson and others in the development of the renormalization group method (see Section 2.6 for more details). This is a systematic procedure for analyzing the scaling properties of rather complex systems. An alternative viewpoint was put forward by Barenblatt [7]. Renormalization has become an essential component of modern physics. Yet from a mathematical viewpoint it still remains rather mysterious. This is quite disturbing since in the end these are mathematical techniques for taking infrared or ultraviolet limits and finding effective models.

Also important is the work of Mori and of Zwanzig, which provides a general formalism for eliminating degrees of freedom in a system [36, 52]. Its importance lies in its generality. It may provide a general principle, using which reduced models can be derived, as well as a starting point for making systematic approximations of the original microscale model [18, 34].

Turning now to algorithms, it is fair to say that multiscale ideas are behind some of the most successful numerical algorithms used today, including the multigrid method and the fast multipole method [14, 30]. These algorithms are designed for general problems, but they rely heavily on the ideas of multiscale decomposition. Adaptive methods, such as adaptive mesh refinement or ODE solvers with adaptive step size control, have also used multiscale thinking in one way or another. In particular, implicit schemes are rather powerful multiscale techniques for a class of ODEs with multiple time scales: they allow us to capture the large-scale features without resolving the small-scale transients.

Ivo Babuska pioneered the study of numerical algorithms for PDEs with multiscale data. Babuska's focus was on finite element methods for elliptic equations with multiscale coefficients. Among other things, Babuska discussed the inadequacies of homogenization theory and the necessity to include at least some small-scale features [3, 4]. Babuska and Osborn proposed the generalized finite element method, which lies at the heart of modern multiscale finite element methods (see Chapter 8 for details) [6]. Engquist considered hyperbolic problems with multiscale data and explored the possibility of capturing the macroscale behavior of the solutions without resolving the small-scale details [26].

As mentioned earlier, there is a very long history of multi-physics modeling using sequential coupling, i.e. precomputing crucial parameters or constitutive relations using microscale models. The importance of concurrent coupling was emphasized in the work of Car and Parrinello [16] as well as in the earlier work of Warshel and Levitt [49] on QM–MM methods (see also [48]). In recent years, we have seen a drastic explosion of activity in this style of work. Indeed, a main purpose of the present volume is to discuss the foundation of this type of multiscale modeling.

References for Chapter 1

[1] F.F. Abraham, J.Q. Broughton, N. Bernstein and E. Kaxiras, "Concurrent coupling of length scales: methodology and application," *Phys. Rev. B*, vol. 60, no. 4, pp. 2391–2402, 1999.

[2] D. Anderson, G. McFadden and A. Wheeler, "Diffuse-interface methods in fluid mechanisms," *Ann. Rev. Fluid Mech.*, vol. 30, pp. 139–165, 1998.

[3] I. Babuska, "Solution of interface by homogenization, I, II, III," *SIAM J. Math. Anal.*, vol. 7, pp. 603–645, 1976; vol. 8, pp. 923–937, 1977.

[4] I. Babuska, "Homogenization and its applications, mathematical and computational problems," in *Numerical Solutions of Partial Differential Equations III*, B. Hubbard, ed., pp. 89–116, Academic Press, 1976.

[5] I. Babuska, G. Caloz and J. Osborn, "Special finite element methods for a class of second order elliptic problems with rough coefficients," *SIAM J. Numer. Anal.*, vol. 31, pp. 945–981, 1994.

[6] I. Babuska and J.E. Osborn, "Generalized finite element methods: their performance and their relation to mixed methods," *SIAM J. Numer. Anal.*, vol. 20, no. 3, pp. 510–536, 1983.

[7] G.I. Barenblatt, *Scaling*, Cambridge University Press, 2003.

[8] J. Bear and Y. Bachmat, *Introduction to Modeling of Transport Phenomena in Porous Media*, Kluwer Academic Publishers, 1990.

[9] C.M. Bender and S.A. Orszag, *Advanced Mathematical Methods for Scientists and Engineers*, McGraw-Hill, 1978.

[10] A. Bensoussan, J.-L. Lions and G.C. Papanicolaou, *Asymptotic Analysis for Periodic Structures*, North-Holland, 1978.

[11] R.B. Bird, R.C. Armstrong and O. Hassager, *Dynamics of Polymeric Liquids, Vol. 1: Fluid Mechanics*, John Wiley, 1987.

[12] R.B. Bird, C.F. Curtiss, R.C. Armstrong and O. Hassager, *Dynamics of Polymeric Liquids, Vol. 2: Kinetic Theory*, John Wiley, 1987.

[13] M. Born and K. Huang, *Dynamical Theory of Crystal Lattices*, Oxford University Press, 1954.

[14] A. Brandt, "Multi-level adaptive solutions to boundary value problems," *Math. Comp.*, vol. 31, no. 138, pp. 333–390, 1977.

[15] H.-J. Bungartz and M. Griebel, "Sparse grids," *Acta Numer.*, vol. 13, pp. 147–269, 2004.

[16] R. Car and M. Parrinello, "Unified approach for molecular dynamics and density-functional theory," *Phys. Rev. Lett.*, vol. 55, no. 22, pp. 2471–2474, 1985.

[17] S. Chen and G.D. Doolen, "Lattice Boltzmann methods for fluid flows," *Ann. Rev. Fluid Mech.*, vol. 30, pp. 329–364, 1998.

[18] A.J. Chorin, O. Hald and R. Kupferman, "Optimal prediction with memory," *Physica D*, vol. 166, nos. 3–4, pp. 239–257, 2002.

[19] E. Clementi and S. F. Reddaway "Global scientific and engineering simulations on scalar, vector and parallel LCAP-type supercomputers, *Phil. Trans. Roy. Soc. London*, Ser. A, vol. 326, pp. 445–470, 1988.

[20] H. Conrad and J. Narayan, "On the grain size softening in nanocrystalline materials," *Scripta Mater.*, vol. 42, no. 11, pp. 1025–1030, 2000.

[21] I. Daubechies, *Ten Lectures on Wavelets*, SIAM, 1992.

[22] P. Dirac, "Quantum mechanics of many-electron systems," *Proc. Roy. Soc. London*, vol. 123, no. 792, 714–733, 1929.

[23] L.J. Durlofsky, "Numerical calculation of equivalent grid block permeability tensors for heterogeneous porous media," *Water Resour. Res.*, vol. 27, pp. 699–708, 1991.

[24] W. E and B. Engquist, "The heterogeneous multi-scale methods," *Comm. Math. Sci.*, vol. 1, pp. 87–133, 2003.

[25] W. E and B. Engquist, "Multiscale modeling and computation," *Not. Amer. Math. Soc.*, vol. 50, no. 9, pp. 1062–1070, 2003.

[26] B. Engquist, "Computation of oscillatory solutions to hyperbolic differential equations," in *Lecture Notes in Mathematics*, vol. 1270, pp. 10–22, 1987.

[27] C. Garcia-Cervera, W. Ren, J. Lu and W. E, "Sequential multiscale modeling using sparse representation," *Comm. Comput. Phys.*, vol. 4, pp. 1025–1033, 2008.

[28] A. Gilbert, S. Guha, P. Indyk, S. Muthukrishnan and M. Strauss, "Near-optimal sparse Fourier representations via sampling," in *Proc. 2002 ACM Symposium on Theory of Computing (STOC)*, pp. 389–398, 2002.

[29] J. Glimm and D.H. Sharp, "Multiscale science," *SIAM News*, October, 1997.

[30] L. Greengard and V. Rokhlin, "A fast algorithm for particle simulations," *J. Comput. Phys.*, vol. 73, pp. 325–348, 1987.

[31] E.O. Hall, "The deformation and ageing of mild steel: III, discussion of results," *Proc. Phys. Soc.*, vol. 64, pp. 747–753, 1951.

[32] B.L. Holian and R. Ravelo, "Fracture simulation using large-scale molecular dynamics", *Phys. Rev. B*, vol. 51, pp. 11 275–11 288, 1995.

[33] J. Kevorkian and J.D. Cole, *Perturbation Methods in Applied Mathematics*, Springer-Verlag, 1981.

[34] X. Li and W. E "Variational boundary conditions for molecular dynamics simulation of crystalline solids at finite temperature: treatment of the thermal bath," *Phys. Rev. B*, vol. 76, no. 10, pp. 104 107–104 129, 2007.

[35] A. Messiah, *Quantum Mechanics*, Dover Publications, 1999.

[36] H. Mori, "Transport, collective motion, and Brownian motion," *Prog. Theor. Phys.*, vol. 33, pp. 423–455, 1965.

[37] G.A. Pavliotis and A.M. Stuart, *Multiscale Methods: Averaging and Homogenization*, Springer-Verlag, 2008.

[38] L.E. Reichl, *A Modern Course in Statistical Physics*, University of Texas Press, 1980.

[39] D. Ruelle, *Statistical Mechanics: Rigorous Results*, W. A. Benjamin, 1969.

[40] O. Runborg, "Mathematical models and numerical methods for high frequency waves," *Comm. Comput. Phys.*, vol. 2, pp. 827–880, 2007.

[41] M. Sahimi, *Flow and Transport in Porous Media and Fractured Rock*, VCH, 1995.

[42] J. Schiotz, F.D. Di Tolla and K.W. Jacobsen, "Softening of nanocrystalline metals at very small grains," *Nature*, vol. 391, pp. 561–563, 1998.

[43] Y. Sone, *Molecular Gas Dynamics*, Birkhäuser, 2007.

[44] H. Spohn, *Large Scale Dynamics of Interacting Particles*, Springer-Verlag, 1991.

[45] S. Torquato, *Random Heterogeneous Materials: Microstructure and Macroscopic Properties*, Springer-Verlag, 2002.

[46] S.H. Tsien, *Lectures on Physical Mechanics* (in Chinese), Science Press, 1962.

[47] C.-Y. Wang, S.-Y. Liu and L.-G. Han, "Electronic structure of impurity (oxygen)-stacking-fault complex in nickel," *Phys. Rev. B*, vol. 41, pp. 1359–1367, 1990.

[48] A. Warshel and M. Karplus, "Calculation of ground and excited state potential surfaces of conjugated molecules. I. Formulation and parametrization," *J. Amer. Chem. Soc.*, vol. 94, pp. 5612–5625, 1972.

[49] A. Warshel and M. Levitt, "Theoretical studies of enzymic reactions," *J. Mol. Biol.*, vol. 103, pp. 227–249, 1976.

[50] K.G. Wilson and J. Kogut, "The renormalization group and the ε expansion", *Phys. Rep.*, vol. 12, pp. 75–200, 1974.

[51] C. Zenger, in *Parallel Algorithms for Partial Differential Equations*, pp. 241–251, Vieweg, 1991.

[52] R. Zwanzig, "Collision of a gas atom with a cold surface," *J. Chem. Phys.*, vol. 32, pp. 1173–1177, 1960.

J. Li and M. Kwauk, "Exploring complex systems in chemical engineering – the multiscale methodology", *Chemical Engineering Science*, vol. 58, pp. 521–535, 2003.

2

Analytical methods

This chapter is a review of the major analytical techniques that have been developed for addressing multiscale problems. Our objective is to illustrate the main ideas and to put things into perspective. It is not our intention to give a thorough treatment of the topics discussed. Nor is it our aim to provide a rigorous treatment, even though in many cases rigorous results are indeed available.

It should be remarked that even though most attention in multiscale modeling has been focused on developing numerical methods, one should not underestimate the usefulness of analytical techniques. They offer much needed insight into the multiscale nature of the problem. In simple cases, they give rise to simplified models. For more complicated problems, they give guidelines for designing and analyzing numerical algorithms.

The problems that we consider below all have a small parameter that is responsible for the multiscale behavior. For analyzing such problems, our basic strategy is to introduce additional variables such that in the new variables the scale of interest becomes $\mathcal{O}(1)$. For example, in boundary layer analysis we introduce an inner variable in which the width of the boundary layer becomes $\mathcal{O}(1)$. We then make an ansatz for the leading-order profile of the multiscale solution, based on our intuition for the problem. If we have the right intuition, we can use asymptotics to refine this intuition so that a better ansatz can be made. Eventually this process allows us to find the leading-order behavior of the multiscale solution. If our intuition is incorrect, the asymptotics will lead us nowhere.

2.1 Matched asymptotics

The method of matched asymptotics is most useful for problems in which the solutions change abruptly in a small, localized, region, usually called the *inner region*. This region can comprise boundary layers, internal layers or vortices in fluids or type-II superconductors. Matched asymptotics is a systematic procedure for obtaining approximate solutions for such problems.

The main reason for distinguishing the inner and the *outer region* is that the dominant features in the two regions are different. To account for the rapid change of the solution in the inner region we introduce a stretched variable, also called the *inner variable*, so that in the new variable the scale of the inner region becomes $\mathcal{O}(1)$. The dominant terms in the inner region can be found by transforming the model to the inner variables. Together with the consideration that the *inner solution* should be matched smoothly to the approximate solution in the outer region, this often allows us to find the leading-order behavior, or the effective models that control the leading order behavior, of the solution in both regions.

2.1.1 A simple advection–diffusion equation

Consider the following simple example:

$$\partial_t u^\varepsilon + \partial_x u^\varepsilon = \varepsilon \partial_x^2 u^\varepsilon \tag{2.1.1}$$

for $x \in [0,1]$, with boundary conditions

$$u^\varepsilon(0,t) = u_0(t), \quad u^\varepsilon(1,t) = u_1(t).$$

Here, as usual, ε is a small parameter representing the small scale. In this problem the highest-order derivative in the differential equation has a small coefficient. If we neglect this term we obtain

$$\partial_t U + \partial_x U = 0. \tag{2.1.2}$$

This is an advection equation. Its solution is called the *outer solution*. One thing to notice about the outer solution is that we are only allowed to impose a boundary condition at $x = 0$. In general the boundary condition for u^ε at $x = 1$ will not be honored by solutions of (2.1.2). This suggests that the solutions to (2.1.1) and (2.1.2) may behave quite differently near $x = 1$. Indeed, we will see that, for the solution of the original problem

(2.1.1), a boundary layer exists at $x = 1$ which serves to connect the value of U near $x = 1$ to the boundary condition of u^ε at $x = 1$.

To see this more clearly, let us change to a stretched variable at $x = 1$ by setting

$$y = \frac{x - 1}{\delta},$$

where δ is the unknown boundary layer thickness. In the stretched variable, our original PDE becomes

$$\partial_t \tilde{u} + \frac{1}{\delta} \partial_y \tilde{u} = \frac{\varepsilon}{\delta^2} \partial_y^2 \tilde{u}. \tag{2.1.3}$$

It is easy to see that to be able to obtain a non-trivial balance between the terms in (2.1.3), we must have $\delta \sim \varepsilon$. Without loss of generality we will take $\delta = \varepsilon$. We then have, to leading order,

$$\partial_y \tilde{u} = \partial_y^2 \tilde{u}.$$

Integrating once, we obtain

$$\tilde{u} = \partial_y \tilde{u} + C_1.$$

To be able to match to the outer solution U as we come out of the boundary layer, i.e. as $y \to -\infty$, we must have

$$\lim_{y \to -\infty} \partial_y \tilde{u} = 0.$$

Hence, we have

$$C_1 = U(1, t).$$

Another integration gives

$$\tilde{u}(y, t) = U(1, t) + (u_1(t) - U(1, t))e^y$$

or, in terms of the original variable,

$$\tilde{u}^\varepsilon(x, t) = U(1, t) + (u_1(t) - U(1, t))e^{(x-1)/\varepsilon}.$$

This gives the leading-order behavior inside the boundary layer.

Here we started with the recognition that a boundary layer exists at $x = 1$. If we had made the mistake of assuming that the boundary layer was located at $x = 0$, we would have obtained the trivial solution $\tilde{u} = u_0(t)$ there, and we would have seen that the boundary condition at $x = 1$ still presents a problem.

To make this procedure more systematic, we should expand both the inner and outer solutions $\tilde{u}$ and U, and we should identify the matching region. See [21] for a more detailed discussion of this methodology.

2.1.2 Boundary layers in incompressible flows

Consider two-dimensional incompressible flow in the half-plane $\Omega = \{\mathbf{x} = (x,y), y \geq 0\}$. We write the governing equation, the Navier–Stokes equation, as

$$\partial_t \mathbf{u}^\varepsilon + (\mathbf{u}^\varepsilon \cdot \nabla)\mathbf{u}^\varepsilon + \nabla p^\varepsilon = \varepsilon \Delta \mathbf{u}^\varepsilon, \qquad (2.1.4)$$

$$\nabla \cdot \mathbf{u}^\varepsilon = 0, \qquad (2.1.5)$$

with the no-slip boundary condition

$$\mathbf{u}^\varepsilon(\mathbf{x}) = 0, \quad \mathbf{x} \in \partial\Omega.$$

Here $\mathbf{u}^\varepsilon = (u^\varepsilon, v^\varepsilon)$ is the velocity field of the fluid and p^ε is the pressure field. In addition to its practical and theoretical importance, this problem is also historically important since it was among the earliest major examples treated using boundary layer theory [35].

If we neglect the viscous term on the right-hand side, we obtain Euler's equation

$$\partial_t \mathbf{u} + (\mathbf{u} \cdot \nabla)\mathbf{u} + \nabla p = 0,$$

$$\nabla \cdot \mathbf{u} = 0.$$

In general, solutions to Euler's equation can only accommodate the boundary condition that there is no normal flow, i.e. $v = 0$. Our basic intuition is that when the viscosity ε is small ($\varepsilon \ll 1$), the fluid behaves mostly as an inviscid fluid that obeys Euler's equation except in a small region, next to the boundary, in which viscous effects are important.

Having guessed that a boundary layer exists at $y = 0$, let us now find out the structure of the solutions inside the boundary layer. As before, we introduce the stretched variable $\tilde{y}$ and perform a change of variable:

$$\tilde{y} = y/\delta \quad \text{and} \quad \tilde{x} = x.$$

The incompressibility condition $\nabla \cdot \mathbf{u}^\varepsilon = 0$ becomes (suppressing the superscript ε)

$$\partial_{\tilde{x}} u + \frac{1}{\delta} \partial_{\tilde{y}} v = 0. \qquad (2.1.6)$$

To balance the terms, the velocity in the y direction should also be of order δ. Therefore, we use

$$\tilde{v} = v/\delta, \quad \tilde{u} = u \quad \text{and} \quad \tilde{p} = p$$

as our new dependent variables. After this rescaling, (2.1.6) becomes

$$\partial_{\tilde{x}} u + \partial_{\tilde{y}} \tilde{v} = 0.$$

Equation (2.1.4) becomes

$$\partial_t \tilde{u} + \tilde{u} \partial_{\tilde{x}} \tilde{u} + \tilde{v} \partial_{\tilde{y}} \tilde{u} + \partial_{\tilde{x}} \tilde{p} = \varepsilon \partial_{\tilde{x}}^2 \tilde{u} + \frac{\varepsilon}{\delta^2} \partial_{\tilde{y}}^2 \tilde{u},$$

$$\partial_t \tilde{v} + \tilde{u} \partial_{\tilde{x}} \tilde{v} + \tilde{v} \partial_{\tilde{y}} \tilde{v} + \frac{1}{\delta^2} \partial_{\tilde{y}} \tilde{p} = \varepsilon \partial_{\tilde{x}}^2 \tilde{v} + \frac{\varepsilon}{\delta^2} \partial_{\tilde{y}}^2 \tilde{v}.$$

In the first equation, to balance the $\mathcal{O}(\varepsilon/\delta^2)$ term with the other terms, we should have $\delta \sim \varepsilon^{1/2}$. This means that the thickness of the boundary layer is $\mathcal{O}(\varepsilon^{1/2})$. We will take $\delta = \varepsilon^{1/2}$. From the second equation, we obtain

$$\partial_{\tilde{y}} p = 0.$$

This says that, to leading order, the pressure does not change across the boundary layer (i.e. in the y direction) and is given by the outer solution, the solution to the Euler's equation. This is the main simplification in the boundary layer.

Retaining only the leading-order terms, we obtain Prandtl's equation governing the leading-order behavior inside the boundary layer (omitting the tildes):

$$\partial_t u + u \partial_x u + v \partial_y u + \partial_x P = \partial_y^2 u,$$
$$\partial_x u + \partial_y v = 0;$$

here we write $P = P(x, t)$ for the pressure to indicate that it is assumed to be given.

Prandtl's boundary layer theory is derived under the assumption that the flow field can be decomposed into an outer region where viscous effects can be neglected and an inner region that lies close to the wall. This assumption is valid for laminar flows. However, it is well known that at high Reynolds number (in our setting, small ε), fluid flow generally becomes turbulent. In that case, the assumption stated above no longer holds and Prandtl's theory ceases to be valid. This does not mean that

Prandtl's theory is without its uses. In fact, one very useful application of Prandtl's equation is in the analysis of boundary layer separation, when the boundary layer detaches from the wall, convecting the vorticity generated at the boundary to the interior of the flow domain and eventually causing the generation of turbulence. This is often attributed to singularity formation in the solutions of Prandtl's equation [20]. In addition, an analogous theory has been developed for turbulent boundary layers [40].

2.1.3 Structure and dynamics of shocks

A viscous shock profile provides one of the simplest examples of an internal layer. We will use the Burgers equation as a simplified model of compressible flow:

$$\partial_t u^\varepsilon + \partial_x \left(\frac{(u^\varepsilon)^2}{2} \right) = \varepsilon \partial_x^2 u^\varepsilon. \qquad (2.1.7)$$

As in the last example, the viscosity ε is assumed to be very small. If we set $\varepsilon = 0$ we obtain the inviscid Burgers equation

$$\partial_t u + \partial_x \left(\frac{u^2}{2} \right) = 0. \qquad (2.1.8)$$

As before, this controls the behavior in the outer region.

First let us consider a simple case with initial data

$$u^\varepsilon(x, 0) = \begin{cases} 1, & x < 0, \\ -1, & x > 0. \end{cases}$$

The solution to (2.1.8) with the same initial data is a stationary shock,

$$u(x, t) = \begin{cases} 1, & x < 0, \\ -1, & x > 0. \end{cases}$$

Obviously, this is not a classical solution. The discontinuity at $x = 0$ is a manifestation of the fact that we have neglected viscous effects. Indeed, if we restore the viscous term then the discontinuity becomes a viscous shock inside which the solution changes rapidly but remains smooth.

To get an understanding of the detailed structure of the shock, let us introduce the stretched variable

$$\tilde{x} = x/\delta,$$

where δ is the width of the shock. In the new variable $\tilde{x}$, (2.1.7) becomes (omitting the superscript ε)

$$\partial_t u + \frac{1}{\delta}\partial_{\tilde{x}}\left(\frac{u^2}{2}\right) = \frac{\varepsilon}{\delta^2}\partial_{\tilde{x}}^2 u.$$

To balance terms, we must have $\delta \sim \varepsilon$. We see that the width of the shock is much smaller than the width of the boundary layer in incompressible flows (which is $\mathcal{O}(\sqrt{\varepsilon})$). This fact has important implications for the numerical solution of compressible and incompressible flows at high Reynolds number.

Let $\delta = \varepsilon$. The leading-order equation now becomes

$$\partial_{\tilde{x}}\left(\frac{u^2}{2}\right) = \partial_{\tilde{x}}^2 u$$

with the boundary condition $u \to \pm 1$ as $\tilde{x} \to \mp\infty$. The solution is given by

$$u = -\tanh(\tilde{x}/2) = -\tanh(x/2\varepsilon).$$

In the general case of a moving shock, assuming that the shock speed is s; then, introducing

$$\tilde{x} = (x - st)/\varepsilon,$$

the original PDE (2.1.7) becomes

$$\partial_t u - \frac{s}{\varepsilon}\partial_{\tilde{x}} u + \frac{1}{\varepsilon}\partial_{\tilde{x}}\left(\frac{u^2}{2}\right) = \frac{1}{\varepsilon}\partial_{\tilde{x}}^2 u.$$

The leading-order equation is then

$$-s\partial_{\tilde{x}} u + \partial_{\tilde{x}}\left(\frac{u^2}{2}\right) = \partial_{\tilde{x}}^2 u.$$

Integrating once, we get

$$-su + \frac{u^2}{2} = \partial_{\tilde{x}} u + C. \tag{2.1.9}$$

Let u_- and u_+ be the values of the inviscid solution at the left- and right-hand sides of the shock respectively. In order to match to the inviscid solution, we need

$$\lim_{\tilde{x}\to\pm\infty} u(\tilde{x}, t) = u_\pm.$$

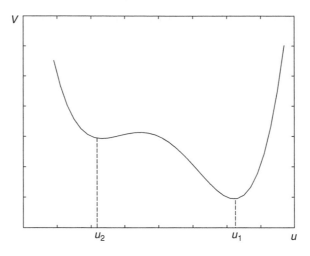

Figure 2.1. A double-well potential.

Therefore, we must have

$$-su_- + \frac{u_-^2}{2} = -su_+ + \frac{u_+^2}{2} = C, \qquad (2.1.10)$$

i.e.

$$s = \frac{u_- + u_+}{2}.$$

This equation determines the speed of the shock. Solving (2.1.9), we obtain the shock profile:

$$u(\tilde{x}, t) = \frac{u_- e^{\tilde{x}u_+/2} - u_+ e^{\tilde{x}u_-/2}}{e^{\tilde{x}u_+/2} - e^{\tilde{x}u_-/2}}.$$

In other words, to leading order, the behavior of u^ε is given by

$$\tilde{u}^\varepsilon = \frac{u_- e^{(x-st)u_+/2\varepsilon} - u_+ e^{(x-st)u_-/2\varepsilon}}{e^{(x-st)u_+/2\varepsilon} - e^{(x-st)u_-/2\varepsilon}}.$$

2.1.4 Transition layers in the Allen–Cahn equation

Next we consider the Allen–Cahn equation, say in two dimensions:

$$\partial_t u^\varepsilon = \varepsilon \Delta u^\varepsilon - \frac{1}{\varepsilon} V'(u^\varepsilon) \qquad (2.1.11)$$

where V, the free energy, is a double-well potential with two local minima, at u_1 and u_2 (see Figure 2.1). This equation describes the dynamics of a scalar order parameter during a first-order phase transition [7].

An inspection of (2.1.11) suggests that the value of u^ε should be very close to u_1 and u_2, so that $V'(u^\varepsilon) \sim 0$, in order for the last term to be balanced by the first two terms. Numerical results confirm that this is indeed the case. Basically, the solution u^ε is very close to u_1 and u_2 except in thin layers in which a transition between u_1 and u_2 takes place. To understand the dynamics of u^ε, we only have to understand the dynamics of these layers. In particular, our outer solution is simply $U = u_1, u_2$.

To understand the dynamics of these layers, we assume that they occupy a thin region around a curve whose configuration at time t is denoted by Γ_t. Let $\varphi(\mathbf{x}, t) = \text{dist}(\mathbf{x}, \Gamma_t)$, the distance between $\mathbf{x}$ and Γ_t. We make the ansatz that

$$u^\varepsilon \sim U_0\left(\frac{\varphi(\mathbf{x}, t)}{\varepsilon}\right) + \varepsilon U_1\left(\frac{\varphi(\mathbf{x}, t)}{\varepsilon}, \mathbf{x}, t\right) + \cdots . \qquad (2.1.12)$$

Substituting this ansatz into (2.1.11) and balancing the leading-order terms in powers of ε, we get

$$\partial_t \varphi U_0'(y) = U_0''(y) - V'(U_0(y)) \qquad (2.1.13)$$

with the boundary condition $\lim_{y \to \pm\infty} U(y) = u_1, u_2$. This is a nonlinear eigenvalue problem, with $\lambda = \partial_t \varphi$ as the eigenvalue. Let us write (2.1.13) as

$$\lambda U_0'(y) = U_0''(y) - V'(U_0(y)). \qquad (2.1.14)$$

If we neglect the term on the left-hand side of (2.1.14), we obtain Newton's equation for an inertial particle in the inverted potential $-V$ (see Figure 2.2). Assume that $u_2 < u_1$. Imagine that such a particle comes down in potential energy from the value $V(u_1)$ at time $y = -\infty$. If there were no friction its potential energy would pass through $V(u_2)$ at some point and then decrease further. This is prevented by the the presence of the friction term $\lambda U_0'$. The value of λ gives just the right amount of friction so that the particle's potential energy reaches $V(u_2)$ as $y \to +\infty$.

Multiplying (2.1.13) by U_0' on both sides and integrating, we obtain

$$\partial_t \varphi \int_{-\infty}^{\infty} (U_0'(y))^2 dy = V(u_1) - V(u_2). \qquad (2.1.15)$$

Since $\partial_t \varphi = \partial_t \varphi / |\nabla \varphi| = v_n$, the normal velocity of Γ_t, we obtain

$$v_n = \frac{V(u_1) - V(u_2)}{\int_{-\infty}^{\infty} (U_0'(y))^2 dy}. \qquad (2.1.16)$$

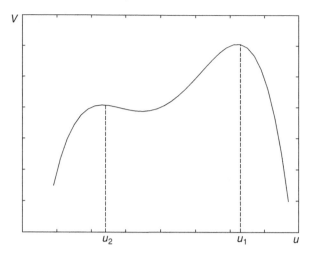

Figure 2.2. Inverted double-well potential that illustrates the solution of (2.1.14).

On this time scale (which is sometimes called the convective time scale, in contrast with the diffusive time scale discussed below), the interface moves to the region of lower free energy, with a constant normal velocity that depends on the difference in the free energy of the two phases u_1 and u_2.

Next we consider the case $V(u_1) = V(u_2)$. To be specific, we will take $V(u) = \frac{1}{4}(1 - u^2)^2$. In this case, on the convective time scale we have $v_n = 0$, i.e. the interface does not move. In order to see the interesting dynamics for the interface we should consider a longer time scale, namely, the diffusive time scale, i.e. we rescale the problem by replacing t by $t\varepsilon$ to obtain

$$\partial_t u^\varepsilon = \Delta u^\varepsilon + \frac{1}{\varepsilon^2} u^\varepsilon (1 - (u^\varepsilon)^2). \qquad (2.1.17)$$

We make the same ansatz as in (2.1.12). The leading-order and the next-order equations (in powers of ε) now become

$$U_0''(y) + (1 - U_0^2)U_0 = 0 \qquad (2.1.18)$$

$$\partial_t \varphi U_0' = U_1'' |\nabla \varphi|^2 + U_0' \Delta \varphi + U_1 (1 - 3U_0^2). \qquad (2.1.19)$$

The solution to (2.1.18) with boundary conditions $\lim_{\varphi \to \pm\infty} U(y) = \pm 1$ is given by

$$U(y) = \tanh \left(\frac{y}{\sqrt{2}} \right).$$

Since $|\nabla\varphi| = 1$, (2.1.19) gives

$$\partial_t\varphi U_0' = U_1'' + U_0'\Delta\varphi + U_1(1 - 3U_0^2).$$

Multiplying both sides by U_0' and then integrating over y, we obtain (note that $\int_{-\infty}^{\infty} U_0'\{U_1'' + U_1(1 - 3U_0^2)\}dy = 0$)

$$\partial_t\varphi = \Delta\varphi.$$

This can also be written as

$$\frac{\partial_t\varphi}{|\nabla\varphi|} = \nabla \cdot \frac{\nabla\varphi}{|\nabla\varphi|}.$$

In terms of Γ_t, this is the same as

$$v_{\mathrm{n}} = \kappa,$$

where κ is the mean curvature of Γ_t.

Examples of this type are discussed in [39].

2.2 The WKB method

To motivate the Wentzel–Kramers–Brillouin (WKB) method, let us look at the ODE

$$\varepsilon^2 \frac{d^2u^\varepsilon}{dx^2} - u^\varepsilon = 0, \tag{2.2.1}$$

with boundary conditions $u^\varepsilon(0) = 0$ and $u^\varepsilon(1) = 1$. It is straightforward to find the exact solution to this problem:

$$u^\varepsilon = Ae^{x/\varepsilon} + Be^{-x/\varepsilon},$$

where A and B are constants determined by the boundary conditions: $A = e^{1/\varepsilon}/(e^{2/\varepsilon} - 1)$ and $B = -e^{1/\varepsilon}/(e^{2/\varepsilon} - 1)$. One thing we learn from this explicit solution is that it contains factors with $1/\varepsilon$ in the exponent. Power series expansions in ε will not be able to capture these terms.

Now consider the problem

$$\varepsilon^2 \frac{d^2u^\varepsilon}{dx^2} - V(x)u^\varepsilon = 0.$$

We assume that $V(x) > 0$ (the case where V changes sign, known as the turning point problem, is more difficult to deal with; see [43]). This equation can no longer be solved explicitly. However, motivated by the form of the explicit solution to (2.2.1), we will look for approximate solutions of the form

$$u^\varepsilon(x) \sim \exp\left(\frac{1}{\varepsilon}(S_0(x) + \varepsilon S_1(x) + \mathcal{O}(\varepsilon^2))\right).$$

Since

$$\frac{d^2 u^\varepsilon}{dx^2} \sim \left(\frac{1}{\varepsilon}S_0'' + S_1''\right)\exp\left(\frac{1}{\varepsilon}S_0 + S_1\right)$$

$$+ \left(\frac{1}{\varepsilon}S_0' + S_1'\right)^2 \exp\left(\frac{1}{\varepsilon}S_0 + S_1\right) + \mathcal{O}(1),$$

we have

$$(\varepsilon S_0'' + \varepsilon^2 S_1'') + (S_0' + \varepsilon S_1')^2 - V(x) = \mathcal{O}(\varepsilon^2).$$

The solution to the leading-order equation

$$(S_0'(x))^2 - V(x) = 0$$

is given by

$$S_0 = \pm \int^x \sqrt{V(y)}\, dy.$$

The next-order equation,

$$S_0'' + 2S_0' S_1' = 0,$$

gives

$$S_1 = -\tfrac{1}{2}\log|S_0'| = -\tfrac{1}{4}\log V.$$

Therefore, to $\mathcal{O}(\varepsilon)$ the solution must be a linear combination of the two solutions obtained above:

$$u(x) \sim A_1 V(x)^{-1/4} \exp\left(\frac{1}{\varepsilon}\int^x \sqrt{V(y)}\, dy\right)$$

$$+ A_2 V(x)^{-1/4} \exp\left(-\frac{1}{\varepsilon}\int^x \sqrt{V(y)}\, dy\right), \qquad (2.2.2)$$

where A_1 and A_2 are constants depending on the boundary conditions and the starting point of the integral in the exponent of (2.2.2).

As another example consider the Schrödinger equation

$$i\varepsilon\partial_t u^\varepsilon = -\tfrac{1}{2}\varepsilon^2 \Delta u^\varepsilon + V(\mathbf{x})u^\varepsilon, \qquad (2.2.3)$$

with initial condition

$$u^\varepsilon(\mathbf{x}, 0) = A(\mathbf{x})e^{iS(\mathbf{x})/\varepsilon}.$$

We will look for approximate solutions in the form:

$$u^\varepsilon(\mathbf{x}, t) \sim A^\varepsilon(\mathbf{x}, t)e^{iS(\mathbf{x},t)/\varepsilon},$$

where $A^\varepsilon(\mathbf{x}, t) \sim A_0(\mathbf{x}, t) + \varepsilon A_1(\mathbf{x}, t) + \cdots$. Substituting into (2.2.3) we obtain the leading- and next-order equations:

$$\partial_t S + \tfrac{1}{2}|\nabla S|^2 + V(\mathbf{x}) = 0,$$
$$\partial_t A_0 + \nabla S \cdot \nabla A_0 + \tfrac{1}{2}A_0 \Delta S = 0.$$

The first equation above is the eikonal equation, which happens to be the usual Hamilton–Jacobi equation in classical mechanics:

$$\partial_t S + H(\mathbf{x}, \nabla S) = 0,$$

where $H(\mathbf{x}, \mathbf{p}) = |\mathbf{p}|^2/2 + V(\mathbf{x})$. The second equation is a transport equation for the leading-order behavior of the amplitude function A_0. This establishes a connection between quantum and classical mechanics via the semiclassical limit $\varepsilon \to 0$.

This asymptotic analysis is valid as long as the solution to the eikonal equation remains smooth. However, singularities may form in the solutions of the eikonal equation, as is the case of caustics, and the asymptotic analysis has to be modified to deal with such a situation [29].

2.3 Averaging methods

In many problems the variables that describe the state of the system can be split into a set of slow variables and a set of fast variables, where we are more interested in the behavior of the slow variables. In situations like this, one can often obtain an effective model for the slow variables by properly averaging the original model over the statistics of the fast variables. We discuss such averaging methods in this section.

2.3.1 Oscillatory problems

Consider the following example:

$$\frac{dx^\varepsilon}{dt} = -\frac{1}{\varepsilon}y^\varepsilon + x^\varepsilon, \qquad (2.3.1)$$

$$\frac{dy^\varepsilon}{dt} = \frac{1}{\varepsilon}x^\varepsilon + y^\varepsilon. \qquad (2.3.2)$$

From now on, for clarity we will sometimes suppress the superscript ε. If we replace (x, y) by the polar coordinates (r, φ), we get

$$\frac{d\varphi}{dt} = \frac{1}{\varepsilon}, \qquad (2.3.3)$$

$$\frac{dr}{dt} = r. \qquad (2.3.4)$$

When $\varepsilon \ll 1$, the dynamics of the angle variable φ is much faster than that of the variable r.

In general, we may use action–angle variables and consider systems of the form

$$\frac{d\varphi}{dt} = \frac{1}{\varepsilon}\omega(I) + f(\varphi, I), \qquad (2.3.5)$$

$$\frac{dI}{dt} = g(\varphi, I). \qquad (2.3.6)$$

Here f and g are assumed to be periodic with respect to the angle variable φ, with period $[-\pi, \pi]$. We also assume that ω does not vanish. In this example φ is the *fast variable* and I is the *slow variable*. Our objective is to derive simplified effective equations for the slow variable. For this purpose, we will make use of the multiscale expansion, which is a very powerful technique that will be discussed in more detail in the next section.

We will look for approximate solutions of (2.3.5) and (2.3.6) in the form

$$\varphi(t) \sim \varphi_0\left(\frac{t}{\varepsilon}\right) + \varepsilon\varphi_1\left(\frac{t}{\varepsilon}, t\right), \qquad (2.3.7)$$

$$I(t) \sim I_0(t) + \varepsilon I_1\left(\frac{t}{\varepsilon}, t\right). \qquad (2.3.8)$$

This says that φ and I are, in principle, functions with two scales but, to leading order, I is a function of only the slow scale. Let $\tau = t/\varepsilon$. To guarantee that the terms φ_0 and I_0 in the above ansatz are indeed the leading-order terms, we will assume that $\varphi_1(\tau, t)$ and $I_1(\tau, t)$ grow sublinearly in τ:

$$\frac{\varphi_1(\tau, t)}{\tau} \to 0, \quad \frac{I_1(\tau, t)}{\tau} \to 0$$

as $\tau \to \infty$. Substituting the ansatz into (2.3.5) and (2.3.6), we obtain the following leading-order equations:

$$\frac{1}{\varepsilon}\left(\frac{d\varphi_0}{d\tau}\left(\frac{t}{\varepsilon}\right) - \omega(I_0(t))\right) + \mathcal{O}(1) = 0,$$

$$\frac{dI_0}{dt}(t) + \frac{\partial I_1}{\partial \tau}\left(\frac{t}{\varepsilon}, t\right) - g\left(\varphi_0\left(\frac{t}{\varepsilon}\right), I_0(t)\right) + \mathcal{O}(\varepsilon) = 0.$$

Our objective is to find $\varphi_0, I_0, \varphi_1, I_1, \ldots$ such that the equations above are satisfied to as high an order in ε as possible. The leading-order equations are

$$\frac{d\varphi_0}{d\tau}\left(\frac{t}{\varepsilon}\right) = \omega(I_0(t)),$$

$$\frac{dI_0}{dt}(t) + \frac{\partial I_1}{\partial \tau}\left(\frac{t}{\varepsilon}, t\right) = g\left(\varphi_0\left(\frac{t}{\varepsilon}\right), I_0(t)\right).$$

To guarantee that these equations hold, we require that the relations

$$\frac{d\varphi_0}{d\tau}(\tau) = \omega(I_0(t)),$$

$$\frac{dI_0}{dt}(t) + \frac{\partial I_1}{\partial \tau}(\tau, t) = g(\varphi_0(\tau), I_0(t))$$

hold for all values of t and τ even though the original problem only requires them to hold when $\tau = t/\varepsilon$.

The solution to the first equation is given by

$$\varphi_0(\tau) = \omega(I_0)\tau + \hat{\varphi}_0. \tag{2.3.9}$$

Averaging the second equation in τ over a large interval $[0, \tau]$ and letting $\tau \to \infty$, we obtain, using the sublinear growth condition,

$$\frac{dI_0}{dt} = \lim_{\tau \to \infty} \frac{1}{\tau} \int_0^\tau g(\varphi_0(\tau'), I_0) \, d\tau' = \frac{1}{2\pi} \int_{[-\pi, \pi]} g(\theta, I_0) \, d\theta.$$

This gives rise to the *averaged system*

$$\frac{dJ}{dt} = G(J), \quad G(J) = \frac{1}{2\pi} \int_{[-\pi, \pi]} g(\theta, J) \, d\theta. \tag{2.3.10}$$

It is a simple matter to show that, for any fixed time T, there exists a constant C such that

$$|I(t) - J(t)| \leq C\varepsilon$$

for $0 \leq t \leq T$ [2]. Therefore, in this case, the dynamics of the slow variable is described accurately by the averaged system.

The story becomes much more complicated when there is more than one degree of freedom. Consider

$$\frac{d\varphi}{dt} = \frac{1}{\varepsilon}\omega(I) + f(\varphi, I), \tag{2.3.11}$$

$$\frac{dI}{dt} = g(\varphi, I), \tag{2.3.12}$$

where $\varphi = (\varphi_1, \ldots, \varphi_d), I = (I_1, \ldots, I_d), f = (f_1, \ldots, f_d)$ and $g = (g_1, \ldots, g_d)$; f and g are periodic functions of φ with period $[-\pi, \pi]^d$. Formally, the "averaging theorem" states that, for small values of ε, the behavior of the action variable is approximately described by the averaged system

$$\frac{dJ}{dt} = G(J), \tag{2.3.13}$$

where

$$G(J) = \frac{1}{(2\pi)^d} \int_{[-\pi,\pi]^d} g(\theta, J)\, d\theta_1 \cdots d\theta_d, \quad \theta = (\theta_1, \ldots, \theta_d). \tag{2.3.14}$$

We have put "averaging theorem" in quotation marks since this description is not always accurate, owing to the possible occurrence of resonance.

To see how (2.3.13) may arise, we again proceed to look for multiscale solutions of the type (2.3.7), (2.3.8); this leads us to

$$\frac{dI_0}{dt} = \lim_{\tau \to \infty} \frac{1}{\tau} \int_0^\tau g(\varphi_0(\tau'), I_0)\, d\tau', \quad \varphi_0(\tau) = \omega(I_0)\tau + \hat{\varphi}_0.$$

Now, to evaluate the right-hand side, we must distinguish two different cases:

(1) The components of $\omega(I_0)$ are rationally dependent, i.e. there exists a vector $\mathbf{q}$ with rational components such that $\mathbf{q} \cdot \omega(I_0) = 0$. This is the resonant case.
(2) The components of $\omega(I_0)$ are rationally independent. In this case, the flow defined by (2.3.9) is ergodic on $[-\pi, \pi]^d$ (periodically extended).

In the second case we obtain

$$\lim_{\tau \to \infty} \frac{1}{\tau} \int_0^\tau g(\varphi_0(\tau'), I_0) \, d\tau' = \frac{1}{(2\pi)^d} \int_{[-\pi,\pi]^d} g(\varphi, I_0) \, d\varphi_1 \cdots d\varphi_d.$$

This gives us the averaged equation (2.3.13). This equation neglects the resonant case. The rationale for doing so is that there are many more rationally independent vectors than rationally dependent ones. Therefore, generically, the system will pass through resonances instead of being locked in them. However, it is easy to see that if the system is indeed locked in a resonance then the "averaging theorem" may fail.

As an example, consider the case when $d = 2$, $\omega(I) = (I_1, I_2)$, $f_1 = f_2 = 0$, $g_1 = 0$ and $g_2 = \cos(\varphi_1 - \varphi_2)$ (see [2]). In this case $G_1 = G_2 = 0$. Let the initial data be given by

$$I_1(0) = I_2(0) = 1, \quad \varphi_1(0) = \varphi_2(0) = 0.$$

For the original system, we have

$$I_1(t) = 1, \quad I_2(t) = 1 + t.$$

However, the solution of the averaged system with the same initial data is given by

$$J_1(t) = J_2(t) = 1.$$

The deviation of I from J comes from the fact that for the original system (2.3.11), $(\varphi_1(t), \varphi_2(t))$ stays on the diagonal in the $\varphi_1\varphi_2$-plane whereas the averaged system assumes that $(\varphi_1(t), \varphi_2(t))$ is ergodic on the periodic cell $[-\pi, \pi]^2$ in the φ-plane.

It should be noted that the violation of the "averaging theorem" does not mean that there is a flaw in the perturbation analysis. Indeed, it was clear from the perturbation analysis that there would be a problem when resonance occurs.

A thorough discussion of this topic can be found in the classic book of Arnold [2].

2.3.2 Stochastic ordinary differential equations

First, a remark about notation. We will not follow standard practice in stochastic analysis, using X_t to denote the value of the stochastic process X at time t and expressing stochastic ODEs in terms of stochastic

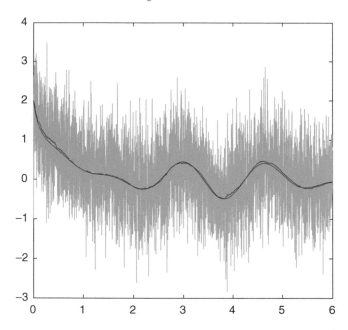

Figure 2.3. The solution of (2.3.15) and (2.3.16) with $x = 2, y = 1$ and $\varepsilon = 10^{-3}$; $X(\cdot)$ and $Y(\cdot)$ are shown in blue and green respectively. Also shown, in red is the solution of the limiting equation (2.3.18) (courtesy of Eric Vanden-Eijnden).

differentials. Instead, we will use the standard notation for functions and derivatives.

Consider

$$\frac{dX}{dt} = f(X, Y, t), \qquad\qquad X(0) = x, \qquad (2.3.15)$$

$$\frac{dY}{dt} = -\frac{1}{\varepsilon}(Y - X) + \frac{1}{\sqrt{\varepsilon}}\dot{W}, \qquad Y(0) = y. \qquad (2.3.16)$$

Here X is the slow variable, Y is the fast variable and $\dot{W}$ is the standard Gaussian white noise. An example of the solution to this problem is shown in Figure 2.3 for the case when $f(x, y, t) = -y^3 + \sin(\pi t) + \cos(\sqrt{2}\pi t)$.

The equation for Y at fixed $X(t) = x$ defines an Ornstein–Uhlenbeck process whose equilibrium density is

$$\rho^*(y; x) = \frac{e^{-(y-x)^2}}{\sqrt{\pi}}.$$

Intuitively, it is quite clear that when ε is small, the X variable in (2.3.16) does not change much before Y reaches local equilibrium (which is set by the local value of X) in a time scale of $\mathcal{O}(\varepsilon)$. This means that the dynamics of the fast variable Y is enslaved to that of the slow variable X. Its behavior can be found approximately by holding the slow variable X fixed (at its instantaneous value) and computing the (conditional) equilibrium distribution for Y. In addition, the effective dynamics for the slow variable is obtained by averaging the right-hand side of (2.3.15) with respect to the conditional equilibrium distribution of Y:

$$\frac{dX}{dt} = F(X,t), \quad F(x,t) = \int_{\mathbb{R}} f(x,y,t)\rho^*(y;x)\,dy. \qquad (2.3.17)$$

For the example given above, it is easy to compute the limiting equation explicitly:

$$\frac{dX}{dt} = -X^3 - \tfrac{3}{2}X + \sin(\pi t) + \cos(\sqrt{2}\pi t), \quad X(0) = x. \qquad (2.3.18)$$

Here the new term, $-\tfrac{3}{2}X$, is due to the noise in (2.3.15). The solution of the limiting equation is shown in Figure 2.3.

The limiting system (2.3.17) can be derived using perturbation analysis on the backward Kolmogorov equation. Let $h = h(x,y)$ be a smooth function, and let

$$u(x,y,t) = \mathbb{E}h(X(t),Y(t))$$

where x,y are the initial values of X and Y: $X(0) = x, Y(0) = y$. The function u satisfies the backward Kolmogorov equation [16]

$$\partial_t u = L_0 u + \frac{1}{\varepsilon}L_1 u, \qquad (2.3.19)$$

with initial condition $u(x,y,0) = h(x,y)$, where

$$L_0 = f(x,y,t)\partial_x, \qquad L_1 = -(y-x)\partial_y + \tfrac{1}{2}\partial_y^2.$$

We look for solutions of (2.3.19) in the form

$$u \sim u_0 + \varepsilon u_1 + \varepsilon^2 u_2 + \cdots.$$

Substituting this ansatz into (2.3.19) and collecting terms of equal orders in ε, we get

$$L_1 u_0 = 0, \qquad (2.3.20)$$

$$L_1 u_1 = \frac{\partial u_0}{\partial t} - L_0 u_0, \qquad (2.3.21)$$

$$L_1 u_2 = \cdots. \qquad (2.3.22)$$

The first equation implies that u_0 is a function of x and t only. The second equation requires a solvability condition:

$$\int \rho(y)(\partial_t u_0 - L_0 u_0)(y)\, dy = 0,$$

where ρ is any solution of the adjoint equation

$$L_1^* \rho = \partial_y((y - x)\rho) + \tfrac{1}{2}\partial_y^2 \rho = 0.$$

It is easy to see that solutions to this equation have the form

$$\rho(y; x) = Ce^{-(y-x)^2},$$

where C is a normalization constant. Therefore, $u_0(x, t)$ must satisfy

$$\partial_t u_0 - \bar{L}_0 u_0 = 0,$$

where

$$\bar{L} = F(x, t)\partial_x$$

and F is given by (2.3.17). This is the backward equation for the dynamics described by (2.3.17).

This analysis can be generalized to the case when the equation for the slow variable also contains noise:

$$\frac{dX}{dt} = f(X, Y, t) + \sigma(X, Y, t)\dot{W}_1, \quad X(0) = x, \qquad (2.3.23)$$

$$\frac{dY}{dt} = -\frac{1}{\varepsilon}(Y - X) + \frac{1}{\sqrt{\varepsilon}}\dot{W}_2, \qquad Y(0) = y. \qquad (2.3.24)$$

Here $\dot{W}_1$ and $\dot{W}_2$ are independent white noises. In this case, the limiting dynamics is given by

$$\frac{dX}{dt} = F(X, t) + \Sigma(X, t)\dot{W}, \qquad (2.3.25)$$

where F is given in (2.3.17) and Σ satisfies

$$\Sigma(x, t)^2 = \int_{\mathbb{R}} \sigma(x, y, t)^2 \rho^*(y; x)\, dy.$$

Results of this type have been obtained in many papers, including [22, 23, 24, 36].

It may happen that the effective dynamics of the slow variable is the trivial equation $dx/dt = 0$. In this case we should consider the dynamics at a longer time scale, say $t = \mathcal{O}(1/\varepsilon)$. To do this we introduce the new variable $\tau = t\varepsilon$ and proceed as before, but now using τ.

Let us look at some simple examples (see [27]). Consider

$$\frac{dX}{dt} = Y + \dot{W}_1,$$

$$\frac{dY}{dt} = -\frac{1}{\varepsilon}Y + \frac{1}{\sqrt{\varepsilon}}\dot{W}_2,$$

where again $\dot{W}_1$ and $\dot{W}_2$ are independent white noises. Notice that, in this case, the fast dynamics is independent of the slow variable and is actually a standard Ornstein–Uhlenbeck process. Therefore, the invariant measure for the fast variable is $N(0, \frac{1}{2})$, a normal distribution with mean 0 and variance $\frac{1}{2}$. Obviously, $F = 0$ in this case. Therefore the effective dynamics for the slow variable X is given by

$$\bar{X}(t) = W_1(t).$$

One can indeed show that $\mathbb{E}|X(t) - \bar{X}(t)|^2 \leq C\varepsilon$.

Next, consider

$$\frac{dX}{dt} = Y\dot{W}_1,$$

$$\frac{dX}{dt} = -\frac{1}{\varepsilon}Y + \frac{1}{\sqrt{\varepsilon}}\dot{W}_2.$$

The fast dynamics is still the Ornstein–Uhlenbeck process. According to (2.3.25), the effective dynamics for the slow variable is given by

$$\bar{X}(t) = \frac{1}{\sqrt{2}}W_1(t).$$

However, it is easy to see that

$$\mathbb{E}|X(t) - \bar{X}(t)|^2 = \mathbb{E}\left|\int_0^t (Y(s) - \tfrac{1}{2})\, dW_1(s)\right|^2$$

$$= \mathbb{E}\int_0^t |Y(s) - \tfrac{1}{2}|^2 ds = \mathcal{O}(1).$$

In this situation we see that the fast dynamics is ergodic but the strong version of the "averaging theorem" fails. This is not a contradiction to the perturbation analysis since the latter only suggests that weak convergence holds, i.e. only the averaged quantities converge.

In general, if we have a system of the type

$$\frac{dX}{dt} = a(X, Y) + \sigma_1(X)\dot{W}_1, \qquad (2.3.26)$$

$$\frac{dY}{dt} = \frac{1}{\varepsilon}b(X, Y) + \frac{1}{\sqrt{\varepsilon}}\sigma_2(X, Y)\dot{W}_2,$$

i.e. where σ_1 does not depend on Y, then one can show that the averaging theorem holds in a strong sense, namely strong convergence can be established. See [22, 23, 27]. Otherwise, one can only expect weak converegence.

2.3.3 Stochastic simulation algorithms

We start with a simple example [12, 13] (see also a related example in [6]). Consider a system that contains multiple copies of a molecule in four different conformations, denoted by S_1, S_2, S_3, S_4. The conformation changes are described by

$$S_1 \underset{a_2}{\overset{a_1}{\rightleftharpoons}} S_2, \qquad S_2 \underset{a_4}{\overset{a_3}{\rightleftharpoons}} S_3, \qquad S_3 \underset{a_6}{\overset{a_5}{\rightleftharpoons}} S_4, \qquad (2.3.27)$$

with transition rates $a_1, \ldots, a_6$ which depend on the number of molecules in each conformation. The state of the system is described by a vector $\mathbf{x} = (x_1, x_2, x_3, x_4)$ where x_i denotes the number of molecules in the conformation S_i. To be specific, let us make the following assumptions:

$$\begin{aligned}
a_1 &= 10^5 x_1, & \mathbf{v}_1 &= (-1, +1, \ 0, \ 0), \\
a_2 &= 10^5 x_2, & \mathbf{v}_2 &= (+1, -1, \ 0, \ 0), \\
a_3 &= x_2, & \mathbf{v}_3 &= (\ 0, -1, +1, \ 0), \\
a_4 &= x_3, & \mathbf{v}_4 &= (\ 0, +1, -1, \ 0), \\
a_5 &= 10^5 x_3, & \mathbf{v}_5 &= (\ 0, \ 0, -1, +1), \\
a_6 &= 10^5 x_4, & \mathbf{v}_6 &= (\ 0, \ 0, +1, -1),
\end{aligned} \qquad (2.3.28)$$

The vectors $\{\mathbf{v}_j\}$ are the stoichiometric vectors (also called the state change vectors): after the jth reaction, the state of the system changes from $\mathbf{x}$ to $\mathbf{x} + \mathbf{v}_j$. For example, if the first reaction occurs then one molecule in the S_1 conformation changes to the S_2 conformation. Hence x_1 is decreased by 1 x_2 is increased by 1 and x_3 and x_4 remain the same.

Each reaction R_j is specified by two objects, the transition (reaction) rate function and the stoichiometric vector:

$$R_j = (a_j(\mathbf{x}), \mathbf{v}_j).$$

If we start the system with roughly the same number of molecules in each conformation and observe how the system evolves, we will see frequent conformation changes between S_1 and S_2 and between S_3 and S_4, with occasional, much more rare conformation changes between S_2 and S_3.

In many cases, we are not interested in the details of the fast processes but only the dynamics of the slow processes. We should be able to simplify the problem and obtain an effective description of only the slow processes. To see this, let us partition the reactions into two sets: slow and fast. The latter $R^{\mathrm{f}} = \{(a^{\mathrm{f}}, \mathbf{v}^{\mathrm{f}})\}$ are

$$
\begin{aligned}
a_1^{\mathrm{f}} &= 10^5 x_1, & \mathbf{v}_1^{\mathrm{f}} &= (-1, +1, \ 0, \ 0), \\
a_2^{\mathrm{f}} &= 10^5 x_2, & \mathbf{v}_2^{\mathrm{f}} &= (+1, -1, \ 0, \ 0), \\
a_5^{\mathrm{f}} &= 10^5 x_3, & \mathbf{v}_5^{\mathrm{f}} &= (\ 0, \ 0, -1, +1), \\
a_6^{\mathrm{f}} &= 10^5 x_4, & \mathbf{v}_6^{\mathrm{f}} &= (\ 0, \ 0, +1, -1).
\end{aligned}
\tag{2.3.29}
$$

The slow reactions $R^{\mathrm{s}} = \{(a^{\mathrm{s}}, \mathbf{v}^{\mathrm{s}})\}$ are

$$
\begin{aligned}
a_3^{\mathrm{s}} &= x_2, & \mathbf{v}_3^{\mathrm{s}} &= (0, -1, +1, 0), \\
a_4^{\mathrm{s}} &= x_3, & \mathbf{v}_4^{\mathrm{s}} &= (0, +1, -1, 0).
\end{aligned}
\tag{2.3.30}
$$

Observe that, during the fast reactions, the variables

$$z_1 = x_1 + x_2, \quad z_2 = x_3 + x_4$$

do not change. They are the slow variables. Intuitively, one can think of the whole process as follows. On the fast times cale, z_1 and z_2 are almost fixed and there is fast exchange between x_1 and x_2, and between x_3 and x_4. This process reaches equilibrium during the fast time scale and the equilibrium is given by a Bernoulli distribution:

$$\mu_{z_1, z_2}(x_1, x_2, x_3, x_4) = \frac{z_1! \ z_2!}{x_1! \ x_2! \ x_3! \ x_4!} \left(\frac{1}{2}\right)^{z_1} \left(\frac{1}{2}\right)^{z_2} \delta_{x_1 + x_2 = z_1} \delta_{x_3 + x_4 = z_2}.$$

On the slow time scale, the values of z_1 and z_2 do change significantly. The effective dynamics on this time scale reduces to a dynamics on the space of $z = (z_1, z_2)$, with effective reaction channels given by

$$\bar{a}_3^s = \langle x_2 \rangle = \frac{x_1 + x_2}{2} = \frac{z_1}{2}, \quad \bar{\mathbf{v}}_3^s = (-1, +1),$$

$$\bar{a}_4^s = \langle x_3 \rangle = \frac{x_3 + x_4}{2} = \frac{z_2}{2}, \quad \bar{\mathbf{v}}_4^s = (+1, -1). \tag{2.3.31}$$

These expressions follow from the general theory that we will describe now.

For the general case, let us assume that we have a total of N species of molecules, denoted by $S_1, \ldots, S_N$. The number of molecules of species S_k is denoted by x_k. The state vector is then given by $\mathbf{x} = (x_1, \ldots, x_N)$. We will denote by $\mathcal{X}$ the state space in which $\mathbf{x}$ exists. Assume that there are M reaction channels, each described by its reaction rate and stoichiometric vector:

$$\mathbf{R}_j = (a_j, \mathbf{v}_j), \qquad \mathbf{R} = \{\mathbf{R}_1, \ldots, \mathbf{R}_M\}.$$

Given the state $\mathbf{x}$, the probabilities of occurrence of the various reactions in an infinitesimal time interval dt are independent; the probability for the reaction $\mathbf{R}_j$ to happen during this time interval is given by $a_j(\mathbf{x}) \, dt$. After reaction $\mathbf{R}_j$, the state of the system changes to $\mathbf{x} + \mathbf{v}_j$. In the chemistry and biology literature, this is often called the stochastic simulation algorithm (SSA) or Gillespie algorithm, named after an algorithm that realizes this process exactly (see [17]).

Let $\mathbf{X}(t)$ be the state variable at time t and denote by $\mathbb{E}_\mathbf{x}$ the expectation conditional on $\mathbf{X}(0) = \mathbf{x}$. Consider the observable $u(\mathbf{x}, t) = \mathbb{E}_\mathbf{x} f(\mathbf{X}(t))$; this observable satisfies the following backward Kolmogorov equation [16]:

$$\frac{\partial u(\mathbf{x}, t)}{\partial t} = \sum_j a_j(\mathbf{x}) \left(u(\mathbf{x} + v_j, t) - u(\mathbf{x}, t) \right) = (Lu)(\mathbf{x}, t). \tag{2.3.32}$$

The operator L is the infinitesimal generator of the Markov process associated with the chemical kinetic system we are considering.

Now we turn to chemical kinetic systems with two disparate time scales. Assume that the rate functions have the following form:

$$a(\mathbf{x}) = \left(a^s(\mathbf{x}), \frac{1}{\varepsilon} a^f(\mathbf{x}) \right),$$

where $\varepsilon \ll 1$ is the ratio of the fast and slow timescales of the system. The corresponding reactions and the associated stoichiometric vectors can be grouped accordingly:

$$\mathbf{R}^{\mathrm{s}} = \{(a^{\mathrm{s}}, \mathbf{v}^{\mathrm{s}})\}, \qquad \mathbf{R}^{\mathrm{f}} = \left\{\left(\frac{1}{\varepsilon}a^{\mathrm{f}}, \mathbf{v}^{\mathrm{f}}\right)\right\}. \qquad (2.3.33)$$

We call the set R^{s} the slow reactions and the set R^{f} the fast reactions.

We now show how to derive the effective dynamics on the $\mathcal{O}(1)$ time scale, following the perturbation theory developed in [24, 36]. For this purpose it is helpful to introduce an auxiliary process, called the *virtual fast process* [6]. This auxiliary process retains the fast reactions only; all slow reactions are turned off, assuming, for simplicity, that the sets of fast and slow reactions do not change over time. Intuitively, each realization of the SSA consists of a sequence of realizations of the virtual fast process, punctuated by occasional firing of the slow reactions. Owing to the time scale separation, with high probability the virtual fast process has enough time to relax to equilibrium before another slow reaction takes place. Therefore it is crucial to characterize the ergodic components, and their associated quasi-equilibrium distributions, for the virtual fast process. In particular, it is of interest to consider parametrizations of these ergodic components. A set of variables that parametrizes these ergodic components can be regarded as the slow variables.

Let v be a function of the state variable $\mathbf{x}$. We say that $v(\mathbf{x})$ is a slow variable if it does not change during the fast reactions, i.e. if, for any $\mathbf{x}$ and any stoichiometric vector $\mathbf{v}^{\mathrm{f}}_j$ associated with a fast reaction, we have

$$v(\mathbf{x} + \mathbf{v}^{\mathrm{f}}_j) = v(\mathbf{x}). \qquad (2.3.34)$$

This is equivalent to saying that the slow variables are conserved quantities for the fast reactions. It is easy to see that $v(\mathbf{x}) = \mathbf{b} \cdot \mathbf{x}$ is a slow variable if and only if

$$\mathbf{b} \cdot \mathbf{v}^{\mathrm{f}}_j = 0 \qquad (2.3.35)$$

for all $\mathbf{v}^{\mathrm{f}}_j$. The set of such vectors forms a linear subspace in R^N. Let $\{\mathbf{b}_1, \ldots, \mathbf{b}_J\}$ be a set of basis vectors of this subspace, and let

$$z_j = \mathbf{b}_j \cdot \mathbf{x}, \quad j = 1, \ldots, J. \qquad (2.3.36)$$

Let

$$\mathbf{z} = (z_1, \ldots, z_J)$$

and denote by $\mathcal{Z}$ the space over which $\mathbf{z}$ is defined. The stochiometric vectors associated with these slow variables are naturally defined as

$$\bar{\mathbf{v}}_i^{\mathrm{s}} = (\mathbf{b}_1 \cdot \mathbf{v}_i^{\mathrm{s}}, \dots, \mathbf{b}_J \cdot \mathbf{v}_i^{\mathrm{s}}), \qquad i = 1, \dots, N_{\mathrm{s}}.$$

We now derive the effective dynamics for the slow time scale using singular perturbation theory. We will assume, for the moment, that $\mathbf{z}$ is a complete set of slow variables, i.e., for each fixed value of the slow variable $\mathbf{z}$, the virtual fast process admits a unique equilibrium distribution $\mu_{\mathbf{z}}(\mathbf{x})$ on the state space $\mathcal{X}$.

Define the projection operator $\mathbf{P}$ by

$$(\mathbf{P}v)(\mathbf{z}) = \sum_{\mathbf{x} \in \mathcal{X}} \mu_{\mathbf{z}}(\mathbf{x}) v(\mathbf{x}).$$

According to this definition, for any $v : \mathcal{X} \to \mathbb{R}$, $\mathbf{P}v$ depends only on the slow variable $\mathbf{z}$, i.e. $\mathbf{P}v : \mathcal{Z} \to \mathbb{R}$.

The backward Kolmogorov equation for the multiscale chemical kinetic system reads

$$\frac{\partial u}{\partial t} = L_0 u + \frac{1}{\varepsilon} L_1 u, \tag{2.3.37}$$

where L_0 and L_1 are the infinitesimal generators associated with the slow and fast reactions, respectively: for any $f : \mathcal{X} \to \mathbb{R}$,

$$(L_0 f)(\mathbf{x}) = \sum_{j=1}^{M_{\mathrm{s}}} a_j^{\mathrm{s}}(\mathbf{x})\big(f(\mathbf{x} + \mathbf{v}_j^{\mathrm{s}}) - f(\mathbf{x})\big),$$

$$(L_1 f)(\mathbf{x}) = \sum_{j=1}^{M_{\mathrm{f}}} a_j^{\mathrm{f}}(\mathbf{x})\big(f(\mathbf{x} + \mathbf{v}_j^{\mathrm{f}}) - f(\mathbf{x})\big),$$

where M_{s} is the number of slow reactions in R^{s} and M_{f} is the number of fast reactions in R^{f}. We look for a solution of (2.3.37) in the form

$$u \sim u_0 + \varepsilon u_1 + \varepsilon^2 u_2 + \cdots. \tag{2.3.38}$$

Inserting this into (2.3.37) and equating the coefficients of equal powers of ε, we arrive as in subsection 2.3.2 at the following hierarchy of equations:

$$L_1 u_0 = 0, \tag{2.3.39}$$

$$L_1 u_1 = \frac{\partial u_0}{\partial t} - L_0 u_0, \tag{2.3.40}$$

$$L_1 u_2 = \cdots. \tag{2.3.41}$$

The first equation implies that u_0 belongs to the null space of L_1, which, by the ergodicity assumption of the fast process, is equivalent to saying that

$$u_0(\mathbf{x}, t) = \bar{U}(\mathbf{b} \cdot \mathbf{x}, t) \equiv \bar{U}(\mathbf{z}, t), \qquad (2.3.42)$$

for some $\bar{U}$ yet to be determined. Inserting (2.3.42) into (2.3.40) gives (using the explicit expression for L_0)

$$
\begin{aligned}
L_1 u_1(\mathbf{x}, t) &= \frac{\partial \bar{U}(\mathbf{b} \cdot \mathbf{x}, t)}{\partial t} - \sum_{i=1}^{M_s} a_i^s(\mathbf{x}) \left(\bar{U}(\mathbf{b} \cdot (\mathbf{x} + \mathbf{v}_i^s), t) - \bar{U}(\mathbf{b} \cdot \mathbf{x}, t) \right) \\
&= \frac{\partial \bar{U}(\mathbf{z}, t)}{\partial t} - \sum_{i=1}^{M_s} a_i^s(\mathbf{x}) \left(\bar{U}(\mathbf{z} + \bar{\mathbf{v}}_i^s, t) - \bar{U}(\mathbf{z}, t) \right).
\end{aligned}
\qquad (2.3.43)
$$

This equation requires a solvability condition, namely that the right-hand side be orthogonal to the null space of the adjoint operator of L_1. By the ergodicity assumption of the fast process, this amounts to requiring that

$$\frac{\partial \bar{U}}{\partial t}(\mathbf{z}, t) = \sum_{i=1}^{M_s} \bar{a}_i^s(\mathbf{z}) \left(\bar{U}(\mathbf{z} + \bar{\mathbf{v}}_i^s, t) - \bar{U}(\mathbf{z}, t) \right) \qquad (2.3.44)$$

where

$$\bar{a}_i^s(\mathbf{z}) = (\mathbf{P} a_i^s)(\mathbf{z}) = \sum_{\mathbf{x} \in \mathcal{X}} a_i^s(\mathbf{x}) \mu_z(\mathbf{x}). \qquad (2.3.45)$$

Equation (2.3.44) describes the effective rates on the slow time scale. The effective reaction kinetics on this time scale are given in terms of the slow variable by

$$\bar{\mathbf{R}} = (\bar{a}^s(\mathbf{z}), \bar{\mathbf{v}}^s). \qquad (2.3.46)$$

It is important to note that the limiting dynamics described in (2.3.44) can be reformulated on the original state space $\mathcal{X}$. Indeed, it is easy to check that (2.3.44) is equivalent to

$$\frac{\partial u_0}{\partial t}(\mathbf{x}, t) = \sum_{i=1}^{M_s} \tilde{a}_i^s(\mathbf{x}) \left(u_0(\mathbf{x} + \mathbf{v}_i^s, t) - u_0(\mathbf{x}, t) \right) \qquad (2.3.47)$$

where $\tilde{a}_i^s(\mathbf{x}) = \bar{a}_i^s(\mathbf{b} \cdot \mathbf{x})$, in the sense that if $u_0(\mathbf{x}, t = 0) = f(\mathbf{b} \cdot \mathbf{x})$ then

$$u_0(\mathbf{x}, t) = \bar{U}(\mathbf{b} \cdot \mathbf{x}, t), \qquad (2.3.48)$$

where $\bar{U}(\mathbf{z}, t)$ solves (2.3.44) with the initial condition $\bar{U}(\mathbf{z}, 0) = f(\mathbf{z})$.

This fact is useful for developing numerical algorithms for this kind of problem [6, 12, 13]. In fact, (2.3.47) provides an effective model that is more general than that considered above. Its validity can be understood intuitively as follows [18]. To find an effective dynamics, in which the fast reactions are averaged out, we ask what happens for the original SSA during a time scale that is very short compared with that of the slow reactions but long compared with that of the fast reactions. Over this time scale, the original SSA essentially functions like the virtual fast process. Therefore, to obtain the effective slow reaction rates, we need to average over the states of the virtual fast process. However, over this time scale the virtual fast process has had sufficient time to relax to its quasi-equilibrium. Hence the effective slow reaction rates can be obtained by averaging the original slow reaction rates over the quasi-equilibrium distribution of the virtual fast process. Note that all these can be formulated over the original state space using the original variables. This is the main content of (2.3.47).

In the large-volume limit, when there are a great many molecules in each species, this stochastic discrete model can be approximated by a deterministic continuous model, in the spirit of the law of large numbers [16]:

$$\frac{dx_j}{dt} = \sum_k \nu_{jk} a_k(\mathbf{x}), \quad j = 1, \ldots, N. \tag{2.3.49}$$

The disparity in the rates results in the stiffness of this system of ODEs. Formally, we can write (2.3.49) as

$$\frac{d\mathbf{x}}{dt} = \frac{1}{\varepsilon} \sum_{j=1}^{M_\mathrm{f}} a_j^\mathrm{f}(\mathbf{x})\mathbf{v}_j^\mathrm{f} + \sum_{j=1}^{M_\mathrm{s}} a_j^\mathrm{s}(\mathbf{x})\mathbf{v}_j^\mathrm{s}. \tag{2.3.50}$$

The averaging theorem suggests that, in the limit $\varepsilon \to 0$, the solution of (2.3.50) converges to the solution of

$$\frac{d\bar{\mathbf{x}}}{dt} = \sum_{j=1}^{M_\mathrm{s}} \bar{a}_j^\mathrm{s}(\mathbf{b} \cdot \bar{\mathbf{x}})\mathbf{v}_j^\mathrm{s}. \tag{2.3.51}$$

Here

$$\bar{a}_j^\mathrm{s}(\mathbf{z}) = \int_{\mathbb{R}^n} a_j^\mathrm{s}(\mathbf{x}) \, d\mu_\mathbf{z}(\mathbf{x}), \tag{2.3.52}$$

where $d\mu_{\mathbf{z}}(\mathbf{x})$ is the equilibrium of the virtual fast process

$$\frac{d\mathbf{x}^f}{dt} = \frac{1}{\varepsilon} \sum_{j=1}^{M_f} a_j^f(\mathbf{x}_t^f) \mathbf{v}_j^f. \tag{2.3.53}$$

For instance, in the large-volume limit, the ODEs corresponding to the simple example treated at the begining of this subsection are

$$\frac{dx_1}{dt} = 10^5(-x_1 + x_2), \tag{2.3.54}$$

$$\frac{dx_2}{dt} = 10^5(x_1 - x_2) - x_2 + x_3, \tag{2.3.55}$$

$$\frac{dx_3}{dt} = 10^5(-x_3 + x_4) - x_3 + x_2, \tag{2.3.56}$$

$$\frac{dx_4}{dt} = 10^5(x_3 - x_4). \tag{2.3.57}$$

In this case, the slow variables are $z_1 = x_1 + x_2$ and $z_2 = x_3 + x_4$, and it is easy to see that the fast process drives the variables toward the fixed point

$$x_1^f = x_2^f = \tfrac{1}{2}z_1, \qquad x_3^f = x_4^f = \tfrac{1}{2}z_2, \tag{2.3.58}$$

meaning that $d\mu_{\mathbf{z}}(\mathbf{x})$ is atomic:

$$d\mu_{\mathbf{z}}(\mathbf{x}) = \delta(x_1 - \tfrac{1}{2}z_1)\delta(x_2 - \tfrac{1}{2}z_1)\delta(x_3 - \tfrac{1}{2}z_2)\delta(x_4 - \tfrac{1}{2}z_2)\, dx_1\, dx_2\, dx_3\, dx_4. \tag{2.3.59}$$

Hence from (2.3.52), the limiting dynamics is

$$\frac{dz_1}{dt} = -x_2 + x_3 = \tfrac{1}{2}(-z_1 + z_2), \tag{2.3.60}$$

$$\frac{dz_2}{dt} = -x_3 + x_2 = \tfrac{1}{2}(z_1 - z_2), \tag{2.3.61}$$

which from (2.3.51), can be written in terms of the original variables as

$$\frac{dx_1}{dt} = 0, \tag{2.3.62}$$

$$\frac{dx_2}{dt} = -x_2 + x_3, \tag{2.3.63}$$

$$\frac{dx_3}{dt} = x_2 - x_3, \tag{2.3.64}$$

$$\frac{dx_4}{dt} = 0. \tag{2.3.65}$$

A helpful modern reference for the averaging methods is [37].

2.4 Multiscale expansions

We now turn to the techniques of multiscale expansion. This is perhaps the most systematic approach for analyzing problems with disparate scales. We begin with simple examples of ODEs, and then turn to PDEs and focus on the homogenization technique.

2.4.1 Removing secular terms

Consider

$$\frac{d^2 y}{dt^2} + y + \varepsilon y = 0, \tag{2.4.1}$$

where $\varepsilon \ll 1$, $y(0) = 1$ and $y'(0) = 0$. The solution to this problem is given by

$$y(t) = \cos(\sqrt{1 + \varepsilon} t) = \cos\left(\left(1 + \frac{\varepsilon}{2} + \cdots\right) t\right)$$

If we omit the term εy, the solution would be $y(t) = \cos t$. This is accurate on the $\mathcal{O}(1)$ time scale but gives $\mathcal{O}(1)$ error on the $\mathcal{O}(1/\varepsilon)$ time scale.

To analyze the behavior of the solutions over the $\mathcal{O}(1/\varepsilon)$ time scale, we use a two-time-scale expansion. Let $\tau = \varepsilon t$ be the slow time variable. Assume that

$$y(t) \sim y_0(t, \varepsilon t) + \varepsilon y_1(t, \varepsilon t) + \varepsilon^2 y_2(t, \varepsilon t). \tag{2.4.2}$$

Substituting into (2.4.1) and letting $\tau = \varepsilon t$, we get

$$\frac{\partial^2 y_0}{\partial t^2} + y_0 + \varepsilon \left(\frac{\partial^2 y_1}{\partial t^2} + y_1 + 2\frac{\partial^2 y_0}{\partial t \partial \tau} + y_0 \right) + \cdots \sim 0. \tag{2.4.3}$$

Our objective is to choose functions $y_0, y_1, \ldots$ such that the above equation is valid to as high an order as possible, when $\tau = \varepsilon t$. Indeed, one idea of multiscale analysis is to require that this equation holds for all values of τ, not just when $\tau = \varepsilon t$. The leading-order equation given by the $\mathcal{O}(1)$ terms in (2.4.3) is

$$\frac{\partial^2 y_0}{\partial t^2}(t, \tau) + y_0(t, \tau) = 0.$$

The solution to this equation can be expressed as $y_0(t, \tau) = C_1(\tau) \cos t + C_2(\tau) \sin t$. The next-order equation is

$$\frac{\partial^2 y_1}{\partial t^2} + y_1 = -2\frac{\partial^2 y_0}{\partial t \partial \tau} - y_0.$$

Substituting the expression for y_0 we get

$$\frac{\partial^2 y_1}{\partial t^2} + y_1 = (2C_1'(\tau) - C_2(\tau)) \sin t - (2C_2'(\tau) + C_1(\tau)) \cos t. \quad (2.4.4)$$

If the coefficients of $\cos t$ and $\sin t$ do not vanish then we will have terms that behave like $t \cos t$ in y_1, with a linear dependence on the fast time t.

However, when making an ansatz involving two time scales, we always assume that the lower-order terms grow sublinearly as a function of the fast time scale. This is the minimum condition necessary to guarantee that the leading-order terms remain dominant over the time interval of interest. The effective equation for the leading-order terms is often found from this condition. Terms that grow linearly or faster are called secular terms. One example was already seen in the previous section, when we derived the averaged equation (2.3.13) as a consequence of the condition that $I_1(\tau, t)$ grows sublinearly as a function of τ.

Coming back to (2.4.4), we require that the coefficients of the terms on the right-hand side must vanish:

$$2C_1'(\tau) - C_2(\tau) = 0,$$
$$2C_2'(\tau) + C_1(\tau) = 0.$$

These lead to

$$C_1(\tau) = A_1 \cos \frac{\tau}{2} + A_2 \sin \frac{\tau}{2},$$
$$C_2(\tau) = -A_1 \sin \frac{\tau}{2} + A_2 \cos \frac{\tau}{2}.$$

Combined with the initial condition, one gets

$$\tilde{y}(t) = \cos \frac{\varepsilon t}{2} \cos t - \sin \frac{\varepsilon t}{2} \sin t = \cos \left(1 + \frac{\varepsilon}{2}\right) t.$$

We see that $\tilde{y}$ approximates the original solution accurately over an $\mathcal{O}(1/\varepsilon)$ time interval. If we are interested in longer time intervals, we have to extend the asymptotics to higher-order terms. This procedure is called the Poincaré–Lighthill–Kuo (PLK) method [33].

The problem treated here is a simple linear equation whose exact solution can be written down easily. However, it is obvious that the methodology presented is general and so can be used for general nonlinear problems.

2.4.2 Homogenization of elliptic equations

Let us consider the problem

$$-\nabla \cdot (a^\varepsilon(\mathbf{x})\nabla u^\varepsilon(\mathbf{x})) = f(\mathbf{x}), \quad \mathbf{x} \in \Omega, \qquad (2.4.5)$$

with the boundary condition $u^\varepsilon|_{\partial\Omega} = 0$. Here $a^\varepsilon(\mathbf{x}) = a(\mathbf{x}, \mathbf{x}/\varepsilon)$ and $a(\mathbf{x}, \mathbf{y})$ is assumed to be periodic with respect to $\mathbf{y}$, with period $\Gamma = [0,1]^d$. We will also assume that a^ε satisfies the ellipticity condition uniformly in ε, namely that there exists a constant α which is independent of ε such that

$$a^\varepsilon(\mathbf{x}) \geq \alpha I,$$

where I is the identity tensor.

Equation (2.4.5) is a standard model for many different problems in physics and engineering, including the mechanical properties of composite materials, heat conduction in heterogeneous media, flow in porous media etc. It has also been used as a standard model for illustrating some of the most important analytical techniques in multiscale analysis such as homogenization theory. The standard reference for the material discussed in this section is [4]. See also [37].

We begin with a formal multiscale analysis, using the ansatz

$$u^\varepsilon(\mathbf{x}) \sim u_0\left(\mathbf{x}, \frac{\mathbf{x}}{\varepsilon}\right) + \varepsilon u_1\left(\mathbf{x}, \frac{\mathbf{x}}{\varepsilon}\right) + \varepsilon^2 u_2\left(\mathbf{x}, \frac{\mathbf{x}}{\varepsilon}\right) + \cdots.$$

Here the functions $u_j(\mathbf{x}, \mathbf{y})$, $j = 0, 1, 2, \ldots$ are assumed to be periodic with period Γ. Using

$$\frac{\partial}{\partial \mathbf{x}} u_0\left(\mathbf{x}, \frac{\mathbf{x}}{\varepsilon}\right) \sim \nabla_{\mathbf{x}} u_0\left(\mathbf{x}, \frac{\mathbf{x}}{\varepsilon}\right) + \frac{1}{\varepsilon}\nabla_{\mathbf{y}} u_0\left(\mathbf{x}, \frac{\mathbf{x}}{\varepsilon}\right) + \cdots,$$

the original PDE (2.4.5) becomes

$$\frac{1}{\varepsilon^2}\nabla_{\mathbf{y}} \cdot \left(a\left(\mathbf{x}, \frac{\mathbf{x}}{\varepsilon}\right)\nabla_{\mathbf{y}} u_0\left(\mathbf{x}, \frac{\mathbf{x}}{\varepsilon}\right)\right) + \frac{1}{\varepsilon}\left[\nabla_{\mathbf{y}} \cdot \left(a\left(\mathbf{x}, \frac{\mathbf{x}}{\varepsilon}\right)\nabla_{\mathbf{x}} u_0\left(\mathbf{x}, \frac{\mathbf{x}}{\varepsilon}\right)\right)\right.$$
$$\left. + \nabla_{\mathbf{x}} \cdot \left(a\nabla_{\mathbf{y}} u_0\left(\mathbf{x}, \frac{\mathbf{x}}{\varepsilon}\right)\right) + \nabla_{\mathbf{y}} \cdot \left(a\nabla_{\mathbf{y}} u_1\left(\mathbf{x}, \frac{\mathbf{x}}{\varepsilon}\right)\right)\right]$$
$$+ \nabla_{\mathbf{y}} \cdot \left(a\left(\mathbf{x}, \frac{\mathbf{x}}{\varepsilon}\right)\nabla_{\mathbf{x}} u_1\left(\mathbf{x}, \frac{\mathbf{x}}{\varepsilon}\right)\right) + \nabla_{\mathbf{x}} \cdot \left(a\left(\mathbf{x}, \frac{\mathbf{x}}{\varepsilon}\right)\nabla_{\mathbf{y}} u_1\left(\mathbf{x}, \frac{\mathbf{x}}{\varepsilon}\right)\right)$$
$$+ \nabla_{\mathbf{y}} \cdot \left(a\left(\mathbf{x}, \frac{\mathbf{x}}{\varepsilon}\right)\nabla_{\mathbf{y}} u_2\left(\mathbf{x}, \frac{\mathbf{x}}{\varepsilon}\right)\right)$$
$$+ \nabla_{\mathbf{x}} \cdot \left(a\left(\mathbf{x}, \frac{\mathbf{x}}{\varepsilon}\right)\nabla_{\mathbf{x}} u_0\left(\mathbf{x}, \frac{\mathbf{x}}{\varepsilon}\right)\right) + \cdots = -f(\mathbf{x}).$$

To leading order, we have

$$-\nabla_{\mathbf{y}} \cdot \left(a\left(\mathbf{x}, \frac{\mathbf{x}}{\varepsilon}\right) \nabla_{\mathbf{y}} u_0\left(\mathbf{x}, \frac{\mathbf{x}}{\varepsilon}\right) \right) = 0.$$

This holds for $\mathbf{y} = \mathbf{x}/\varepsilon$. We will require that it holds for *all* values of $\mathbf{x}$ and $\mathbf{y}$:

$$-\nabla_{\mathbf{y}} \cdot \left(a(\mathbf{x}, \mathbf{y}) \nabla_{\mathbf{y}} u_0(\mathbf{x}, \mathbf{y}) \right) = 0.$$

From this, we conclude that

$$\nabla_{\mathbf{y}} u_0 = 0, \quad u_0 = u_0(\mathbf{x}).$$

The next-order equation is given by

$$-\nabla_{\mathbf{y}} \cdot (a(\mathbf{x}, \mathbf{y}) \nabla_{\mathbf{y}} u_1(\mathbf{x}, \mathbf{y})) = \nabla_{\mathbf{y}} \cdot \left(a(\mathbf{x}, \mathbf{y}) \nabla_{\mathbf{x}} u_0(\mathbf{x}) \right).$$

This should be viewed as a PDE for u_1; the term $\nabla_{\mathbf{x}} u_0$ enters only as a constant coefficient. To find solutions to this equation, it helps to introduce the *cell problem*:

$$-\nabla_{\mathbf{y}} \cdot (a(\mathbf{x}, \mathbf{y}) \nabla_{\mathbf{y}} \xi_j) = \nabla_{\mathbf{y}} \cdot (a(\mathbf{x}, \mathbf{y}) \cdot \mathbf{e}_j), \qquad (2.4.6)$$

where $\mathbf{e}_j$ is the standard jth basis vector for the Euclidean space $\mathbb{R}^d$. Equation (2.4.6) is solved with a periodic boundary condition. Clearly, as the solution to this cell problem, ξ_j is a function of $\mathbf{y}$. But it also depends on $\mathbf{x}$ through the dependence of a on $\mathbf{x}$. We can then express u_1 as

$$u_1(\mathbf{x}, \mathbf{y}) = \xi(\mathbf{x}, \mathbf{y}) \cdot \nabla_{\mathbf{x}} u_0(\mathbf{x})$$

where $\xi = (\xi_1, \dots, \xi_d)$.

We now turn to the $\mathcal{O}(1)$ equation:

$$\begin{aligned} \nabla_{\mathbf{y}} \cdot (a(\mathbf{x}, \mathbf{y}) \nabla_{\mathbf{y}} u_2(\mathbf{x}, \mathbf{y})) &+ \nabla_{\mathbf{y}} \cdot (a(\mathbf{x}, \mathbf{y}) \nabla_{\mathbf{x}} u_1(\mathbf{x}, \mathbf{y})) \\ + \nabla_{\mathbf{x}} \cdot (a(\mathbf{x}, \mathbf{y}) \nabla_{\mathbf{y}} u_1(\mathbf{x}, \mathbf{y})) &+ \nabla_{\mathbf{x}} \cdot (a(\mathbf{x}, \mathbf{y}) \nabla_{\mathbf{x}} u_0(\mathbf{x})) = -f(\mathbf{x}). \quad (2.4.7) \end{aligned}$$

This is viewed as an equation for u_2. If we are to find solutions that are periodic in $\mathbf{y}$ then u_0 and u_1 must satisfy some solvability condition. To identify this condition, let us define the averaging operator (with respect to the fast variable $\mathbf{y}$):

$$\langle f \rangle = \int_{\Gamma = [0,1]^d} f(\mathbf{y}) \, d\mathbf{y} = \bar{f}.$$

Applying this operator to both sides of (2.4.7), we obtain

$$\nabla_{\mathbf{x}} \cdot (\langle a(\mathbf{x},\mathbf{y})\nabla_{\mathbf{y}} u_1(\mathbf{x},\mathbf{y})\rangle) + \nabla_{\mathbf{x}} \cdot (\langle a(\mathbf{x},\mathbf{y})\rangle \nabla_{\mathbf{x}} u_0(\mathbf{x})) = -f(\mathbf{x}).$$

Since

$$a(\mathbf{x},\mathbf{y})\nabla_{\mathbf{y}} u_1(\mathbf{x},\mathbf{y}) = a(\mathbf{x},\mathbf{y})\nabla_{\mathbf{y}}(\xi(\mathbf{x},\mathbf{y})\nabla_{\mathbf{x}} u_0(\mathbf{x})),$$

we have

$$\nabla_{\mathbf{x}} \cdot (\langle a\nabla_{\mathbf{y}}\xi + a\mathbf{I}\rangle \nabla_{\mathbf{x}} u_0) = -f.$$

Defining

$$A(\mathbf{x}) = \langle a(\mathbf{x},\mathbf{y})(\mathbf{I} + \nabla_{\mathbf{y}}\xi(\mathbf{x},\mathbf{y}))\rangle,$$

we arrive at a homogenized equation for u_0:

$$-\nabla \cdot (A(\mathbf{x})\nabla u_0(\mathbf{x})) = f(\mathbf{x}).$$

Let us look at a simple one-dimensional example. In this case, the cell problem becomes

$$-\partial_y (a(x,y)\partial_y\xi) = \partial_y a.$$

Integrating once, we get

$$a(x,y)(1 + \partial_y\xi) = C_0.$$

Dividing by a and applying the averaging operator on both sides, we get

$$1 = \left\langle \frac{C_0}{a} \right\rangle = C_0 \left\langle \frac{1}{a} \right\rangle = C_0 \int_0^1 \frac{1}{a(x,y)} \, dy.$$

Hence, we obtain

$$A(x) = C_0 = \left\langle \frac{1}{a} \right\rangle^{-1}.$$

This is the well-known result that the coefficient for the homogenized problem is the harmonic average, rather than the arithmetic average, of the original coefficient.

The same result can also be derived by applying multiscale test functions in a weak formulation of the original PDE. This has the advantage that it can be easily made rigorous. More importantly, it illustrates that we should think about multiscale problems through test functions. The basic idea is to use multiscale test functions to probe the behavior of the multiscale solutions at different scales. The basis of this procedure is the following result of Nguetseng [34]. Early applications of this idea can be found in [1, 11].

Lemma *Let $\{u^\varepsilon\}$ be a sequence of functions, uniformly bounded in $L^2(\Omega)$, with $\|u^\varepsilon\|_{L^2} \leq C$ for some constant C. Then there exists a subsequence $\{u^{\varepsilon_j}\}$, and a function $u \in L^2(\Omega \times \Gamma), \Gamma = [0,1]^d$, such that, as $\varepsilon_j \to 0$,*

$$\int u^{\varepsilon_j}(\mathbf{x})\varphi\left(\mathbf{x},\frac{\mathbf{x}}{\varepsilon}\right) d\mathbf{x} \to \iint u(\mathbf{x},\mathbf{y})\varphi(\mathbf{x},\mathbf{y})d\mathbf{x}d\mathbf{y}$$

for any function $\varphi \in C(\Omega \times \Gamma)$ which is periodic in the second variable with period Γ.

To apply this result, we rewrite the PDE $-\nabla \cdot (a^\varepsilon(\mathbf{x})\nabla u^\varepsilon(\mathbf{x})) = f(\mathbf{x})$ as

$$\mathbf{w}^\varepsilon = \nabla u^\varepsilon, \quad -\nabla \cdot (a^\varepsilon(\mathbf{x})\mathbf{w}^\varepsilon(\mathbf{x})) = f(\mathbf{x}),$$

or, in a weak form,

$$\int_\Omega a^\varepsilon(\mathbf{x})\mathbf{w}^\varepsilon(\mathbf{x})\nabla\varphi^\varepsilon(\mathbf{x})\,d\mathbf{x} = \int_\Omega f(\mathbf{x})\varphi^\varepsilon(\mathbf{x})\,d\mathbf{x}.$$

Applying a test function of the form $\varphi^\varepsilon(\mathbf{x}) = \varphi_0(\mathbf{x}) + \varepsilon\varphi_1(\mathbf{x},\mathbf{x}/\varepsilon)$, we obtain

$$\int a\left(\mathbf{x},\frac{\mathbf{x}}{\varepsilon}\right)\mathbf{w}^\varepsilon(\mathbf{x})\left(\nabla_{\mathbf{x}}\varphi_0(\mathbf{x}) + \nabla_{\mathbf{y}}\varphi_1\left(\mathbf{x},\frac{\mathbf{x}}{\varepsilon}\right) + \varepsilon\nabla_{\mathbf{x}}\varphi_1\left(\mathbf{x},\frac{\mathbf{x}}{\varepsilon}\right)\right) d\mathbf{x}$$

$$= \int f(\mathbf{x})\left(\varphi_0(\mathbf{x}) + \varepsilon\varphi_1\left(\mathbf{x},\frac{\mathbf{x}}{\varepsilon}\right)\right) d\mathbf{x}.$$

Using the previous lemma, we can conclude that there exists a function $\mathbf{w} \in L^2(\Omega \times \Gamma)$ such that

$$\iint a(\mathbf{x},\mathbf{y})\mathbf{w}(\mathbf{x},\mathbf{y})\left(\nabla_{\mathbf{x}}\varphi_0(\mathbf{x}) + \nabla_{\mathbf{y}}\varphi_1(\mathbf{x},\mathbf{y})\right) d\mathbf{x}\,d\mathbf{y} = \int f(\mathbf{x})\varphi_0(\mathbf{x})\,d\mathbf{x}.$$

Setting $\varphi_1 = 0$ in this equation, we get

$$-\nabla_{\mathbf{x}} \cdot \langle a\mathbf{w}\rangle = f. \tag{2.4.8}$$

If instead we let $\varphi_0 = 0$, we obtain

$$-\nabla_{\mathbf{y}} \cdot (a\mathbf{w}) = 0. \tag{2.4.9}$$

Since $\mathbf{w}^\varepsilon = \nabla u^\varepsilon$, we have

$$\int \mathbf{w}^\varepsilon(\mathbf{x})\varphi^\varepsilon(\mathbf{x})\,d\mathbf{x} = -\int u^\varepsilon(\mathbf{x})\nabla \cdot \varphi^\varepsilon(\mathbf{x})\,d\mathbf{x}.$$

Applying the same kind of test function as above, we conclude that there exists a function $u \in L^2(\Omega \times \Gamma)$ such that

$$\iint \mathbf{w}(\mathbf{x}, \mathbf{y})\varphi_0(\mathbf{x})\,d\mathbf{x}\,d\mathbf{y} = -\iint u(\mathbf{x}, \mathbf{y})\big(\nabla_\mathbf{x}\varphi_0(\mathbf{x}) + \nabla_\mathbf{y}\varphi_1(\mathbf{x}, \mathbf{y})\big)\,d\mathbf{x}\,d\mathbf{y}.$$
$$(2.4.10)$$

Letting $\varphi_0 = 0$, we get

$$\iint u(\mathbf{x}, \mathbf{y})\nabla_\mathbf{y}\varphi_1(\mathbf{x}, \mathbf{y})\,d\mathbf{x}\,d\mathbf{y} = 0$$

for any φ_1. Hence we have

$$\nabla_\mathbf{y} u = 0,$$

i.e. $u(\mathbf{x}, \mathbf{y}) = u_0(\mathbf{x})$. Letting $\varphi_1 = 0$ in (2.4.10), we get

$$\langle \mathbf{w} \rangle = -\nabla_\mathbf{x}\langle u \rangle = -\nabla_\mathbf{x} u_0.$$

Applying the same argument to the equation $\nabla \times \mathbf{w}^\varepsilon = 0$, we conclude that there exists a $u_1 \in H^1(\Omega \times \Gamma)$ such that

$$\mathbf{w}(\mathbf{x}, \mathbf{y}) = \nabla_\mathbf{y} u_1(\mathbf{x}, \mathbf{y}) + \nabla_\mathbf{x} u_0(\mathbf{x}).$$

Substituting into (2.4.8) and (2.4.9), we obtain

$$\begin{cases} -\nabla_\mathbf{y} \cdot (a(\nabla_\mathbf{y} u_1 + \nabla_\mathbf{x} u_0)) = 0, \\ -\nabla_\mathbf{x} \cdot \big\langle a(\nabla_\mathbf{y} u_1 + \nabla_\mathbf{x} u_0)\big\rangle = f \end{cases}$$

It is easy to check that this is equivalent to the homogenized equation obtained earlier.

2.4.3 Homogenization of the Hamilton–Jacobi equations

Consider the following problem:

$$\partial_t u^\varepsilon + H\left(\frac{\mathbf{x}}{\varepsilon}, \nabla u^\varepsilon\right) = 0, \qquad (2.4.11)$$

where $H(\mathbf{y}, \mathbf{p})$ is assumed to be periodic in $\mathbf{y}$ with period $\Gamma = [0, 1]^d$. Such a problem arises, for example, as a model for interface propagation in heterogeneous media, where u^ε is the level set function of the interface, $H(\mathbf{x}/\varepsilon, \nabla\varphi) = c(\mathbf{x}/\varepsilon)|\nabla\varphi|$, and $c(\cdot)$ is the given normal velocity.

The homogenization of (2.4.11) was considered in the pioneering, but unpublished, paper by Lions, Papanicolaou and Varadhan [26] (see also [25]). Here we will proceed to get a quick understanding of the main features of this problem using formal asymptotics.

We begin with the ansatz

$$u^\varepsilon(\mathbf{x}, t) \sim u_0(\mathbf{x}, t) + \varepsilon u_1(\mathbf{x}, \mathbf{x}/\varepsilon, t) + \cdots,$$

where $u_1(\mathbf{x}, \mathbf{y}, t)$ is periodic in $\mathbf{y}$. The leading-order asymptotic equation is given by

$$\partial_t u_0 + H(\mathbf{y}, \nabla_\mathbf{x} u_0 + \nabla_\mathbf{y} u_1) = 0.$$

To understand this problem, we rewrite it as

$$H(\mathbf{y}, \nabla_\mathbf{x} u_0 + \nabla_\mathbf{y} u_1) = -\partial_t u_0. \qquad (2.4.12)$$

This should be viewed as the cell problem for u_1, with $\mathbf{p} = \nabla_x u_0$ as the parameter. Note that the right-hand side is independent of $\mathbf{y}$. It should be viewed as a function of the parameter $\nabla_\mathbf{x} u_0$. In fact, the following result asserts that it is a well-defined function of $\nabla_\mathbf{x} u_0$ [26].

Lemma *Assume that $H(\mathbf{x}, \mathbf{p}) \to \infty$ as $|\mathbf{p}| \to \infty$ uniformly in $\mathbf{x}$. Then, for any $\mathbf{p} \in \mathbb{R}^d$, there exists a unique constant λ such that the cell problem*

$$H(\mathbf{y}, \mathbf{p} + \nabla_\mathbf{y} v) = \lambda \qquad (2.4.13)$$

has a periodic viscosity solution v.

We refer to [25] for the definition of viscosity solutions.

The unique value of λ mentioned in the lemma will be denoted as $\bar{H}(\mathbf{p})$, where $\bar{H}$ is the effective Hamiltonian.

Going back to (2.4.12), we obtain the homogenized equation by letting $\mathbf{p} = \nabla_\mathbf{x} u_0$:

$$\bar{H}(\nabla_\mathbf{x} u_0) = \lambda = -\partial_t u_0$$

or

$$\partial_t u_0 + \bar{H}(\nabla_\mathbf{x} u_0) = 0.$$

As a first example, let us consider the one-dimensional "Hamiltonian" $H(x, p) = |p|^2/2 - V(x)$. (In classical mechanics, this would be called the Lagrangian. Here we are following standard practice in the theory of Hamilton–Jacobi equations and calling it the Hamiltonian [25]). The potential V is assumed to be periodic with period 1 and is normalized so that $\min_y V(y) = 0$. The effective Hamiltonian $\bar{H}(p)$ can be obtained by solving the cell problem (2.4.13):

$$\tfrac{1}{2}|p + \partial_y v|^2 - V(y) = \lambda. \tag{2.4.14}$$

Obviously, $\lambda \geq 0$. Let us first consider the case when (2.4.14) has a smooth solution. Smooth solutions have to satisfy

$$p + \partial_y v(y) = \pm\sqrt{2}\sqrt{V(y) + \lambda}. \tag{2.4.15}$$

Thus

$$p = \pm\sqrt{2} \int_0^1 \sqrt{V(y) + \lambda}\, dy.$$

Therefore we can expect to have smooth solutions only when

$$|p| \geq \sqrt{2} \int_0^1 \sqrt{V(y)}\, dy.$$

When

$$|p| \leq \sqrt{2} \int_0^1 \sqrt{V(y)}\, dy,$$

the solution to the cell problem cannot be smooth, i.e. it has to jump between the $+$ and $-$ branches of (2.4.15). Being a viscosity solution means that $u = \partial_y v$ has to satisfy the entropy condition, which in this case says that the limit of u from the left cannot be smaller than the limit of u from the right:

$$u(y_0+) \leq u(y_0-),$$

where y_0 is any location where a jump occurs. In order to retain periodicity, the two branches have to intersect. This implies that

$$\lambda = 0.$$

Indeed, it can be easily verified that one can always choose appropriate jump locations to find u such that

$$p + u(y) = \pm\sqrt{2}\sqrt{V(y)}$$

and the entropy condition is satisfied.

In summary, we have

$$\bar{H}(p) = \begin{cases} 0, & |p| \leq \sqrt{2} \int_0^1 \sqrt{V}\, dy, \\ \lambda, & |p| = \sqrt{2} \int_0^1 \sqrt{V + \lambda}\, dy, \quad \lambda \geq 0. \end{cases}$$

Note that, unlike H, the effective Hamiltonian $\bar{H}(p)$ is no longer quadratic in p.

As the second example, let us consider the case when $H(\mathbf{y}, \mathbf{p}) = c(\mathbf{y})|\mathbf{p}|$. The homogenized Hamiltonian is then $\bar{H}(\mathbf{p}) = c(\mathbf{y})|\nabla_{\mathbf{y}} v + \mathbf{p}|$, where v is a periodic viscosity solution to the cell problem. In this case, the original Hamiltonian is a homogeneous function of degree 1 in $\mathbf{p}$. It is easy to see that the homogenized Hamiltonian inherits this property, i.e. $\bar{H}(\lambda \mathbf{p}) = \lambda \bar{H}(\mathbf{p})$ for $\lambda > 0$. In fact, let $v_\lambda = \lambda v$; then $\lambda \bar{H}(\mathbf{p}) = c(\mathbf{y})|\nabla_{\mathbf{y}} v_\lambda + \lambda \mathbf{p}|$ and v_λ is clearly periodic.

In one dimension, if we assume that c never vanishes we have

$$\partial_y v + p = \frac{\bar{H}(p)}{c(y)},$$

since the right-hand side does not change sign. Integrating both sides, we arrive at

$$\bar{H}(p) = \left(\int_0^1 \frac{1}{c(y)}\, dy \right)^{-1} |p|.$$

If $c(y)$ vanishes at some point then clearly $\bar{H}(p) = 0$.

2.4.4 Flows in porous media

Next, we consider the homogenization of flows in a porous medium, assuming that the microstructure of the medium is periodic. Let Ω_ε be the flow domain, and Ω_ε^c be the solid matrix. Then $\Omega_\varepsilon = \varepsilon \cup_{\mathbf{z} \in \mathcal{Z}} (\Omega + \mathbf{z})$, the union of the periodic extension of Ω rescaled by a factor of ε. In principle, one should start with the Navier–Stokes equations in the flow domain with the no-slip boundary condition at $\partial \Omega_\varepsilon$ and then study the asymptotics as $\varepsilon \to 0$. However, since the velocity is small, we can omit the dynamic terms in the Navier–Stokes equation and use Stokes' equation:

$$-\nu \Delta \mathbf{u}^\varepsilon + \nabla p^\varepsilon = 0 \quad \text{in } \Omega_\varepsilon,$$
$$\nabla \cdot \mathbf{u}^\varepsilon = 0 \quad \text{in } \Omega_\varepsilon,$$
$$\mathbf{u}^\varepsilon = 0 \quad \text{on } \partial \Omega_\varepsilon.$$

To study the asymptotic behavior, we make the ansatz

$$\mathbf{u}^{\varepsilon}(\mathbf{x}) \sim \varepsilon^2 \mathbf{u}_0(\mathbf{x}, \mathbf{x}/\varepsilon) + \varepsilon^3 \mathbf{u}_1(\mathbf{x}, \mathbf{x}/\varepsilon) + \cdots,$$
$$p^{\varepsilon}(\mathbf{x}) \sim p_0(\mathbf{x}, \mathbf{x}/\varepsilon) + \varepsilon p_1(\mathbf{x}, \mathbf{x}/\varepsilon) + \cdots.$$

The leading-order equation is $\nabla_{\mathbf{y}} p_0 = 0$; hence $p_0 = p_0(\mathbf{x})$ is independent of $\mathbf{y}$. The next-order equation is

$$\begin{aligned}
-\nu \Delta_{\mathbf{y}} u_0 + \nabla_{\mathbf{y}} p_1 &= -\nabla_{\mathbf{x}} p_0 \quad &\text{in } \Omega, \\
\nabla_{\mathbf{y}} u_0 &= 0 \quad &\text{in } \Omega, \\
u_0 &= 0 \quad &\text{on } \partial\Omega.
\end{aligned}$$

Let (χ_j, q_j), $j = 1, 2$, be the solution of

$$\begin{aligned}
-\nu \Delta_{\mathbf{y}} \chi_j + \nabla_{\mathbf{y}} q_j &= \mathbf{e}_j \quad &\text{in } \Omega, \\
\nabla_{\mathbf{y}} \chi_j &= 0 \quad &\text{in } \Omega, \\
\chi_j &= 0 \quad &\text{on } \partial\Omega,
\end{aligned}$$

where $\mathbf{e}_1 = (1, 0)^{\mathrm{T}}$ and $\mathbf{e}_2 = (0, 1)^{\mathrm{T}}$. Given $\nabla_{\mathbf{x}} p_0$, the solution to the next-order equation is given by $u_0(\mathbf{x}, \mathbf{y}) = -(\chi_1, \chi_2)\nabla_{\mathbf{x}} p_0$. The average velocity is

$$\bar{u}(\mathbf{x}) = -\left(\int_{\Omega} (\chi_1, \chi_2) d\mathbf{y} \right) \nabla_{\mathbf{x}} p_0(\mathbf{x}).$$

This is Darcy's law for flows in a porous medium; $K = \int (\chi_1, \chi_2) d\mathbf{y}$ is the effective permeability tensor.

2.5 Scaling and self-similar solutions

So far we have focused on problems with disparate scales and discussed techniques that can be used to eliminate the small scales and obtain simplified effective models for the large-scale behavior of the problem. Another class of problems for which simplified models can be obtained comprises problems that exhibit self-similar behavior. We will begin with a discussion of self-similar solutions in this section. In the next section, we turn to the renormalization group method, which is a systematic way of extracting effective models in systems that exhibit self-similar behavior at small scales.

2.5.1 Dimensional analysis

Dimensional analysis is a very simple and elegant tool for making a quick guess about the important features of a problem. The most impressive example of dimensional analysis is Kolmogorov's prediction about the small-scale structure in a fully developed turbulent flow. The quantity of interest there is the (kinetic) energy density at wave number k, denoted as $E(k)$. The behavior of $E(k)$ varies according to the regime. In the large-scale regime, i.e. scales comparable with the size of the system the behavior of $E(k)$ is non-universal and depends on how energy is supplied to the system and on the geometry of the system etc. In the dissipation regime, viscous effects dominate. Between the large-scale and the dissipation regimes lies the so-called inertial regime. Here, the conventional picture is that energy is simply transported from large to small scales without being dissipated. It is widely expected, and partly confirmed by experimental results [15, 31], that the behavior of turbulent flows in this inertial range is universal and exhibits some kind of self-similarity.

To find the self-similar behavior, Kolmogorov made the basic assumption that the only important quantity for the inertial regime is the mean energy dissipation rate,

$$\bar{\varepsilon} = \nu \left\langle |\nabla \mathbf{u}|^2 \right\rangle,$$

where ν is the dynamic viscosity; the bracket denotes the ensemble average of the quantity inside. To find the behavior of $E(k)$, which is assumed to depend on k and $\bar{\varepsilon}$ only, let us look at the dimensions of the quantities $E(k)$, k and $\bar{\varepsilon}$. We will use L to denote length, T to denote time, and $[\cdot]$ to denote the dimension of the quantity inside the bracket. Obviously, $[k] = 1/L$. Since $[\int E(k)\, dk] = L^2/T^2$, we have $[E(k)]/L = L^2/T^2$. Hence $[E(k)] = L^3/T^2$. Also,

$$[\bar{\varepsilon}] = [\nu][\nabla \mathbf{u}]^2 = (L^2/T)(1/T)^2 = L^2/T^3.$$

Assume that $E(k) = \bar{\varepsilon}^\alpha k^\beta$. It is easy to see that to be dimensionally consistent we must have $\alpha = 2/3$ and $\beta = -5/3$. Therefore, we conclude that

$$E(k) \propto \bar{\varepsilon}^{2/3} k^{-5/3}.$$

Experimental results are remarkably close to the behavior predicted by this simple relation. However, as was pointed out by Landau, it is unlikely that this is the real story: the dissipation field $\nu|\nabla \mathbf{u}|^2$ is a highly inter-mittent quantity. Its contribution to the flow field is dominated by intense events in a very small part of the whole physical domain. Therefore we do not expect the mean-field picture put forward by Kolmogorov to be entirely accurate. Indeed there is evidence suggesting that the small-scale structure in a turbulent flow exhibits multi-fractal statistics:

$$\langle |\mathbf{u}(\mathbf{x}+\mathbf{r}) - \mathbf{u}(\mathbf{x})|^p \rangle \sim |\mathbf{r}|^{\zeta_p},$$

where ζ_p is a nonlinear function of p. In contrast, Kolmogorov's argument would give $\zeta_p = \frac{1}{3}p$. To this day, finding ζ_p remains one of the main challenges of turbulence theory. Many phenomenological models have been proposed, but we still lack an understanding of models based on first principles (i.e. the Navier–Stokes equation) [15].

2.5.2 Self-similar solutions of PDEs

Consider the Barenblatt equation, which arises from modeling gas flow in a porous medium [3],

$$\partial_t u = \partial_x^2 u^2, \qquad (2.5.1)$$

with initial condition $u(x,0) = \delta(x)$. We look for a self-similar solution that satisfies the relation

$$u(x,t) = \lambda^\alpha u(\lambda x, \lambda^\beta t) \qquad (2.5.2)$$

for all $\lambda > 0$. At $t = 0$, we have

$$\delta(x) = \lambda^\alpha \delta(\lambda x).$$

Therefore $\alpha = 1$. Substituting the ansatz (2.5.2) into the PDE (2.5.1), we obtain

$$\lambda^{\beta+1}\partial_t u = \lambda^4 \partial_x^2 u^2.$$

Hence $\beta = 3$. This means that

$$u(x,t) = \lambda u(\lambda x, \lambda^3 t) = \frac{1}{t^{1/3}} U\left(\frac{x}{t^{1/3}}\right),$$

where $U(y) = u(y,1)$ and we have used $\lambda = 1/t^{1/3}$. The original PDE for u now becomes an ODE for U:

$$t^{-4/3}(U^2)'' = -\tfrac{1}{3}t^{-4/3}U - \tfrac{1}{3}t^{-5/3}xU',$$

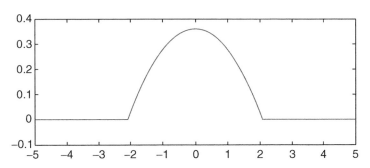

Figure 2.4. The profile of the solution to the Barenblatt equation at $t = 1$.

or, if we write $y = x/t^{1/3}$,

$$(U^2)'' = -\tfrac{1}{3}U - \tfrac{1}{3}yU'.$$

Integrating once, we get

$$(U^2)' = -\tfrac{1}{3}yU + C.$$

As $y \to \pm\infty$, we expect that $U \to 0$. Hence $C = 0$. Integrating one more time, we get

$$U(y) = \begin{cases} U_0 - \tfrac{1}{12}y^2, & U_0 > \tfrac{1}{12}y^2, \\ 0 & \text{otherwise.} \end{cases}$$

The constant U_0 can be determined by conservation of mass:

$$\int U(y)dy = \int u(y,1)dy = 1.$$

This gives $U_0 = 3^{1/3}/4$. The solution is shown in Figure 2.4. What is interesting about this solution is that it has finite speed of propagation. This is possible due to the degeneracy of the diffusion term, and is consistent with our intuition about gas flow in a porous medium.

Next, we study an example that could model a viscous rarefaction wave:

$$\partial_t u + u^2 \partial_x u = \partial_x^2 u, \qquad (2.5.3)$$

with initial condition

$$u(x,0) = \begin{cases} 0, & x \le 0, \\ 1, & x > 0. \end{cases} \qquad (2.5.4)$$

Analytical methods

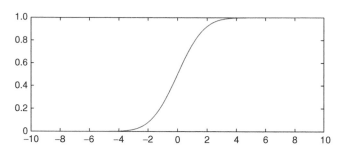

Figure 2.5. The profile of the solution to (2.5.5) at $t = 1$.

It is impossible to find a self-similar solution to this problem. In fact, we will show that the solution has three different regimes.

Let us first study the effect of the diffusion and convection terms seperately. Keeping only the diffusion term, (2.5.3) becomes

$$\partial_t u = \partial_x^2 u. \tag{2.5.5}$$

For this problem with the initial condition (2.5.4) we can find a self-similar solution. Assume that

$$u(x, t) = \lambda^\alpha u(\lambda x, \lambda^\beta t).$$

Since the initial condition is invariant under scaling, $\alpha = 0$. Substituting into (2.5.5) gives $\beta = 2$. Therefore,

$$u(x, t) = u(\lambda x, \lambda^2 t) = U\left(\frac{x}{t^{1/2}}\right),$$

where $U(y) = u(y, 1)$. Moreover, U satisfies the ODE

$$U'' = -\frac{1}{2} y U'.$$

The solution to this equation with the prescribed initial condition is given by

$$U(y) = \frac{1}{2\sqrt{\pi}} \int_{-\infty}^{y} e^{-z^2/4} \, \mathrm{d}z.$$

As shown in Figure 2.5, the effect of the diffusion term is to smear out the discontinuity in the initial data.

If we keep only the convection term then (2.5.3) becomes

$$\partial_t u + u^2 \partial_x u = 0. \tag{2.5.6}$$

By the same argument as before we can find a self-similar solution that satisfies

$$u(x,t) = u(\lambda x, \lambda t) = U\left(\frac{x}{t}\right).$$

The quantity U satisfies the equation

$$-yU' + U^2U' = 0.$$

The solution is

$$U = \begin{cases} 0, & y \leq 0, \\ \sqrt{y}, & 0 \leq y \leq 1, \\ 1, & y \geq 1. \end{cases}$$

Going back to the original problem, it is easy to see that for small t diffusion dominates and for large t convection dominates. This can also be shown by a scaling argument. Assume that

$$u(x,t) = u(\lambda x, \lambda^\beta t).$$

Substituting into the original PDE, we get

$$\lambda^\beta \partial_t u + \lambda u^2 \partial_x u = \lambda^2 \partial_{xx} u.$$

When $\lambda \ll 1$, corresponding to large times, the term $\lambda u^2 u_x$ is more important and hence the convective behavior is dominant. When $\lambda \gg 1$, corresponding to small times, the viscous behavior dominates.

What happens at intermediate times? Using the ansatz

$$u(x,t) = \lambda^\alpha u(\lambda x, \lambda^\beta t),$$

we get $\alpha = 1/2$ and $\beta = 2$ by balancing the convective and diffusive terms. Hence

$$u(x,t) \sim \frac{1}{t^{1/4}} U\left(\frac{x}{t^{1/2}}\right).$$

As before, we obtain an equation for U:

$$U'' = -\tfrac{1}{4}U - \tfrac{1}{2}yU' + U^2U'.$$

This can be solved numerically to give the behavior at intermediate times.

In summary, at early times the dynamics of the solution is dominated by the smoothing out of the discontinuity of the initial data. At large times the dynamics is dominated by the convective behavior induced by the different asymptotic behaviors of the initial data for large $|x|$. The self-similar profile at intermediate times serves to connect these two types of behavior.

2.6 Renormalization group analysis

The renormalization group method is a very powerful tool for analyzing the coarse-grained behavior of complex systems. Its main idea is to define a flow in the space of models or model parameters induced by a coarse-graining operator, in which the scale plays the role of time, and to analyze the large-scale behavior of the original model with the help of this flow. The main components of the renormalization group analysis are

(1) the renormalization transformation, i.e. a coarse-graining operator that acts on the state variables;
(2) the induced transformation on the model, or on the model parameters, which defines the renormalization group flow;
(3) analysis of the fixed points for this transformation, as well as the behavior near the fixed points.

It is rarely the case that one can carry out step 2 described above exactly and explicitly. Very often we have to use drastic approximations, as we will see in the example discussed below. The hope is that these approximations preserve the universality class of the model with which we started. Unfortunately, there is no a priori guarantee that this will be the case.

We will illustrate the main ideas of renormalization group analysis using a classical model in statistical physics, the Ising model.

2.6.1 The Ising model and critical exponents

Consider the two-dimensional Ising model on a triangular lattice $\mathcal{L}$ (see for example the classical textbook on statistical physics by Reichl [38]). Each site is assigned a spin variable s which takes two possible values ± 1. The state of the system is specified by the spin configuration over the lattice. For each configuration $\{s_i\}$, we associate a Hamiltonian

$$\tilde{H}\{s_i\} = -J \sum_{\langle i,j \rangle} s_i s_j, \qquad (2.6.1)$$

where $\langle i, j \rangle$ means that i and j are nearest neighbors, and $J > 0$ is the coupling constant. We will denote by $K = J/(k_B T)$ the non-dimensionalized coupling strength. The probability of a configuration is given by

$$P\{s_i\} = \frac{1}{Z}e^{-\tilde{H}\{s_i\}/k_B T}, \tag{2.6.2}$$

where the partition function Z is defined as follows:

$$Z = Z(N, K) = \sum_{\{s_j\}} e^{-\tilde{H}\{s_j\}/k_B T}.$$

The sum is carried out over all possible configurations. To simplify the notation, we set $H\{s_i\} = K\sum_{\langle i,j \rangle} s_i s_j$, so that

$$P\{s_i\} = \frac{e^{H\{s_i\}}}{\sum_{\{s_j\}} e^{H\{s_j\}}}.$$

Statistical averages $\langle \cdot \rangle$ are defined with respect to this probability density. The free energy is given by the large-volume limit

$$\zeta(K) = -k_B T \lim_{N \to \infty} \frac{1}{N} \ln Z(N, K).$$

This is a standard model for ferromagnetic materials. The expectation value of a single spin $\langle s \rangle$ measures the macroscopic magnetization of the system.

Strictly speaking, we should consider finite lattices and take the thermodynamic limit as the lattice goes to $\mathbb{Z}^d$. It is well known that this model has a phase transition in the thermodynamic limit: there exists a critical temperature T^* such that if $T \geq T^*$ then the thermodynamic limit is unique, i.e. it does not depend on how we take the limit and what boundary condition we use for the finite lattices. In addition, we have $\langle s \rangle = 0$ in the thermodynamic limit. However, if $T < T^*$ then the limit is no longer unique and does depend on the boundary condition used for the finite lattices. In particular, it may happen that $\langle s \rangle$ does not vanish in the thermodynamic limit. Intuitively, from (2.6.1) the energy-minimizing configurations are configurations for which all the spins are aligned, i.e. s_j is 1 or -1 for all $j \in \mathcal{L}$. Therefore, at low temperatures the spins prefer to be aligned, giving rise to a non-zero average spin. At high temperatures the spins prefer to be random, from entropic considerations, giving rise to a zero average. However, the fact that there is a sharp transition at some finite temperature is a rather non-trivial fact [38].

It is also of interest to consider the correlations between the spins at different sites. There is strong evidence that, when $T \neq T^*$,

$$\langle s_i s_j \rangle - \langle s_i \rangle \langle s_j \rangle \sim e^{-r/\xi}, \qquad r = |i - j|.$$

Here ξ is called the correlation length and depends on T. It is an important intrinsic length scale for the system. The critical point T^* is characterized by the divergence of ξ:

$$\xi \rightarrow \infty \quad \text{as } T \rightarrow T^*.$$

In fact, from various considerations, one expects that

$$\xi \sim (T - T^*)^{-\nu},$$

where $\nu > 0$ is the critical exponent for the correlation length.

Heuristically, correlated spins tend to form clusters in which all spins have the same sign, and so ξ is of the order of the largest diameter of the finite volume clusters. This is illustrated in Figure 2.6. What is most interesting is the situation at the critical temperature T^*, which exhibits self-similarity: the picture looks essentially the same at different scales or with different resolutions. This motivates us to seek statistically self-similar behavior at $T = T^*$.

Given the above background, our goal is to compute T^* (or equivalently K^*) and ν. The basic idea is to divide the spins into blocks and regard the blocks as coarse-grained spins. This corresponds to considering the problem at a coarser resolution. One way of defining the spin of a block is to use the majority rule:

$$s_I = \text{sgn}\left(\sum_{i \in I} s_i\right),$$

where I indexes the blocks. Here we assume that the number of spins within every block is odd. Denote the size of a block by L. Then each block has roughly L^d spins, where d is the space dimension. Assume for simplicity that the Hamiltonian for the coarse-grained problem is given *approximately* by

$$H\{s_I\} = -J_L \sum_{|I-J|=1} s_I s_J = -\frac{K_L}{k_B T} \sum_{|I-J|=1} s_I s_J.$$

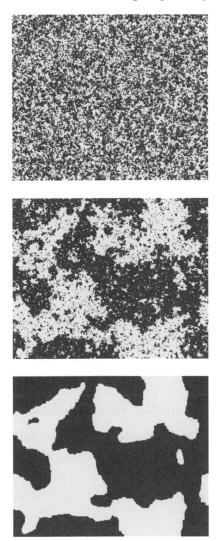

Figure 2.6. Typical spin configurations for the two-dimensional Ising model at low temperatures, the critical temperature and high temperatures (produced by Hao Shen).

Then the coarse-graining operator for the spin configuration induces a transformation in the space of the model parameters:

$$K_L = \mathcal{R}(K).$$

This is the renormalization transformation.

The information we are looking for can be obtained from this transformation. In particular, we have the following.

(1) The critical temperature is given by the fixed point of this transformation: $K^* = \mathcal{R}(K^*)$.

(2) It is easy to see that

$$\xi(\mathcal{R}(K)) = L^{-1}\xi(K).$$

Therefore at the fixed point of $\mathcal{R}$, $\xi = \infty$.

(3) The critical exponent ν can be found by linearizing $\mathcal{R}$ at its fixed point. For $|K - K^*| \ll 1$,

$$K_L = \mathcal{R}(K) = K^* + \lambda(K - K^*) + \cdots,$$

where $\lambda = (d\mathcal{R}/dK)(K^*)$; therefore

$$\frac{1}{L} = \frac{\xi(K_L)}{\xi(K)} = \frac{|T_L - T^*|^{-\nu}}{|T - T^*|^{-\nu}} = \frac{|K^* - K_L|^{-\nu}}{|K^* - K|^{-\nu}} = \lambda^{-\nu}.$$

This gives a prediction of the value of ν.

2.6.2 *An illustration of the renormalization transformation*

It is useful to illustrate the basic ideas of renormalization group analysis using the one-dimensional Ising model. Consider a one-dimensional lattice with N sites. At the jth site we assign a spin variable s_j, which takes two possible values, ± 1. The Hamiltonian for this model is given by

$$H = -J \sum_{|i-j|=1} s_i s_j.$$

The renormalization group analysis proceeds as follows. First, let us define a coarse-graining operator. For example, starting with a system with N sites (we assume N to be even), we integrate out the spin variables on the even sites and take only the odd ones, $\{s_1, s_3, \ldots, s_{N-1}\}$, as the variables of our system.

Next we define the renormalization group flow. The transformation

$$\{s_1, s_2, \ldots, s_N\} \rightarrow \{s_1, s_3, \ldots, s_{N-1}\}$$

induces a transformation in the space of the model Hamiltonians. After summing over s_2, s_4 etc., the partition function becomes

$$Z(N, K) = \sum_{\{s_j\}} \exp(K(s_1 s_2 + s_2 s_3 + \cdots))$$

$$= \sum_{s_1, s_3, \cdots} \left[\exp(K(s_1 + s_3)) + \exp(-K(s_1 + s_3)) \right]$$

$$\times \left[\exp(K(s_3 + s_5)) + \exp(-K(s_3 + s_5)) \right] \cdots .$$

We may view this new form of the partition function as the partition function of a new Hamiltonian on a lattice with $N/2$ sites. This means that we would like to find a constant K_L (with $L = 2$) such that

$$Z(N, K) = \sum_{s_1, s_3, \cdots} \exp(K_L(s_1 s_3 + s_3 s_5 + \cdots)) \, f(K)^{N/2}. \qquad (2.6.3)$$

Here the last term is a scaling factor that has to be introduced to compensate for the reduction in the configuration space. For (2.6.3) to hold it suffices to require that

$$\exp(K(s_1 + s_3)) + \exp(-K(s_1 + s_3)) = f(K) \exp(K_L s_1 s_3) \qquad (2.6.4)$$

for all choices of s_1 and s_3. The solution to these equations is given by

$$K_L = \tfrac{1}{2} \ln \cosh 2K, \quad f(K) = 2 \cosh^{1/2} 2K. \qquad (2.6.5)$$

The first equation defines the renormalization group flow in the space of the model parameters, here given by the value of K. We then have

$$Z(N, K) = f(K)^{N/2} Z(N/2, K_L). \qquad (2.6.6)$$

The renormalized Hamiltonian is

$$H_L = -J_L \sum_{|i-j|=2} s_i s_j, \qquad (2.6.7)$$

where $J_L = k_B T K_L$. From (2.6.6), we obtain

$$-\zeta(K) = \tfrac{1}{2} \ln f(K) - \tfrac{1}{2} \zeta(K_L).$$

From (2.6.5), if we know the value of ζ at K then we can obtain the value of ζ at K_L.

To analyze the dynamics of the renormalization group flow and the change in the free energy along the flow, we may proceed as follows. When $K = 0$, the partition function is simply $Z(N, 0) = 2^N$. Therefore for some small value of K, say 0.01, we may set $Z(N, 0.01) \approx 2^N$. We then use the

Analytical methods

Table 2.1. *Accuracy of the renormalization group*
approximation to the free energy of the
one-dimensional Ising model

| | $\zeta(K)$ | |
| | | |
K	Renormalization group	Exact
0.01	ln 2	0.693 197
0.100 334	0.698 147	0.698 172
0.327 447	0.745 814	0.745 827
0.636 247	0.883 204	0.883 210
0.972 710	1.106 299	1.106 302
1.316 710	1.386 078	1.386 080
1.662 637	1.697 968	1.697 968
2.009 049	2.026 876	2.026 877
2.355 582	2.364 536	2.364 537
2.702 146	2.706 633	2.706 634

renormalization transformation (2.6.5) iteratively to find the free energy at other values of K. For the one-dimensional Ising model, this approximation is indeed quite good [28]. See Table 2.1 (which is reprinted with permission from [28]. Copyright 1978, American Association of Physics Teachers).

In this example, the renormalized Hamiltonian (2.6.7) is in exactly the same form as the Hamiltonian we began with, i.e. we were able to find values of $f(K)$ and K_L such that (2.6.4) is satisfied for all values of s_1 and s_3 and no approximations are needed. This is rarely the case in general. Instead, one has to make approximations in order to maintain some feasibility for the renormalization group analysis, as the next example shows.

2.6.3 Renormalization group analysis of the two-dimensional Ising model

We will now look at the Ising model on a two-dimensional triangular lattice with lattice constant a [38]. To coarse-grain the model we divide the lattice into a triangular lattice of triangular blocks, indexed by I, having three vertices. See Figure 2.7. We define a block spin $S_I = \text{sgn}(s_1 + s_2 + s_3)$, where s_1, s_2, s_3 are the spins on the vertices of the Ith

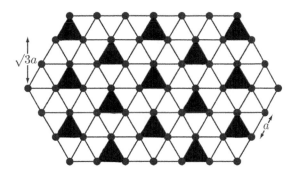

Figure 2.7. Coarse-graining the Ising model on a triangular lattice.

block, ordered in some particular way. After coarse-graining, the lattice constant changes from a to $\sqrt{3}a$. When $S_I = 1$, we have four possible spin configurations for s_1, s_2 and s_3, which we denote by $\sigma_I = 1, 2, 3, 4$: they are $++-$, $+-+$, $-++$ and $+++$, where we use $+$ for spin up and $-$ for spin down. These are the internal variables over which we will integrate. To renormalize, note that we can write H as a function of the variables $\{S_I, \sigma_I\}$, namely as $H = H(K, \{S_I, \sigma_I\})$. Integrating over the variables $\{\sigma_I\}$, we can write $Z(K, N)$ in the form $\sum_{S_I} \exp(-H(K_L, \{S_I\}))$ by taking

$$H(K_L, \{S_I\}) = -\log \sum_{\{\sigma_I\}} \exp(-H(K, \{S_I, \sigma_I\})).$$

It would be useful if $H(K_L, \{S_I\})$ had the same form as the original Hamiltonian, say if $H = -J_L \sum_{\langle I,J \rangle} S_I S_J$ for some new parameter J_L. However, this is impossible unless we make some approximations.

The starting point is

$$P\{S_I\} = \sum_{\{s_i|S_I\}} P\{s_i\}, \qquad (2.6.8)$$

where $\sum_{\{s_i|S_I\}}$ means a sum over all configurations $\{s_i\}$ that give rise to $\{S_I\}$ after coarse graining. For later use, we mention the obvious relation

$$\sum_{\{s_i\}} = \sum_{\{S_I\}} \sum_{\{s_i|S_I\}} \qquad (2.6.9)$$

and the elementary but useful factorization

$$\sum_{x_1} \cdots \sum_{x_n} f_1(x_1) \cdots f_n(x_n) = \left(\sum_{x_1} f_1(x_1) \right) \cdots \left(\sum_{x_n} f_n(x_n) \right). \qquad (2.6.10)$$

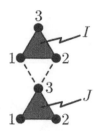

Figure 2.8. Interaction between two block spins.

Rewriting (2.6.8) we have

$$\frac{\exp(K_L \sum S_I S_J)}{\sum_{\{S_I\}} \exp\left(K_L \sum S_I S_J\right)} = \sum_{\{s_i|S_I\}} \frac{\exp(K \sum s_i s_j)}{\sum_{\{s_i\}} \exp(K \sum s_i s_j)}. \qquad (2.6.11)$$

In this equation, $K\sum_{|i-j|=1} s_i s_j$ consists of two parts: one that corresponds to the internal block, including all terms for which s_i and s_j are in the same block, and another, interblock, part that includes all terms for which s_i and s_j are in two different blocks. See Figure 2.8. We denote these parts by H_0 and H_1 respectively.

Claim 2.1 *For (2.6.11) to hold, it is sufficient that*

$$\exp\left(K_L \sum S_I S_J\right) = \frac{\sum_{\{s_i|S_I\}} \exp\left(K \sum s_i s_j\right)}{\sum_{\{s_i|S_I\}} \exp\left(-H_0/(k_B T)\right)}. \qquad (2.6.12)$$

Proof Note that

$$\sum_{\{s_i|S_I\}} \exp\left(-H_0/(k_B T)\right) = \prod_{I=1}^{M}\left(\sum_{\sigma_I} \exp\left(K(S_I^1 S_I^2 + S_I^2 S_I^3 + S_I^3 S_I^1)\right)\right)$$
$$= (e^{3K} + 3e^{-K})^M, \qquad (2.6.13)$$

where $S_I^i, i=1,2,3$, are the internal spins of the Ith block and M is the number of blocks. Note also that the right-hand side of (2.6.13) does not depend on $\{S_I\}$. Summing over $\{S_I\}$ on both sides of (2.6.12), we get

$$\sum_{\{S_I\}} \exp\left(K_L \sum S_I S_J\right) = \frac{\sum_{\{s_i\}} \exp\left(K \sum s_i s_j\right)}{\sum_{\{s_i|S_I\}} \exp\left(-H_0/(k_B T)\right)},$$

where we have used (2.6.9). The ratio of (2.6.12) and the last equation gives (2.6.11). □

Next define the weighted average

$$\langle A \rangle = \frac{\sum_{\{s_i|S_I\}} \exp\left(-H_0/(k_B T)A\right)}{\sum_{\{s_i|S_I\}} \exp\left(-H_0/(k_B T)\right)};$$

then (2.6.12) can be written as

$$\exp\left(K_L \sum S_I S_J\right) = \langle \exp\left(-H_1/(k_B T)\right)\rangle$$

$$= \exp\left(-\langle H_1 \rangle/(k_B T) - \frac{1}{2}(\langle H_1^2 \rangle - \langle H_1 \rangle^2)/(k_B T)\right) + \cdots.$$

We will make the approximation

$$\langle \exp -H_1/(k_B T)\rangle \approx \exp -\langle H_1 \rangle/(k_B T). \qquad (2.6.14)$$

This is a drastic approximation. Now compare the exponents:

$$-K_L \sum_{|I-J|=1} S_I S_J = \langle H_1 \rangle/(k_B T) = -\sum_{|I-J|=1} K\left(\langle S_I^1 S_J^3 \rangle + \langle S_I^2 S_J^3 \rangle\right),$$

where $S_I^i, S_J^j, i, j = 1, 2, 3$, refer to Figure 2.8.

Using (2.6.10) again, we have

$$\langle S_I^1 S_J^3 \rangle = \frac{\sum_{\{s_i|S_I\}} \exp\left(-H_0/(k_B T)\right) S_I^1 S_J^3}{\sum_{\{s_i|S_I\}} \exp\left(-H_0/(k_B T)\right)}$$

$$= \frac{\sum_{\sigma_I} \exp\left(K(S_I^1 S_I^2 + S_I^2 S_I^3 + S_I^3 S_I^1)\right) S_I^1}{\sum_{\sigma_I} \exp\left(K(S_I^1 S_I^2 + S_I^2 S_I^3 + S_I^3 S_I^1)\right)}$$

$$\times \frac{\sum_{\sigma_J} \exp\left(K(S_J^1 S_J^2 + S_J^2 S_J^3 + S_J^3 S_J^1)\right) S_J^3}{\sum_{\sigma_J} \exp\left(K(S_J^1 S_J^2 + S_J^2 S_J^3 + S_J^3 S_J^1)\right)}.$$

We get a result that is similar to (2.6.13):

$$\sum_{\sigma_I} \exp\left(K(S_I^1 S_I^2 + S_I^2 S_I^3 + S_I^3 S_I^1)\right) S_I^1 = S_I(e^{3K} + e^{-K}).$$

Therefore, we have

$$K_L \sum_{|I-J|=1} S_I S_J = \sum_{|I-J|=1} 2K \left(\frac{e^{3K} + e^{-K}}{e^{3K} + 3e^{-K}}\right)^2 S_I S_J,$$

in other words

$$K_L = \mathcal{R}(K) = 2K \left(\frac{e^{3K} + e^{-K}}{e^{3K} + 3e^{-K}} \right)^2.$$

This is the renormalization group flow under the approximation (2.6.14).

The nontrivial fixed point of this map is $K^* = \frac{1}{4}\ln(1 + 2\sqrt{2}) \approx 0.3356$, from which we obtain $\nu = \ln L / \ln \mathcal{R}'(K^*) \approx 1.134$. As a comparison, note that Onsager's exact solution for the two-dimensional Ising model on a triangular lattice gives $K^* = \frac{1}{4}\ln 3 \approx 0.2747$ and $\nu = 1$.

In principle we may try to improve upon the approximation (2.6.14). In practice this is not easy, since the computation quickly becomes very involved.

2.6.4 A PDE example

Consider the example

$$\begin{cases} \partial_t u_1 = \partial_x^2 u_1 - u_1 u_2^2, \\ \partial_t u_2 = D\partial_x^2 u_2 + u_1 u_2^2, \end{cases} \qquad (2.6.15)$$

with initial data that decays sufficiently fast at infinity. This example was analyzed in [5]. Our interest here is to see how simple-minded renormalization group methods can be used for this problem.

First let us define the renormalization transformation. Let $L > 1$ be fixed, and let (u_1, u_2) be the solution of (2.6.15). Define

$$\begin{cases} \tilde{u}_1(x, t) = L^{\alpha_1} u_1(Lx, L^2 t), \\ \tilde{u}_2(x, t) = L^{\alpha_2} u_2(Lx, L^2 t). \end{cases} \qquad (2.6.16)$$

Next, we define the renormalization group flow, now in the space of nonlinear operators. It is easy to see that $(\tilde{u}_1, \tilde{u}_2)$ satisfies

$$\begin{cases} \partial_t \tilde{u}_1 = \partial_x^2 \tilde{u}_1 - L^{2(1-\alpha_2)} \tilde{u}_1 \tilde{u}_2^2, \\ \partial_t \tilde{u}_2 = D\partial_x^2 \tilde{u}_2 + L^{2-(\alpha_1+\alpha_2)} \tilde{u}_1 \tilde{u}_2^2. \end{cases} \qquad (2.6.17)$$

Therefore, under the renormalization transformation (2.6.16), the operator in the PDE (2.6.15) acts as follows:

$$\begin{pmatrix} \partial_x^2 u_1 - u_1 u_2^2 \\ D\partial_x^2 u_2 + u_1 u_2^2 \end{pmatrix} \longrightarrow \begin{pmatrix} \partial_x^2 u_1 - L^{2(1-\alpha_2)} u_1 u_2^2 \\ D\partial_x^2 u_2 + L^{2-(\alpha_1+\alpha_2)} u_1 u_2^2 \end{pmatrix}. \qquad (2.6.18)$$

This suggests that we should consider the class of operators with the form

$$\mathcal{L}\begin{pmatrix} u_1 \\ u_2 \end{pmatrix} = \begin{pmatrix} \partial_x^2 u_1 + c_1 u_1 u_2^2 \\ D\partial_x^2 u_2 + c_2 u_1 u_2^2 \end{pmatrix}. \tag{2.6.19}$$

Indeed, under the renormalization transformation (2.6.16), such an operator has the action

$$\tilde{\mathcal{L}}\begin{pmatrix} u_1 \\ u_2 \end{pmatrix} = \begin{pmatrix} \partial_x^2 u_1 + c_1 L^{2(1-\alpha_2)} u_1 u_2^2 \\ D\partial_x^2 u_2 + c_2 L^{2-(\alpha_1+\alpha_2)} u_1 u_2^2 \end{pmatrix} \tag{2.6.20}$$

and so is still in the same class of operators. Therefore our renormalization group flow is defined by

$$c_1 \to c_1 L^{2(1-\alpha_2)}, \quad c_2 \to c_2 L^{2-(\alpha_1+\alpha_2)}. \tag{2.6.21}$$

We are interested in the fixed points of the map (2.6.16), (2.6.21). The fixed points of (2.6.21) should satisfy

$$c_1^* = c_1^* L^{2(1-\alpha_2)}, \quad c_2^* = c_2^* L^{2-(\alpha_1+\alpha_2)}. \tag{2.6.22}$$

It is beyond the scope of this book to carry out a complete analysis of all the fixed points. But one interesting set of fixed points identified by Bricmont *et al.* is given by [5]

$$c_1^* = -1, \quad c_2^* = 0, \quad \alpha_2 = 1, \quad \alpha_1 > 1. \tag{2.6.23}$$

These fixed points correspond to the PDE

$$\begin{cases} \partial_t u_1 = \partial_x^2 u_1 - u_1 u_2^2, \\ \partial_t u_2 = D\partial_x^2 u_2. \end{cases} \tag{2.6.24}$$

The final step in our consideration of (2.6.15) is to study the linearization of the renormalization group flow (2.6.21) around this fixed point PDE. We refer to [5] for details. Let us just note that this PDE has solutions that are fixed points of the map (2.6.16). The value of α_1 can be found by looking for solutions of (2.6.24) that are also fixed points of the renormalization transformation (2.6.16). Such solutions must take the form

$$u_1(x,t) = t^{-\alpha_1/2} u_1\left(\frac{x}{\sqrt{t}}\right), \quad u_2(x,t) = t^{-1/2} u_2\left(\frac{x}{\sqrt{t}}\right).$$

It is easy to see that

$$u_2(\xi) = Ae^{-\xi^2/4}$$

and that u_1 satisfies

$$-\frac{d^2 u_1}{d\xi^2} - \frac{1}{2}\xi \frac{du_1}{d\xi} - \frac{\alpha_1}{2} u_1 + u_2^2 u_1 = 0.$$

For small values of A, α_1 can be calculated perturbatively and has the form [5]

$$\alpha_1 = 1 + \frac{A^2}{4\pi \sqrt{2D+1}} + \mathcal{O}(A^4).$$

2.7 The Mori–Zwanzig formalism

So far we have discussed two special situations in which simplified effective models can be obtained using asymptotic analysis. The first is the case when the system has disparate scales. The second occurs when the system has self-similar behavior at different scales. A natural question is: what happens in more general situations? To address this question, we turn to the Mori–Zwanzig formalism.

The Mori–Zwanzig formalism is a strategy for integrating out a subset of variables in a problem, which is precisely what we would like to do when performing coarse-graining [32, 44]. What is remarkable about the Mori–Zwanzig formalism is that it applies to general nonlinear systems. However, at some level it is also a tautology, since it simply represents the solution for a subset of the variables (those to be eliminated) as a functional of the other variables (those to be kept) and then substitutes back into the (projected form of the) original equation. In this way, some variables are formally eliminated; however, the remaining model is often far from being explicit. Nevertheless, it is still a very good starting point for model reduction or coarse-graining, for the following reasons.

(1) It is a general strategy.
(2) It suggests the form of the reduced model. For example, if we start with the equations in molecular dynamics then the model obtained after elimination of some variables can be viewed as a generalized Langevin equation, with noise and memory terms.
(3) It provides the starting point for making further approximations. From a practical viewpoint, this is the most important application of the Mori–Zwanzig formalism. For example, the abstract generalized Langevin equation, which arises from the application

of the Mori–Zwanzig formalism to molecular dynamics models, can sometimes be approximated by a much simpler Langevin equation.

We start with a simple example to illustrate the basic idea. Consider a linear system

$$\frac{dp}{dt} = A_{11}p + A_{12}q,$$

$$\frac{dq}{dt} = A_{21}p + A_{22}q.$$

Our objective is to obtain a closed model for p, which is the quantity in which we are really interested. To this end we solve the second equation for q, viewing p as a parameter:

$$q(t) = e^{A_{22}t}q(0) + \int_0^t e^{A_{22}(t-\tau)}A_{21}p(\tau)d\tau.$$

Substituting this into the first equation, we obtain

$$\frac{dp}{dt} = A_{11}p + A_{12}\int_0^t e^{A_{22}(t-\tau)}A_{21}p(\tau)d\tau + A_{12}e^{A_{22}t}q(0)$$

$$= A_{11}p + \int_0^t K(t-\tau)p(\tau)d\tau + f(t),$$

where $K(t) = A_{12}e^{A_{22}t}A_{21}$ is the *memory kernel* and $f(t) = A_{12}e^{A_{22}t}q(0)$ is the so-called *noise term*, if we think of $q(0)$ as a random variable. Note that it does indeed make sense to think of $q(0)$ as a random variable since we are not really interested in the details of q.

Our next example is the original model considered by Zwanzig [44]. Consider a system with the Hamiltonian:

$$H(x, y, \mathbf{p}, \mathbf{q}) = \tfrac{1}{2}y^2 + U(x) + \sum_{k=1}^{N}\left(\tfrac{1}{2}p_k^2 + \tfrac{1}{2}\omega_k^2\left(q_k - \frac{\gamma_k}{\omega_k^2}x\right)^2\right).$$

One may think of x, y as the variables that describe the phase space of a distinguished particle, while $\{(q_k, p_k)\}$ describes other particles that provide the thermal bath. We will use the Mori–Zwanzig formalism to integrate out the variables $\{(q_k, p_k)\}$. The Hamilton equations are

$$\dot{x} = y,$$

$$\dot{y} = -U'(x) + \sum_k \gamma_k \left(q_k - \frac{\gamma_k}{\omega_k^2} x \right),$$

$$\dot{q}_k = p_k,$$

$$\dot{p}_k = -\omega_k^2 q_k + \gamma_k x.$$

Integrating the equations for q_k and p_k, we get

$$q_k(t) = q_k(0) \cos \omega_k t + \frac{p_k(0)}{\omega_k} \sin \omega_k t - \frac{\gamma_k^2}{\omega_k} \int_0^t \sin(\omega_k(t - \tau)) x(\tau) d\tau.$$

Therefore, the equations for x and y become

$$\dot{x} = y,$$

$$\dot{y} = -U'(x) - \sum_k \frac{\gamma_k^2}{\omega_k^2} \int_0^t \cos(\omega_k(t - \tau)) y(\tau) d\tau$$

$$+ \sum_k \left(\gamma_k \left(q_k(0) - \frac{\gamma_k}{\omega_k^2} x(0) \right) \cos \omega_k t + \frac{\gamma_k p_k(0)}{\omega_k} \sin \omega_k t \right).$$

Again, this has the form of a generalized Langevin equation with memory and noise terms. Under suitable assumptions on the initial values $q_k(0)$ and $p_k(0)$, $k = 1, \ldots, N$, the noise term will be Gaussian.

Our next example is a general procedure for eliminating variables in linear elliptic equations. Consider a linear differential equation

$$Lu = f. \tag{2.7.1}$$

Here L is some nice operator and f is the forcing function. Let H be a Hilbert space which contains both u and f, and let H_0 be a closed subspace of H containing only the coarse-scale component of the functions in H. Let P be the projection operator to H_0 and let $Q = I - P$: functions in H can be decomposed into the form

$$v = v_0 + v', \qquad v \in H, \quad v_0 \in H_0, \quad v' \in H_0^\perp. \tag{2.7.2}$$

Here $H_0^\perp$ is the orthogonal complement of H_0, and v_0 and v' are respectively the coarse- and fine-scale parts of v:

$$v_0 = Pv, \quad v' = Qv. \tag{2.7.3}$$

We can now write (2.7.1) equivalently as

$$PLPu + PLQu = Pf, \tag{2.7.4}$$

$$QLPu + QLQu = Qf. \tag{2.7.5}$$

Assuming that QLQ is invertible on $H_0^\perp$, we can write an abstract equation for $u_0 = Pu$ [14]:

$$\bar{L}u_0 = \bar{L}Pu = \bar{f}, \tag{2.7.6}$$

where

$$\bar{L} = PLP - PLQ(QLQ)^{-1}QLP, \qquad \bar{f} = Pf - PLQ(QLQ)^{-1}Qf. \tag{2.7.7}$$

This is an abstract upscaling procedure, and (2.7.6) is an abstract form of an upscaled equation. Note that even if the original equation were local, say an elliptic differential equation, (2.7.6) is in general nonlocal. Its effectiveness depends on the choice of H_0 and whether the operator $\bar{L}$ can be simplified. We will return to this issue in Chapter 8, when we discuss the upscaling of elliptic differential equations with multiscale coefficients.

We now explain the general Mori–Zwanzig formalism. The main steps are as follows [8].

(1) Divide the variables into two sets: a set of retained variables and a set of variables to be eliminated.
(2) Given the time history of the retained variables and the initial condition for the variables to be eliminated, one can find the solution for the variables to be eliminated.
(3) Substituting these solutions into the equation for the retained variables, one obtains the generalized Langevin equation for the retained variables, if we regard the initial condition for the eliminated variables as being random.

We now show how these steps can be implemented [9].

Consider the general nonlinear system

$$\frac{d\mathbf{x}}{dt} = \mathbf{f}(\mathbf{x}). \tag{2.7.8}$$

Let φ_0 be a function of $\mathbf{x}$. Define the Liouville operator L:

$$(L\varphi_0)(\mathbf{x}) = (\mathbf{f} \cdot \nabla\varphi_0)(\mathbf{x}).$$

Let $\varphi(t) = \varphi_0(\mathbf{x}(t))$, where $\mathbf{x}(t)$ is the solution of (2.7.8) at time t. Using semigroup notation, we can write $\varphi(t)$ as follows:

$$\varphi(t) = e^{tL}\varphi_0.$$

Now let $\mathbf{x} = (\mathbf{p}, \mathbf{q})$, where $\mathbf{q}$ is the variable to be eliminated and $\mathbf{p}$ is the retained variable (see for example [9]). We will consider a setting where the initial conditions of $\mathbf{q}$ are random. Define the projection operator as a conditional expectation:

$$Pg = \mathbb{E}(g|\mathbf{q}(0)),$$

where g is a function of $\mathbf{x}$, and let $Q = I - P$. Let φ be a function of $\mathbf{p}$ only; then

$$\frac{d}{dt}\varphi(t) = e^{tL}L\varphi(0) = e^{tL}PL\varphi(0) + e^{tL}QL\varphi(0).$$

Using Dyson's formula,

$$e^{tL} = e^{tQL} + \int_0^t e^{(t-s)L}PLe^{sQL}\,ds,$$

we get

$$\frac{d}{dt}\varphi(t) = e^{tL}PL\varphi(0) + \int_0^t e^{(t-s)L}K(s)\,ds + R(t), \qquad (2.7.9)$$

where

$$R(t) = e^{tQL}QL\varphi(0), \quad K(t) = PLR(t).$$

This is the desired model for the retained variables. As one can see, (2.7.9) is not in the form of an equation for $\mathbf{p}$ directly. Instead, it is an equation for any function of $\mathbf{p}$.

The term $R(t)$ is usually regarded as a noise term since it is related to the initial conditions of $\mathbf{q}$, which are random. The term involving $K(t)$ is a memory term and is often dissipative in nature, as can be seen from the examples discussed earlier. Therefore (2.7.9) is often regarded as a generalized Langevin equation.

2.8 Notes

This chapter is a crash course in the analytical techniques used in multi-scale problems. Many textbooks have been written on some topics discussed here. Our treatment is certainly not thorough, but we have attempted to cover the basic issues and ideas.

The techniques discussed here are largely considered to be mathematically well understood except for the renormalization group method and the Mori–Zwanzig formalism. While many physicists are at ease using renormalization group analysis, a mathematical understanding of this technique is still in its infancy. Our treatment of the two-dimensional Ising model has largely followed that of [38]. One can see clearly that drastic approximations are necessary in order to obtain numbers that can be compared with experiment, or, in this case, with the exact solution of Onsager.

For some progress on the renormalization group analysis of one-dimensional models and Feigenbaum universality, we refer to [30].

The issue with the Mori–Zwanzig formalism is quite different. Here the question is: what should we do with it? This formalism is exact, but the resulting model is often extremely complicated. Clearly we need to make approximations but what is not clear is how one should make them and how accurate they are. We will return to this issue at the end of the book.

References for Chapter 2

[1] G. Allaire, "Homogenization and two-scale convergence," *SIAM J. Math. Anal.*, vol. 23, no. 6, pp. 1482–1518, 1992.

[2] V.I. Arnold, *Geometrical Methods in the Theory of Ordinary Differential Equations*, Springer-Verlag, New York, 1983.

[3] G.I. Barenblatt, *Scaling*, Cambridge University Press, 2003.

[4] A. Bensoussan, J.-L. Lions and G.C. Papanicolaou, *Asymptotic Analysis for Periodic Structures*, Studies in Mathematics and Its Applications, vol. 5, 1978.

[5] J. Bricmont, A. Kupiainen and J. Xin, "Global large time self-similarity of a thermal-diffusive combustion system with critical nonlinearity," *J. Diff. Equations*, vol. 130, pp. 9–35, 1996.

[6] Y. Cao, D. Gillespie and L. Petzold, "Multiscale stochastic simulation algorithm with stochastic partial equilibrium assumption for chemically reacting systems," *J. Comput. Phys.*, vol. 206, pp. 395–411, 2005.

[7] P. Chaikin and T. Lubensky, *Principles of Condensed Matter Physics*, Cambridge University Press, 1995.

[8] A.J. Chorin and O.H. Hald, *Stochastic Tools in Mathematics and Science*, 2nd edition, Springer-Verlag, 2009.

[9] A.J. Chorin, O.H. Hald and R. Kupferman, "Optimal prediction and the Mori–Zwanzig representation of irreversible processes," *Proc. Nat. Acad. Sci. USA*, vol. 97, pp. 2968–2973, 2000.

[10] J. Creswick and N. Morrison, "On the dynamics of quantized vortices," *Phys. Lett. A*, vol. 76, pp. 267–268, 1980.

[11] W. E, "Homogenization of linear and nonlinear transport equations," *Comm. Pure Appl. Math.*, vol. 45, pp. 301–326, 1992.

[12] W. E, D. Liu and E. Vanden-Eijnden, "Nested stochastic simulation algorithm for chemical kinetic systems with disparate rates," *J. Chem. Phys.*, vol. 123, pp. 194 107–194 115, 2005.

[13] W. E, D. Liu and E. Vanden-Eijnden, "Nested stochastic simulation algorithms for chemical kinetic systems with multiple time scales," *J. Comput. Phys.*, vol. 221, pp. 158–180, 2007.

[14] B. Engquist and O. Runborg, "Wavelet-based numerical homogenization with applications," in *Multiscale and Multiresolution Methods: Theory and Applications, Lecture Notes in Computer Science and Engineering*, T.J. Barth *et al.*, eds., vol. 20, pp. 97–148, Springer-Verlag, 2002.

[15] U. Frisch, *Turbulence: The Legacy of A.N. Kolmogorov*, University Press, 1995.

[16] C.W. Gardiner, *Handbook of Stochastic Methods*, 2nd edition, Springer-Verlag, 1997.

[17] D. Gillespie, "A general method for numerically simulating the stochastic time evolution of coupled chemical reactions," *J. Comput. Phys.*, vol. 22, pp. 403–434, 1976.

[18] D. Gillespie, L.R. Petzold and Y. Cao, "Comment on 'Nested stochastic simulation algorithm for chemical kinetic systems with disparate rates' [12], *J. Chem. Phys.*, vol. 126, pp. 137 101–137 105, 2007.

[19] N. Goldenfeld, *Lectures on Phase Transitions and the Renormalization Group*, Perseus Books, 1992.

[20] S. Goldstein, "On laminar boundary layer flow near a point of separation," *Quart. J. Mech. Appl. Math.*, vol. 1, pp. 43–69, 1948.

[21] J. Kevorkian and J.D. Cole, *Perturbation Methods in Applied Mathematics*, Springer-Verlag, 1981.

[22] R.Z. Khasminskii, *Stochastic Stability of Differential Equations*, Monographs and Textbooks on Mechanics of Solids and Fluids, vol. 7, *Mechanics and Analysis*, Sijthoff and Noordhoff, 1980.

[23] R.Z. Khasminskii, "A limit theorem for the solutions of differential equations with random right-hand sides," *Theory Prob. Appl.*, vol. 11, pp. 390–406, 1966.

[24] T.G. Kurtz, "A limit theorem for perturbed operator semigroups with applications for random evolutions," *J. Funct. Anal.*, vol. 12, pp. 55–67, 1973.

[25] P.-L. Lions, *Generalized Solutions of Hamilton–Jacobi Equations*, Pitman Advanced Publishing Program, 1982.

[26] P.-L. Lions, G.C. Papanicolaou and S.R.S. Varadhan, "Homogenization of Hamilton–Jacobi equations," unpublished.

[27] D. Liu, "Analysis of multiscale methods for stochastic dynamical systems with multiple time scales," preprint.

[28] H.J. Maris and L.P. Kadanoff, "Teaching the renormalization group," *Amer. J. Phys.*, vol. 46, pp. 652–657, 1978.

[29] V.P. Maslov and M.V. Fedoriuk, *Semi-classical Approximations in Quantum Mechanics*, Reidel, 1981.

[30] C. McMullen, *Complex Dynamics and Renormalization*, Princeton University Press, 1994.

[31] A.S. Monin and A.M. Yaglom, *Statistical Fluid Mechanics: Mechanics of Turbulence*, MIT Press, 1971.

[32] H. Mori, "Transport, collective motion, and Brownian motion," *Progr. Theor. Phys.*, vol. 33, pp. 423–455, 1965.

[33] A.H. Nayfeh, *Perturbation Methods*, John Wiley, 1973.

[34] G. Nguetseng, "A general convergence result for a functional related to the theory of homogenization," *SIAM J. Math. Anal.*, vol. 20, no. 3, pp. 608–623, 1989.

[35] K. Nickel, "Prandtl's boundary layer theory from the viewpoint of a mathematician," *Ann. Rev. Fluid Mech.*, vol. 5, pp. 405–428, 1974.

[36] G.C. Papanicolaou, "Introduction to the asymptotic analysis of stochastic differential equations,", in *Lectures in Applied Mathematics*, R.C. DiPrima, ed., vol. 16, pp. 109–149, 1977.

[37] G.A. Pavliotis and A.M. Stuart, *Multiscale Methods: Averaging and Homogenization*, Springer-Verlag, 2008.

[38] L.E. Reichl, *A Modern Course in Statistical Physics*, University of Texas Press, 1980.

[39] J. Rubinstein, J.B. Keller and P. Sternberg, "Fast reaction, slow diffusion and curve shortening," *SIAM J. Appl. Math.*, vol. 49, 116–133, 1989.

[40] H. Schlicting, *Boundary Layer Theory*, 4th edition, McGraw-Hill, 1960.

[41] N.G. Van Kampen, *Stochastic Processes in Physics and Chemistry*, 3rd edition, Elsevier, 2007.

[42] E. Vanden-Eijnden, unpublished lecture notes.

[43] R.B. White, *Asymptotic Analysis of Differential Equations*, Imperial College Press, 2005.

[44] R. Zwanzig, "Collision of a gas atom with a cold surface," *J. Chem. Phys.*, vol. 32, pp. 1173–1177, 1960.

3
Classical multiscale algorithms

In this chapter we discuss some of the classical algorithms that have been developed using multiscale ideas. Unlike the algorithms that we consider in the second half of this volume, the methods presented here are developed for the purpose of resolving the fine-scale details of solutions. For this reason, the best we can hope for from the viewpoint of the computational complexity is linear scaling. This should be contrasted with the sublinear scaling algorithms that we will discuss in the second half of this volume.

We have chosen the following examples:

(1) the multigrid method;
(2) the fast multipole method;
(3) domain decomposition methods;
(4) adaptive mesh refinement;
(5) multi-resolution representation.

Much has been written on these topics, including some helpful monographs (see for example, [1, 13, 37]). It is *not* our intention to discuss them in great detail. Our focus will be on the main ideas and the general aspects that are relevant to the overall theme of this volume.

3.1 Multigrid method

Although its basic ideas were initiated in the 1960s [22], the multigrid method only became a practical tool in the late 1970s. Initially, it was developed as a tool for solving the algebraic equations that arise from discretizing PDEs [11, 27]. Later, it was extended to other contexts such as Monte Carlo methods and molecular dynamics [12, 24]. In particular, Brandt proposed to extend the multigrid method as a tool for capturing the macroscale behavior of multiscale or multi-physics problems [12], a topic that we will take up in Chapter 6. Here we will focus on the classical multigrid method.

Consider the following problem:

$$-\triangle u(\mathbf{x}) = f(\mathbf{x}), \quad \mathbf{x} \in \Omega, \tag{3.1.1}$$

with the boundary condition $u|_{\partial\Omega} = 0$. The standard procedure for solving such a PDE is as follows.

(1) First discretize the PDE using, for example, finite difference or finite element methods. This gives rise to a linear system of equations of the form

$$A_h u_h = b_h \tag{3.1.2}$$

where h is the typical grid size used in the discretization.

(2) Solve the linear system (3.1.2) using some direct or iterative method.

The rate of convergence of an iterative method depends on the *condition number* of the matrix A_h, defined as $\kappa(A_h) = ||A_h||_2 ||A_h^{-1}||_2$. Here $||A_h||_2$ is the l^2 norm of the matrix A_h. In most cases the condition number of A_h scales as follows:

$$\kappa(A_h) \sim h^{-2}. \tag{3.1.3}$$

This suggests that as h becomes smaller, iterative methods converge more slowly. It is easy to understand why this has to be the case. The finite difference or finite element operators are local, but the solution operator (the Green's function for (3.1.1)) is global. If one makes an error at one corner of the domain, it affects the solution at all other places. If the iteration method is local, as most iterative methods are, in order to remove such an error one has to run the propagator, i.e. the iteration, at least $\mathcal{O}(1/h)$ times. Of course, this argument would not work if the iterative procedure itself were global, which is the case when global bases are used to represent numerical solutions. Indeed, as we discuss later in this chapter, it is possible to construct hierarchical bases for finite element methods for which the condition number of A_h is drastically reduced [42].

An important observation is that the slow convergence is caused mostly by the error in the large-scale components of the solution. Such large-scale components can also be accurately represented on coarser grids. Since it is less costly to eliminate the errors on coarser grids, we should be able to construct much more efficient iterative algorithms by correcting the large-scale components of the error on coarser grids. This is the main intuition behind the multigrid method.

In order to turn this intuition into real algorithms, we have to know how to move the numerical solutions across different grids and how to set up iterations on different grids. This is indeed the central issue of the multigrid method. Our presentation here is very much influenced by the tutorial of Briggs *et al.* [13], which gives a very clear introduction to this topic.

To illustrate the details of the algorithm, it is useful to consider the one-dimensional case of (3.1.1). When $\Omega = [0, 1]$, (3.1.1) becomes

$$-\frac{d^2 u}{dx^2} = f(x), \quad x \in [0, 1],$$
$$u(0) = u(1) = 0. \tag{3.1.4}$$

We consider a finite difference discretization on a grid with grid points $x_j = jh$, $j = 1, \ldots, n - 1$, with $h = 1/n$. Using a standard second-order finite difference formula, we obtain from (3.1.4)

$$-\frac{u_{j+1} - 2u_j + u_{j-1}}{h^2} = f(x_j), \quad j = 1, \ldots, n - 1.$$

This can be written as

$$A_h u_h = f_h, \tag{3.1.5}$$

where

$$A_h = \frac{1}{h^2} \begin{pmatrix} 2 & -1 & & & \\ -1 & 2 & -1 & & \\ & \ddots & \ddots & \ddots & \\ & & \ddots & 2 & -1 \\ & & & -1 & 2 \end{pmatrix}$$

is a $(n - 1) \times (n - 1)$ tridiagonal matrix, $u_h = (u_1, \ldots, u_{n-1})^{\mathrm{T}}$ and $f_h = (f(x_1), \ldots, f(x_{n-1}))^{\mathrm{T}}$.

What makes this example particularly useful is that the eigenvalues and eigenvectors of A_h can be computed explicitly using discrete Fourier analysis. Here are the results.

(1) The eigenvalues are

$$\lambda_k = \frac{4}{h^2} \sin^2 \frac{k\pi h}{2}, \quad k = 1, \ldots, n-1. \qquad (3.1.6)$$

(2) The eigenvector w_k^h corresponding to the eigenvalue λ_k is given by

$$w_k^h(j) = \sin \pi k x_j = \sin \pi k j / n. \qquad (3.1.7)$$

The condition number is

$$\kappa(A_h) = \frac{\max|\lambda_k|}{\min|\lambda_k|} = \frac{\sin^2 \pi(1-h)/2}{\sin^2 \pi h/2} = \mathcal{O}(h^{-2}).$$

We now look at what happens when *iterative methods* are used to solve the linear system (3.1.5). The simplest iterative algorithms can be written in the form

$$u^{n+1} = Bu^n + g,$$

where B is the iteration matrix and g is a constant vector. Let u^* be the exact solution of (3.1.5), and define the error vector $e^n = u^n - u^*$. If we express e^0 as a linear combination of the eigenvectors $\{v_k\}$ of B,

$$e^0 = \sum_k c_k v_k,$$

then

$$e^n = \sum_k c_k \left(\mu_k\right)^n v_k$$

where μ_k is the eigenvalue of B corresponding to the eigenvector v_k. Hence

$$\frac{\|e^n\|}{\|e^0\|} \sim (\rho(B))^n$$

where $\rho(B) = \max\{|\mu_k|\}$ is the spectral radius of B. The convergence rate is determined by the spectrum of B.

Let D be the diagonal part of A_h and L be the lower subdiagonal part of $-A_h$. Then the symmetric matrix A_h can be rewritten as $A_h = D - L - L^{\mathsf{T}}$. Some of the simplest iterative algorithms are listed below.

(1) *Jacobi iteration*:

$$u^{n+1} = D^{-1}(L + L^{\mathrm{T}})u^n + D^{-1}f_h.$$

The Jacobi iteration matrix is given by

$$B_{\mathrm{J}} = D^{-1}(L + L^{\mathrm{T}}).$$

Its eigenvectors are the same as those of A_h. Its eigenvalues are $\mu_k^J = 1 - 2\sin^2(\pi k h/2)$, $k = 1, \ldots, n-1$.

(2) *Gauss–Seidel iteration*:

$$u^{n+1} = (D - L)^{-1}L^{\mathrm{T}}u^n + (D - L)^{-1}f_h.$$

Thus, the Gauss–Seidel iteration matrix is

$$B_{\mathrm{GS}} = (D - L)^{-1}L^{\mathrm{T}}.$$

(3) *Relaxation methods*:
   Starting with

$$u^{n+1} = Bu^n + g,$$

we can construct the relaxed iteration scheme

$$u^{n+1} = (1 - \omega)u^n + \omega(Bu^n + g).$$

The relaxed iteration matrix is given by

$$B_\omega = (1 - \omega)I + \omega B.$$

One can choose an optimal value of ω that minimizes the spectral radius of B_ω. The best-known such relaxation method is the successive overrelaxation method studied by David Young; see [41].

For the simple one-dimensional example discussed earlier, the spectral radius of B_{J} is

$$\rho(B_{\mathrm{J}}) = \mu_1^J = 1 - 2\sin^2(\pi h/2) \sim 1 - \pi^2 h^2/2.$$

Hence

$$\log \rho(B_{\mathrm{J}}) \sim \pi^2 h^2/2.$$

To reduce the size of the error e^n to a prescribed tolerance ε, the number of iterations needed is roughly

$$N \sim \log \varepsilon / \log \rho(B_{\mathrm{J}}) \sim \mathcal{O}(h^{-2}).$$

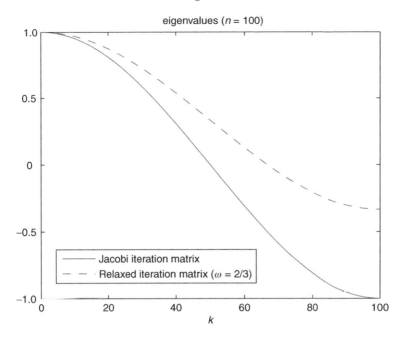

Figure 3.1. The eigenvalues as a function of k for a Jacobi matrix and a relaxed Jacobi matrix with $\omega = 2/3$ (courtesy of Xiang Zhou).

For the Gauss–Seidel iteration (see [13]),

$$\rho(B_{\mathrm{GS}}) = \cos^2(\pi h) \sim 1 - \pi^2 h^2,$$
$$\log \rho(B_{\mathrm{GS}}) \sim \pi^2 h^2.$$

Comparing with the Jacobi iteration, we have

$$\log \rho(B_{\mathrm{GS}}) \sim 2 \log \rho(B_{\mathrm{J}}).$$

Therefore, Gauss–Seidel iteration is about twice as fast as Jacobi iteration.

For relaxed Jacobi iteration, the eigenvalues are

$$\mu_k(B_\omega^J) = 1 - 2\omega \sin^2(\pi k h/2)$$

(see Figure 3.1). Take $\omega = 2/3$; then $\mu_k(B_\omega^J) < 1/2$ for $k > n/2$, i.e., the damping factor for high frequencies is less than half that for the relaxed iteration method with $\omega = 2/3$. This is better than Jacobi iteration. But, for the low-frequency components the damping factor is still of order $1 - \mathcal{O}(h^2)$, which is comparable with that of Jacobi iteration.

Note that the difficult components are the low-frequency components. Indeed for this reason, the iteration procedure is also called *smoothing*.

the numerical solution becomes smoother after a few iteration steps since the high-frequency components are effectively damped out. The low-frequency components can be accurately represented on coarser grids, on which the iterations converge faster. This is the key observation behind the multigrid method. To turn this observation into an algorithm, we need to know the following.

(1) How do we transfer functions between different grids?
(2) What are the iterative procedures on each different grid?

To see how things work, let us first consider a *two-grid* algorithm, on grids Ω_h and Ω_{2h} that correspond to grid sizes h and $2h$, respectively. The iteration scheme on each grid can be either the relaxation method or the Gauss–Seidel method. To transfer between functions defined on Ω_h and Ω_{2h}, we will make use of *restriction (or projection) operators*, which map functions defined on fine grids to functions on coarser grids:

$$I_h^{2h} : \Omega_h \to \Omega_{2h},$$

and *interpolation (or prolongation) operators*, which map functions defined on coarse grids to functions on finer grids:

$$I_{2h}^h : \Omega_{2h} \to \Omega_h.$$

With these operators, the two-grid algorithm can be expressed in the following form.

(1) Run a few smoothing steps on Ω_h using some simple iteration algorithms such as the Gauss–Seidel or the relaxation method.
(2) Calculate the residual $r_h = A_h u_h - f_h$.
(3) Project to a coarse grid: $r_{2h} = I_h^{2h} r_h$.
(4) Solve on the coarse grid: $A_{2h} e_{2h} = r_{2h}$.
(5) Interpolate to the fine grid: $u_h \leftarrow u_h - I_{2h}^h e_{2h}$.

It remains to specify the intergrid operators I_h^{2h}, I_{2h}^h and the operator A_{2h} on the coarse grid. In principle A_{2h} should be the effective operator on Ω_{2h}. In some situations, e.g. problems of the type treated in Chapter 8 or molecular dynamics models, it can be quite nontrivial to find a good A_{2h}. However, for the problem considered here one may simply choose A_{2h} to be the corresponding matrix obtained by discretizing (3.1.1) on the coarse grid with mesh size $2h$. One may also choose

$$A_{2h} = I_h^{2h} A_h I_{2h}^h.$$

To define I_h^{2h}, let v_h be any function defined on the grid Ω_h. We may simply define $I_h^{2h} v_h$ as the function on Ω_{2h} obtained by evaluating v_h on Ω_{2h}:

$$v_{2h}(j) := I_h^{2h}(v_h)(j) = v_h(2j), \quad j = 1, 2, \ldots, n/2 - 1.$$

In matrix form we have

$$
\begin{pmatrix} v_{2h}(1) \\ v_{2h}(2) \\ \vdots \\ v_{2h}(\frac{1}{2}n - 1) \\ v_{2h}(\frac{1}{2}n) \end{pmatrix}
=
\begin{pmatrix}
0 & 1 & 0 & 0 & 0 & \cdots & 0 & 0 \\
0 & 0 & 0 & 1 & 0 & \cdots & 0 & 0 \\
& & & \ddots & & \ddots & & \\
0 & 0 & 0 & 0 & 0 & \cdots & 0 & 1
\end{pmatrix}
\begin{pmatrix} v_h(1) \\ v_h(2) \\ \vdots \\ v_h(n-1) \\ v_h(n) \end{pmatrix}.
$$

We may also define $I_h^{2h} v_h$ by averaging over neighboring grid points. For example, we can average over the nearest neighbors with weights $(\frac{1}{4}, \frac{1}{2}, \frac{1}{4})$:

$$v_{2h}(j) := I_h^{2h}(v_h)(j) = \tfrac{1}{4} v_h(2j-1) + \tfrac{1}{2} v_h(2j) + \tfrac{1}{4} v_h(2j+1).$$

In matrix form, this is

$$
\begin{pmatrix} v_{2h}(1) \\ v_{2h}(2) \\ \vdots \\ v_{2h}(\frac{1}{2}n - 1) \\ v_{2h}(\frac{1}{2}n) \end{pmatrix}
=
\begin{pmatrix}
\frac{1}{4} & \frac{1}{2} & \frac{1}{4} & 0 & 0 & \cdots & & 0 \\
0 & 0 & \frac{1}{4} & \frac{1}{2} & \frac{1}{4} & 0 & \cdots & 0 \\
& & & \ddots & & \ddots & & \\
0 & 0 & 0 & 0 & 0 & \cdots & 0 & 1
\end{pmatrix}
\begin{pmatrix} v_h(1) \\ v_h(2) \\ \vdots \\ v_h(n-1) \\ v_h(n) \end{pmatrix}.
$$

$$(3.1.8)$$

For I_{2h}^h, we may define it to be (a constant multiple of) the adjoint operator (or transpose) of I_h^{2h}. For example, if I_{2h}^h is given by (3.1.8) then we may define I_h^{2h} as

$$
\begin{pmatrix} v_h(1) \\ v_h(2) \\ \vdots \\ v_h(n-1) \\ v_h(n) \end{pmatrix}
=
\begin{pmatrix}
\frac{1}{2} & 0 & 0 & \cdots & 0 \\
1 & 0 & 0 & \cdots & 0 \\
\frac{1}{2} & \frac{1}{2} & 0 & \cdots & 0 \\
0 & 1 & 0 & \cdots & 0 \\
0 & \frac{1}{2} & \frac{1}{2} & \cdots & 0 \\
& & & \ddots & \\
0 & 0 & 0 & \cdots & 1
\end{pmatrix}
\begin{pmatrix} v_{2h}(1) \\ v_{2h}(2) \\ \vdots \\ v_{2h}(\frac{1}{2}n - 1) \\ v_{2h}(\frac{1}{2}n) \end{pmatrix}.
\qquad (3.1.9)
$$

This simply means that, on the coarse-grid points, $v_h = I_{2h}^h v_{2h}$ takes the same values as v_{2h}. On the remaining grid points the value of v_h

is obtained through averaging the values on the neighboring coarse-grid points. In this case, $I_h^{2h} = 2(I_{2h}^h)^{\mathrm{T}}$.

Instead of averaging, one can also use linear interpolation to obtain the values of v_h on the remaining grid points.

To understand the convergence properties of such a procedure, note that we can write the multigrid iteration matrix as

$$B_{\mathrm{MG}} = (I - I_{2h}^h A_{2h}^{-1} I_h^{2h} A_h)(B_{J,\omega})^m,$$

where $B_{J,\omega} = (1 - \omega)I + \omega B_J$ and m is the number of smoothing steps on Ω_h. Here we have assumed that the relaxed Jacobi method is used as the smoothing operator. Even though the $\{w_h^k\}$ defined in (3.1.7) are no longer eigenvectors of B_{MG}, the subspace spanned by the pair $(w_h^k, w_h^{k'})$, $k' = n - k$, is still an invariant subspace of the matrix B_{MG}. In fact, it is easy to see that if I_{2h}^h and I_h^{2h} are defined as in (3.1.8) and (3.1.9) then we have

$$B_{MG}w_h^k = \left(1 - 2\omega \sin^2 \frac{k\pi h}{2}\right)^m \sin^2 \frac{k\pi h}{2}\left(w_h^k + w_h^{k'}\right), \quad (3.1.10)$$

$$B_{MG}w_h^{k'} = \left(1 - 2\omega \sin^2 \frac{k'\pi h}{2}\right)^m \cos^2 \frac{k\pi h}{2}\left(w_h^k + w_h^{k'}\right), \quad (3.1.11)$$

$k = 1, \ldots, n/2$. Take $\omega = 2/3$. For the high-frequency components corresponding to k', the damping factor is less than $1/2$, as indicated in Figure 3.1. For the low-frequency components corresponding to k, the factor $\sin^2(k\pi h/2)$, which is less than $1/2$, helps to reduce the damping factor to less than $1/2$. Therefore, as long as $m \geq 1$, the damping factor is uniformly less than $1/2$ for all frequencies; see Figure 3.2.

This two-grid algorithm can be easily generalized so that more grids, $\Omega_{4h}, \Omega_{8h}, \ldots$, are used. Instead of solving the equation $A_{2h}e_{2h} = r_{2h}$ exactly, we apply the same two-grid algorithm to this equation and proceed recursively. In this situation it is important to specify an overall strategy for moving from grid to grid. We may either follow a *V-cycle* or a *W-cycle*. In a *V*-cycle one starts from the fine grid Ω_h and then moves to progressively coarser grids until the coarsest grid is reached; the procedure is then reversed. In a *W*-cycle one may reverse the direction before the finest or coarsest grid is reached [13].

In this example, the mechanism responsible for the effectiveness of the multigrid method originates in the smoothing effect of the iterative

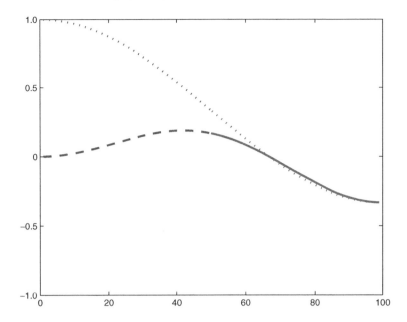

Figure 3.2. The damping factors in (3.1.10) and (3.1.11). Broken line, the low-frequency components $k = 1, \ldots, n/2$; solid line, the high-frequency components $k' = n - k$. The result from the relaxed Jacobi iteration method (the dotted curve, the same as in Figure 3.1) is also shown for comparison (courtesy of Xiang Zhou).

algorithms on each grid. One should not get the impression, however, that this is the only way in which this method can be effective.

The multigrid method has been extended to many other situations such as steady-state calculations of transonic flows [14] and Monte Carlo sampling methods [24] (see [12]). In these cases, the explanation given above is no longer sufficient. At present a unified general understanding of the mechanisms behind the effectiveness of the multigrid method is still lacking.

3.2 Fast summation methods

We now turn to fast summation methods. The general problem treated by such methods can be formulated as follows: given a kernel $K(\mathbf{x}, \mathbf{y})$ and a set of sources located at $\{\mathbf{x}_i\}$ with weights $\{w_j\}$, as well as a set of target locations $\{\mathbf{y}_k\}$, we want to compute

$$u_k = \sum_{j=1}^{N} w_j K(\mathbf{x}_j, \mathbf{y}_k) \qquad (3.2.1)$$

for $k = 1, \ldots, M$. Let us assume that M and N are comparable in size. Some examples of kernels include:

(1) the gravitational and Coulombic potential, for which $K(\mathbf{x}, \mathbf{y}) = \dfrac{1}{|\mathbf{x} - \mathbf{y}|}$;

(2) the gravitational force, for which $K(\mathbf{x}, \mathbf{y}) = \dfrac{\mathbf{x} - \mathbf{y}}{|\mathbf{x} - \mathbf{y}|^3}$;

(3) the heat kernel, for which $K(\mathbf{x}, \mathbf{y}) = e^{-|\mathbf{x} - \mathbf{y}|^2 / 4T}$.

In what follows, will focus on *analysis-based fast algorithms*, of which the fast multipole method (FMM) is the primary example. These algorithms are based on the analytical properties of the kernels involved in the problem. They should be contrasted with the other class of fast algorithms, such as the fast Fourier transform (FFT), based on the algebraic properties of the summands.

If we calculate (3.2.1) by direct summation, the cost is $\mathcal{O}(N^2)$. The tree code, which was a precursor of FMM, brings the cost down to $\mathcal{O}(N \log N)$. The use of FMM brings the cost down further to $\mathcal{O}(N)$ for arbitrary precision. However, it should be pointed out that the importance of FMM is not just in removing the logarithmic factor in the cost; it establishes a new framework for designing hierarchical algorithms based on analytical approximations, making it possible to develop fast summation algorithms beyond the use of multipole expansions.

3.2.1 Low-rank kernels

Assume that the kernel is in tensor product form, say $K(\mathbf{x}, \mathbf{y}) = K_1(\mathbf{x}) K_2(\mathbf{y})$; then, to evaluate

$$u_k = \sum_{j=1}^{N} w_j K_1(\mathbf{x}_j) K_2(\mathbf{y}_k)$$

for a set of sources $\{\mathbf{x}_j\}$ and targets $\{\mathbf{y}_k\}$, we may simply compute

$$A = \sum_{j=1}^{N} w_j K_1(\mathbf{x}_j)$$

and use

$$u_k = A K_2(\mathbf{y}_k).$$

This requires only $\mathcal{O}(M + N)$ operations.

Tensor product kernels are rank-1 kernels. Kernels of practical interest are rarely in tensor product form, however. Fortunately, some important ones can be very accurately approximated by the sum of a small number of kernels in tensor product form:

$$K(\mathbf{x}, \mathbf{y}) \approx \sum_{n=1}^{p} K_{n,1}(\mathbf{x}) K_{n,2}(\mathbf{y}). \tag{3.2.2}$$

In this case, we can use

$$u_k \approx \sum_{j=1}^{N} w_j \sum_{n=1}^{p} K_{n,1}(\mathbf{x}_k) K_{n,2}(\mathbf{y}_j)$$

$$= \sum_{n=1}^{p} M_n K_{n,1}(\mathbf{x}_k),$$

where the $M_n = \sum_{j=1}^{N} w_j K_{n,2}(\mathbf{y}_j)$ are the *moments*. The cost for computing the moments is $\mathcal{O}(Np)$. The additional cost for computing all the u_k is $\mathcal{O}(Mp)$. Thus the total computational cost is $\mathcal{O}(Mp + Np)$.

Expression (3.2.2) means that the kernel K is very well approximated by a kernel whose rank is no greater than p. The Gauss transform is one such example.

Consider the two-dimensional Gauss transform [26]

$$u(\mathbf{y}) = \sum_{j=1}^{N} w_j e^{-|\mathbf{y} - \mathbf{x}_j|^2 / 4T}.$$

A key observation is the following Taylor series expansion for one-dimensional heat kernels:

$$e^{-(x - x_0)^2} = \sum_{n=0}^{\infty} \frac{x_0^n}{n!} h_n(x),$$

where $h_n(x)$ is the nth-order Hermite function,

$$h_n(x) = (-1)^n \frac{d^n}{dx^n} \left(e^{-x^2} \right).$$

In two dimensions we have, for any $\mathbf{c} \in \mathbb{R}^2$,

$$e^{-|\mathbf{x} - \mathbf{x}_i|^2 / 4T} = \sum_{n_1, n_2 = 0}^{\infty} \Phi_{n_1, n_2}(\mathbf{x} - \mathbf{c}) \Psi_{n_1, n_2}(\mathbf{x}_i - \mathbf{c}), \tag{3.2.3}$$

where $\mathbf{x} = (x, y)$ and

$$\Phi_{n_1, n_2}(\mathbf{x}) = h_{n_1}\left(\frac{x}{\sqrt{4T}}\right) h_{n_2}\left(\frac{y}{\sqrt{4T}}\right),$$

$$\Psi_{n_1, n_2}(\mathbf{x}) = \frac{1}{n_1! n_2!}\left(\frac{x}{\sqrt{4T}}\right)^{n_1}\left(\frac{y}{\sqrt{4T}}\right)^{n_2}.$$

More importantly, if $\|\mathbf{x}_i - \mathbf{c}\| \leq \sqrt{T}$ then we have [26]

$$|e^{-|\mathbf{x}-\mathbf{x}_i|^2/4T} - \sum_{n_1, n_2=0}^{p} \Phi_{n_1, n_2}(\mathbf{x} - \mathbf{c})\Psi_{n_1, n_2}(\mathbf{x}_i - \mathbf{c})| \leq \left(\frac{1}{p!}\right)\left(\frac{1}{8}\right)^p.$$

$$(3.2.4)$$

The right-hand side decays very quickly as p increases. This means that the heat kernel is very well approximated by *low-rank kernels*.

Assume first that all the sources $\{\mathbf{x}_i\}$ are located in a square centered at $\mathbf{c}$ with side $\sqrt{T}$. Then

$$u_k = \sum_{j=1}^{N} w_j e^{-|\mathbf{y}_k - \mathbf{x}_j|^2/4T}$$

can be approximated by

$$u_k \approx \sum_{j=1}^{N} w_j \sum_{n_1, n_2=0}^{p} \Phi_{n_1, n_2}(\mathbf{y}_k - \mathbf{c})\Psi_{n_1, n_2}(\mathbf{x}_j - \mathbf{c}).$$

One can then use the ideas discussed earlier.

In the general case, let us cover all the sources and targets by a bounding box. If the box has side length larger than $\sqrt{T}$, we divide the bounding box into smaller boxes of size $\sqrt{T}$. Owing to the very rapid exponential decay of the heat kernel, to calculate u_k for each target $\mathbf{y}_k$ we need only sum over the contribution from the sources which lie in the $(2n+1)^2$ neighboring boxes:

$$u_k = \sum_{j} w_j e^{-|\mathbf{y}_k - \mathbf{x}_j|/4T} = \sum_{\mathbf{x}_j \in \mathcal{N}_k^n} w_j e^{-|\mathbf{y}_k - \mathbf{x}_j|/4T} + \mathcal{O}(e^{-n^2/4}). \quad (3.2.5)$$

Here n is a parameter that can be used to control the accuracy of the algorithm, $\mathcal{N}_k^n$ is the collection of the boxes within a distance $n\sqrt{T}$ from the box in which the target $\mathbf{y}_k$ lies ($|\mathcal{N}_k^n| = (2n+1)^2$ in two dimensions).

Using (3.2.4), we can compute the moments for each box B:

$$M^B_{n_1,n_2} = \sum_{\mathbf{x}_j \in B} w_j \Psi_{n_1,n_2}(\mathbf{x}_j - \mathbf{c}_B),$$

where $\mathbf{c}_B$ is the center of the box B. The cost of computing the moments for all the small boxes is Np^2. Since the moments are independent of the targets they can be precomputed. Now, for each target $\mathbf{y}_k$, (3.2.5) can be calculated approximately by using

$$u_k \approx \sum_{B \in \mathcal{N}^n_k} \sum_{n_1,n_2=0}^{p} M^B_{n_1,n_2} \Phi_{n_1,n_2}(\mathbf{y}_k - \mathbf{c}_B).$$

The computational cost for evaluating this sum is $(2n+1)^2 p^2 M$. Thus the total cost is proportional to $M + N$.

3.2.2 Hierarchical algorithms

We will restrict ourselves to two dimensions, for simplicity, and consider a kernel given by

$$K(\mathbf{x}, \mathbf{x}_0) = -\log(\|\mathbf{x} - \mathbf{x}_0\|). \tag{3.2.6}$$

In this case it is much more convenient to use complex variables. Identifying $\mathbf{x} = (x, y)$ with the complex number $z = x + yi$, and $\mathbf{x}_0 = (x_0, y_0)$ with the complex number $z_0 = x_0 + y_0 i$, we can then write (3.2.6) as

$$K(\mathbf{x}, \mathbf{x}_0) = \text{Re}(-\log(z - z_0)).$$

We will view the analytic function $\log(z - z_0)$ as the potential due to a unit point charge located at z_0.

Consider now a point charge of strength q, located at z_0. It is well known that, for any z with $|z| > |z_0|$,

$$\log(z - z_0) = \log z - \sum_{k=1}^{\infty} \frac{1}{k} \left(\frac{z_0}{z}\right)^k. \tag{3.2.7}$$

This is the Taylor expansion in z_0, or Laurent expansion in z. If there are m charges of strengths $\{q_i, i = 1, \ldots, m\}$ located at the points $\{z_i, i = 1, \ldots, m\}$ respectively, such that $\max_i |z_i| < r$, then, according to (3.2.7), if $|z| > r$ then the potential $\phi(z) = \sum_i q_i \log(z - z_i)$ induced by the

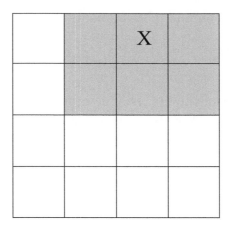

Figure 3.3. The first step of the tree code. The interactions between particles in box X and the white boxes can be computed via multipole expansions. The interactions with the nearest neighbor boxes (gray) are not computed yet.

charges is given by

$$\phi(z) = Q \log z + \sum_{k=1}^{\infty} \frac{a_k}{z^k}, \qquad (3.2.8)$$

where

$$Q = \sum_i q_i, \quad a_k = -\frac{1}{k} \sum_i q_i z_i^k.$$

This is called the multipole expansion of ϕ. The coefficients $\{a_k\}$ are generally referred to as the *moments*. If we truncate this series after p terms, the error incurred is bounded by [7, 25]

$$\left| \phi(z) - Q \log(z) - \sum_{k=1}^{p} \frac{a_k}{z^k} \right| \le \left(\frac{A}{p+1} \right) \left(\frac{1}{c-1} \right) \left(\frac{1}{c} \right)^p, \qquad (3.2.9)$$

where $c = |z/r|$ and $A = \sum_i |q_i|$. Note that the error bound depends on the target z through the parameter c. If we choose $c \ge 2$ then the error is bounded by $A/2^p$. If we want the error to be smaller than δ, we need to use $p = -\log_2 \delta$ terms in the expansion (3.2.8).

This observation together with a hierarchical strategy is the basis for the *tree code* developed by Appel [2] and Barnes and Hut [6].

Suppose that the charges are distributed in a square and that this square is divided into $2^l \times 2^l$ boxes; see Figure 3.3 for $l = 2$. Two boxes are nearest neighbors if they share a boundary point and are said to be well separated if they are not nearest neighbors. For example, the gray boxes in Figure 3.3 are nearest neighbors of the box X and the white boxes are well separated from X. We want to evaluate $\phi(z) = \sum_{i=1}^{M} q_i \log(z - z_i)$ at a location z which lies in, say, X. Denote the collection of boxes well separated from box X by $\mathcal{W}(z)$. For z_i in a box belonging to $\mathcal{W}(z)$, the *multipole expansion* (3.2.8) gives

$$\sum_{z_i \in \mathcal{W}(z)} q_i \log(z - z_i)$$

$$\times \sum_{B \in \mathcal{W}(z)} \sum_{z_i \in B} \left(\left(\sum_i q_i \right) \log(z - z_{0,B}) - \sum_{k=1}^{p} \frac{q_i (z_i - z_{0,B})^k}{k(z - z_{0,B})^k} \right), \quad (3.2.10)$$

where $z_{0,B}$ is the center of box B. Let

$$A_{k,B} = -\sum_{z_i \in B} \frac{q_i (z_i - z_{0,B})^k}{k}, \quad q_B = \sum_{\{i : z_i \in B\}} q_i.$$

Then the use of the multipole expansion for contributions from the well-separated boxes $\mathcal{W}(z)$ leads to the following:

$$\phi(z) = \sum_{i=1}^{M} q_i \log(z - z_i) \qquad (3.2.11)$$

$$\approx \sum_{B \in \mathcal{N}(z)} \sum_{z_i \in B} q_i \log(z - z_i)$$

$$+ \sum_{B \in \mathcal{W}(z)} \left(q_B \log(z - z_{0,B}) - \sum_{k=1}^{p} \frac{A_{k,B}}{(z - z_{0,B})^k} \right),$$

where $\mathcal{N}(z)$ is the collection of nearest neighbor boxes of the box containing z. The error made in these approximations can be estimated explicitly using (3.2.9).

Next, we apply the same idea to calculate the interaction between charges in the box X and its nearest neighbors,

$$\sum_{B \in \mathcal{N}(z)} \sum_{z_i \in B} q_i \log(z - z_i),$$

by subdividing each box into four smaller boxes. To do this systematically, we build a tree on hierarchy of boxes. This can be done inductively. At level 0 we have the entire computational domain. The

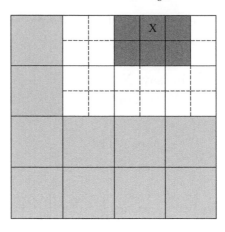

Figure 3.4. The second step of the tree code. After refinement, $l = 3$. The interactions between particles in box X (note that, at the current level, this box X is one of the children of the box X in Figure 3.3) and the light gray boxes were calculated. The white boxes are well separated from box X, which is at level $l = 3$. Therefore these interactions can be calculated using a multipole expansion again. But the nearest neighbor interactions (dark gray) are not computed at this level.

level-$(l+1)$ boxes are obtained recursively from the level-l boxes by subdividing each box into four equal parts. This yields a natural (quad-)tree structure, where for each box, the four boxes obtained by subdividing it are called its children. If we assume that the charges are fairly homogeneously distributed, then we need roughly $\log_4 N$ levels to obtain N boxes at the finest scale. On average, each smallest box contains roughly one charge. We will denote by L the number of levels, i.e. $l = L$ is the finest level.

With each box X we associate an *interaction list*, defined in the following way. Denote the parent of X as $\tilde{X}$. Consider the children of the nearest neighbors of $\tilde{X}$. The interaction list of X is defined as the subset of this collection of boxes which are not the nearest neighbors of X. Obviously boxes in the interaction list of X are at the same level as X. For example, in Figure 3.4 the white boxes are in the interaction list of the box X. A multipole expansion is used to calculate the potential inside the box X due to the charges in the boxes belonging to the interaction list of X.

Given a target z, let $B^l(z)$ be the level-l box that contains z. Let $\mathcal{N}^l(z)$ be the collection of the nearest neighbor boxes of $B^l(z)$ and let $\mathcal{L}^l(z)$ be the interaction list of $B^l(z)$. It is easy to see that

$$\sum_{B \in \mathcal{N}^l(z)} \sum_{z_i \in B} q_i \log(z - z_i) = \sum_{B \in \mathcal{L}^{l+1}(z)} \sum_{z_i \in B} q_i \log(z - z_i)$$

$$+ \sum_{B \in \mathcal{N}^{l+1}(z)} \sum_{z_i \in B} q_i \log(z - z_i). \quad (3.2.12)$$

The calculation of the first term on the right-hand side can be done using the multipole expansion. For the second term, we can use (3.2.12) recursively with l replaced by $l + 1$. These recursive steps can be continued until the finest level is reached. At that point, one can use direct summation to compute the potential due to the charges in the nearest neighbor boxes. In other words, we have, for all z,

$$\sum_i q_i \log(z - z_i) = \sum_{l < L} \sum_{B \in \mathcal{L}^l(z)} \sum_{z_i \in B} q_i \log(z - z_i) + \sum_{B \in \mathcal{N}^L(z)} \sum_{z_i \in B} q_i \log(z - z_i).$$

Note that the coefficients in the multipole expansions are independent of the targets and so can be precomputed. The total cost for precomputing the moments for the boxes in all $\mathcal{O}(\log N)$ levels is approximately $Np \log N$. For each target in a given box, $27p$ operations are required to compute the interaction with the boxes in the interaction list at each refinement level. A total of $27p \log N$ operations is needed for each target. At the finest level, there is on average one particle in each nearest neighbor box, therefore eight operations are required for each target. Hence, the total cost is approximately

$$28Np \log N + 8N.$$

3.2.3 The fast multipole method

In order to obtain a linear scaling algorithm, let us examine the origin of the $\mathcal{O}(N \log N)$ cost. First, the computation of the moments in order to build the multipole expansion for all boxes at *all* levels requires $pN \log N$ operations. This can be avoided if we know how to build efficiently the multipole expansion for a box from the multipole expansions of its four children. The main issue is that the multipole expansions for the four children boxes are made about the center of each of the children boxes, whereas the multipole expansion for the parent box should be made about the center of the parent box. The following simple lemma tells us how to translate a multipole expansion made about to one center to a multipole expansion made about a different center.

Lemma (Translation of multipole expansion) *Suppose that*

$$\sum_{i=1}^{m} q_i \log(z - z_i) = a_0 \log(z - z_0) + \sum_{k=1}^{\infty} \frac{a_k}{(z - z_0)^k}$$

is a multipole expansion of the potential due to a set of m charges of strengths $\{q_i, i = 1, \ldots, m\}$ such that $\max_i |z_0 - z_i| < R$. If z satisfies $|z| < R + |z_0|$, $p > 1$, then

$$\left| \sum_{i=1}^{m} \log(z - z_i) - a_0 \log z - \sum_{l=1}^{p} \frac{b_l}{z^l} \right| \leq \left(\frac{A}{1 - |c|} \right) |c|^{p+1}$$

where $c = (|z_0| + R)/z$, $A = \sum_{i=1}^{m} |q_i|$ and

$$b_l = -\frac{a_0 z_0^l}{l} + \sum_{k=1}^{l} a_k z_0^{l-k} \binom{l - 1}{k - 1}.$$

This lemma tells us how to convert a multipole expansion centered at z_0 to a multipole expansion centered at 0.

Using this lemma, we can develop an *upward pass* procedure to build multipole expansions for all the boxes, starting from the multipole expansion for the boxes at the finest level. This reduces the cost of forming multipole expansions from $\mathcal{O}(N \log N)$ to $\mathcal{O}(pN + 4p^2 \log N)$.

The second source of the $\mathcal{O}(N \log N)$ cost in the tree code comes from the fact that the value of the potential is evaluated separately for each target and that the calculation for each target costs $\mathcal{O}(\log N)$ since the interaction list at all levels has to be visited and there are $\mathcal{O}(\log N)$ interaction lists for each target. The fast multipole method overcomes this problem by introducing local expansions.

For any given box X at some refinement level, we will set up an expansion for the potential at *any position* in box X due to all the sources located outside the nearest neighbors of X. This expansion is called the *local expansion*.

Lemma *Suppose that*

$$\sum_{i=1}^{m} \log(z - z_i) = a_0 \log(z - z_0) + \sum_{k=1}^{\infty} \frac{a_k}{(z - z_0)^k}$$

is a multipole expansion of the potential due to a set of m charges of strengths $\{q_i, i = 1, \ldots, m\}$ such that $\max_i |z_0 - z_i| < R$ and $|z_0| > (c+1)R$. Then, if z is such that $|z| < R$ and if $p > 1$,

$$\left| \sum_{i=1}^{m} \log(z - z_i) - \sum_{l=0}^{p} b_l z^l \right| \leq \frac{A(4e(p+c)(c+1) + c^2)}{c(c-1)} \left(\frac{1}{c} \right)^{p+1},$$

where

$$b_0 = a_0 \log(-z_0) + \sum_{k=1}^{\infty} \frac{a_k}{z_0^k} (-1)^k$$

and

$$b_l = -\frac{a_0}{l z_0^l} + \frac{1}{z_0^l} \sum_{k=1}^{\infty} \frac{a_k}{z_0^k} \binom{l+k-1}{k-1} (-1)^k.$$

This lemma allows us to convert multipole expansions into local expansions. Note that in contrast with the multipole expansions based on the centers of the boxes in which the sources lie, the local expansions are based on the centers of the boxes in which the targets lie.

Once we have the local expansions for all the boxes, the final step is to distribute the contributions of the sources in each box to the targets in its interaction list. For this purpose, we need to translate a local expansion with respect to one center to a local expansion with respect to a different center, i.e. we need to convert an expansion of the form $\sum_{k=0}^{p} a_k (z - z_0)^k$ to one of the form $\sum_{k=0}^{p} b_k z^k$. This can be done using the standard Horner algorithm.

We are ready to describe the main steps in the fast multipole method (FMM).

(1) *Initialization.* Build $\log_4(N/s)$ levels of boxes, so that there are, on average, s particles per box at the finest level.

(2) *Upward pass.* Beginning at the finest level, create the multipole expansions of the potential for each box due to the sources in this box. The expansions for all boxes at coarser levels are then formed by the merging procedure described in the translation of multipole expansion lemma above. The cost is $\mathcal{O}(Np + p^2 N/s)$.

(3) *Downward pass.* Beginning at the coarsest level, convert the multipole expansion of each box to a local expansion about the centers of the boxes in its interaction list.

These local expansions are shifted to their children's level, using the Horner algorithm described above. Therefore, at each level the local expansion for each box receives contributions from two sources: one inherited from the parent box and others from sources in its interaction list.

At the finest level we have local expansions for each finest-scale box, which account for contributions from charges that lie outside the nearest neighbors of each box. These local expansions are then evaluated at each target. Contributions from neighboring finest-level boxes are computed directly.

The cost for precomputing the local expansions is $\mathcal{O}(p^2 N/s)$; the cost for evaluating the local expansions at the targets is $\mathcal{O}(pN/s)$. Thus the total cost in this step is $\mathcal{O}(p^2 N/s)$. The cost for the direct summation over the nearest neighbors at the finest level is $9(N/s)s^2 = 9Ns$ since s^2 operations are required to calculate the interaction between one box and its neighbors.

It should be noted that while the multipole expansion is oriented towards the sources, the local expansion is oriented toward the targets. In addition, when performing the final summation the focus of the $\mathcal{O}(N \log N)$ algorithm is on the targets: for a given target the contributions from the sources in its interaction list are summed from scale to scale. In FMM, however, the focus is on the sources. At each scale, for each box, one asks what the contributions are of the sources in that box to the boxes in its interaction list. Previously we were examining the interaction list of the targets. Now we are examining the interaction list of the sources. This dual viewpoint is very important for generalizing analysis-based fast summation algorithms to other settings.

3.3 Adaptive mesh refinement

Adaptivity is a central issue in modeling and computation and has numerous aspects. One can adaptively choose models, discretization methods, stopping criteria, time-step size, and mesh size. Well-known examples of adaptive numerical techniques include:

(1) adaptive numerical quadrature, in which new quadrature points are added on the basis of some local error indicators [15];
(2) ODE solvers with adaptive time-step size selection [15];
(3) adaptive mesh refinement for solving PDEs [1].

Here we will focus on adaptive mesh refinement for solving PDEs. This has been developed along two different lines.

(1) The adaptive finite element methods pioneered by Babuska and Rheinboldt [3, 4] are based on rigorous a posteriori error estimates.

(2) Adaptive finite difference or finite volume methods for nonlinear time-dependent PDEs such as those that arise in gas dynamics were pioneered by Berger, Collela, Oliger *et al.* [8]. In this case it is much harder to develop rigorous *a posteriori* error estimates. Thus, quantities such as the gradients of the numerical solutions are often used as the error indicators.

Recently, both strategies have been extended to include the possibility of adaptively selecting physical models [1, 23], a topic to which we will return in Chapter 7.

3.3.1 A posteriori *error estimates and local error indicators*

The first thing we need for an adaptive mesh refinement scheme is some local error indicators that are easily computable. This is the purpose of developing a posteriori error estimates. In numerical analysis, we are usually more familiar with a priori error estimates, which are typically in the form

$$\|u - u_h\|_1 \leq Ch\|u\|_2. \tag{3.3.1}$$

Here $\|v\|_1$ and $\|v\|_2$ are the H^1 and H^2 norms, respectively, of the function v. Expression (3.3.1) tells us that the convergence rate is first order. But as a quantitative measure of the error, it has limited value since we do not know the norm of u that appears on the right-hand side of (3.3.1). In addition, (3.3.1) gives an estimate of the global error, not the local error around each element.

The purpose of a posteriori error estimates is to find easily computable quantities $\{e_K(u_h)\}$ that give an estimate of the error near the element K:

$$\|u - u_h\|_{1,K} \sim e_K(u_h).$$

This kind of work was pioneered by Babuska and Rheinboldt [3, 4]. A comprehensive treatment of the subject is presented in [1].

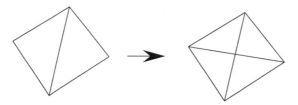

Figure 3.5. Refining the elements

To give a typical example of such a posteriori error estimates, let us consider the second-order elliptic equation in two dimensions:

$$Lu(\mathbf{x}) = -\nabla \cdot (a(\mathbf{x})\nabla u(\mathbf{x})) = f(\mathbf{x}), \quad \mathbf{x} \in \Omega,$$

with the Dirichlet boundary condition $u(\mathbf{x}) = 0, \mathbf{x} \in \partial\Omega$. Consider a piecewise linear finite element method on a regular mesh and denote by u_h the finite element solution. For each element K, define the residual $r_h = Lu_h - f$. For each edge Γ that is shared by a pair of elements K and K', define $J_\Gamma(u_h)$ to be the jump of $\mathbf{n} \cdot (a\nabla u_h)$ across that edge, where $\mathbf{n}$ is the normal to the edge (the sign of $\mathbf{n}$ does not matter).
Define

$$\varepsilon(u_h) = \left(\sum_K h_K^2 \int_K |r_h|^2 d\mathbf{x} + \sum_\Gamma h_\Gamma \int_\Gamma |J_\Gamma(u_h)|^2 ds \right)^{1/2}. \quad (3.3.2)$$

Here the first sum is over all elements and h_K is the diameter of the element K. The second sum is over all interior edges and h_Γ is the length of the edge Γ. It can be shown [1, 21, 38] under various conditions that

$$C_1 \varepsilon(u_h) \leq \|u - u_h\|_1 \leq C_2 \varepsilon(u_h),$$

where C_1 and C_2 are constants. Moreover, each term inside the expression of $\varepsilon(u_h)$ is a measure of the local error near that element. Therefore, these quantities can be used as a criterion for mesh refinement.

Other types of *a posteriori* error estimators are discussed in [1] (see also [5, 43]).

In some cases our interest is not in obtaining an accurate approximation of the whole solution, but rather in some quantities associated with the solution. For example, when modeling the flow past obstacles we might be most interested in the total drag or lift on the body. In this case, we might want to refine further in places, say near the body, where the accuracy of the approximate solution is more important for the quantities in which we are interested. The procedure outlined above has been extended to this case [1]. In such a "goal-oriented adaptive mesh refinement scheme," the error indicators are usually obtained by solving an adjoint problem.

3.3.2 The moving mesh method

Next we discuss briefly an alternative procedure for adapting the mesh, the moving mesh method. In this way of redistributing the grid points, more computational effort is directed at places where it is needed, e.g. places where the solution has large gradients. The moving mesh method have its origin in the moving finite element method [32, 17]. Current versions of the moving mesh method have largely followed the framework proposed by Winslow, in which the mesh is regulated using a *monitor function* [39], which is a carefully designed function of the numerical solution. The quality of the numerical mesh is improved by minimizing the monitor function.

There are two slightly different ways of implementing the moving mesh method. One is to introduce an auxiliary computational domain on which the discretization is carried out, usually a regular grid. The computational domain and the physical domain are related by a mapping, which is obtained adaptively using the monitor function. This approach can be regarded as an adaptive (or dynamic) change of variable. One very elegant way of implementing this is to formulate the original PDE in the computational domain and to supplement the resulting PDE with an additional equation for the mesh [29, 33, 34].

Denote by α the variable in the computational domain and $\mathbf{x}$ the variable in the physical domain. They are related by the function $\alpha = \alpha(\mathbf{x})$. We are looking for the function α such that a regular grid in the α domain is mapped to the desired grid in the physical domain, with grid points clustered in the areas of interest. This can be done using a variational formulation with the functional

$$E(\alpha) = \int \frac{1}{w(\mathbf{x})} |\nabla_{\mathbf{x}} \alpha|^2 \, d\mathbf{x}. \tag{3.3.3}$$

Here $w = w(\mathbf{x})$ is the monitor function. The Euler–Lagrange equation for the minimization of this functional is

$$\nabla \cdot \left(\frac{1}{w(\mathbf{x})} \nabla_{\mathbf{x}} \alpha \right) = 0.$$

In one dimension this reduces to

$$w(x) \frac{dx}{d\alpha} = C.$$

Assume that we have a uniform grid in the α domain, with $\alpha_n = n\Delta\alpha$, $n = 1, \ldots, N$. Let $\alpha_n = \alpha(x_n)$. Integrating over $[\alpha_{n-1}, \alpha_n]$ gives

$$\int_{x_{n-1}}^{x_n} w(x)\, dx = C\Delta\alpha,$$

i.e. $\Delta x\, w(x) \approx \int_{x_{n-1}}^{x_n} w(x)\, dx$ is approximately a constant. The grid points are concentrated in the regions where the monitor function w is large.

The other approach is to work directly with the original physical domain and change the mesh from time to time using the monitor function. At each step, one first minimizes the functional (3.3.3) to find the new mesh. This is followed by an interpolation step to provide the values of the numerical solution on the new mesh. This approach has been explored extensively in [29] and subsequent works.

In cases when a particularly truly high-quality mesh is required, for example when the solution is singular, an iterative procedure is used to improve the quality of the mesh [34].

The advantage of the first approach is that one works with a standard regular grid on the computational domain. There are no complications coming from, for example, numerical interpolation. The disadvantage is that by moving away from the physical domain the structure of the original PDE, such as its conservation form, might be lost. The second approach, however, is exactly complementary to the first.

Finding the right monitor function is a crucial component of the moving mesh method. Commonly used monitor functions are often in the form

$$w(\mathbf{x}) = \sqrt{1 + u(\mathbf{x})^2 + |\nabla u(\mathbf{x})|^2},$$

where u is the numerical solution.

Compared with the adaptive mesh refinement strategies discussed earlier, the moving mesh method has the advantage that essentially it works with a regular and smooth mesh. Therefore it has the nice properties of a regular, smooth mesh, such as superconvergence. However, the moving mesh method is in general less flexible since its control of the mesh is rather indirect, being through the monitor function.

3.4 Domain decomposition methods

The intuitive idea behind the domain decomposition method (DDM) is quite simple: the computational domain is divided into smaller subdomains and a computational strategy is established that is based on solving the given problem on each subdomain and making sure that the solutions on different subdomains match. There are several advantages of such a strategy.

(1) It is very well suited for parallel computation.
(2) It can be used as a localization strategy, as in electronic structure analysis [40].
(3) According to the nature of the solution on the different subdomains, one may use different models or different algorithms. For this reason, DDM has been used extensively as a platform for developing algorithms with multi-physics models.

The main challenge in DDM is to make sure that the solutions on different subdomains match each other. To discuss how this is done, we will distinguish two different cases, when the subdomains do or do not overlap. See Figure 3.6.

3.4.1 Nonoverlapping domain decomposition methods

We first discuss the case when the subdomains do not overlap. We will use the example

$$\begin{cases} -\triangle u(\mathbf{x}) = f(\mathbf{x}), & \mathbf{x} \in \Omega \\ u(\mathbf{x}) \quad = 0, & \mathbf{x} \in \partial\Omega \end{cases} \tag{3.4.1}$$

to illustrate the main issues.

After discretization on Ω, say using a finite difference method, we obtain a linear system:

$$Au = b. \tag{3.4.2}$$

To solve this linear problem, we divide the domain Ω into a union of two

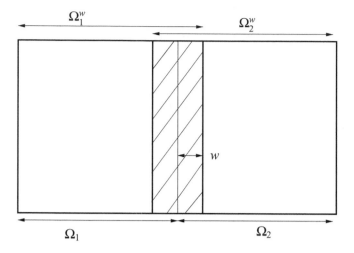

Figure 3.6. Illustration of overlapping domain decomposition (upper labels) and nonoverlapping domain decomposition (lower labels) (courtesy of Xiang Zhou).

non-overlapping subdomains Ω_1 and Ω_2; see Figure 3.6. We denote by u_1 and u_2 the solutions in Ω_1 and Ω_2 respectively, and by w the solution on the interface between Ω_1 and Ω_2. The width of the interfacial region is chosen so that the components u_1 and u_2 do not interact directly, i.e. if we let $u = (u_1 \quad u_2 \quad w)^{\mathrm{T}}$ then (3.4.2) can be written as

$$\begin{pmatrix} A_{11} & 0 & A_{13} \\ 0 & A_{22} & A_{23} \\ A_{31} & A_{32} & A_{33} \end{pmatrix} \begin{pmatrix} u_1 \\ u_2 \\ w \end{pmatrix} = \begin{pmatrix} b_1 \\ b_2 \\ b_3 \end{pmatrix}. \tag{3.4.3}$$

Eliminating u_1 and u_2, we obtain

$$Cw = d, \tag{3.4.4}$$

where $d = b_3 - A_{31}A_{11}^{-1}b_1 - A_{32}A_{22}^{-1}b_2$. The matrix C is the Schur complement of A_{33}:

$$C = A_{33} - A_{31}A_{11}^{-1}A_{13} - A_{32}A_{22}^{-1}A_{23}.$$

It is usually a full matrix and so computing it explicitly is costly. Fortunately there are many numerical algorithms for solving (3.4.4) which only require the action of C on a given vector to be evaluated, not C itself. One example of such an algorithm is the the conjugate gradient method. From (3.4.1), we see that evaluating Cv for a given vector v amounts to solving the linear systems of equations on the subdomains Ω_1 and Ω_2:

$$Cv = A_{33}v - A_{31}v_1 - A_{32}v_2,$$

where v_1 and v_2 are solutions of

$$A_{11}v_1 = A_{13}v, \quad A_{22}v_2 = A_{23}v.$$

These two equations can be solved independently, thereby decoupling the solutions on the two domains Ω_1 and Ω_2. The same can be said about evaluating d.

To understand the nature of C, it helps to introduce the Dirichlet-to-Neumann operator (DtN) for a domain Ω and a subset of the boundary $\partial\Omega$; it is denotd by Γ. Consider the problem

$$-\triangle u(\mathbf{x}) = f(\mathbf{x}), \quad \mathbf{x} \in \Omega, \tag{3.4.5}$$
$$u(\mathbf{x}) = 0, \quad \mathbf{x} \in \partial\Omega\backslash\Gamma. \tag{3.4.6}$$

This is the same as (3.4.1) except that the boundary condition is specified only on $\partial\Omega\backslash\Gamma$, so the problem is not fully determined. Now, given a function w defined on Γ we can obtain a unique solution of (3.4.5) using the additional condition that $u = w$ on Γ. Denote this solution by U_w. The Dirichlet-to-Neumann operator acting on w is defined as the normal derivative of U_w on Γ. Hence it is a map between functions defined on Γ.

Consider the special case when $\Omega = \mathbb{R}^+ = \{(x_1, x_2), x_2 > 0\}$ and $\Gamma = \partial\Omega = \mathbb{R}^1$. Given w on $\partial\Omega = \mathbb{R}^1$, we can solve (3.4.5) with $f = 0$ and w as the boundary condition $u = w$ on $\partial\Omega$ using a Fourier transform in the x_1 direction:

$$\hat{U}_w(k_1, x_2) = e^{-|k_1||x_2|}\hat{w}(k_1),$$

where k_1 is the variable for the Fourier transform in the x_1 direction and $\hat{f}$ denotes the Fourier transform of f. Hence, on the Fourier side, we have

$$\frac{\partial}{\partial x_2}\hat{U}_w(k_1, x_2)|_{x_2=0} = |k_1|\hat{w}(k_1).$$

From this we see that the Fourier symbol of the Dirichlet-to-Neumann operator is $|k_1|$. This means in particular that it is effectively a first-order operator.

Now we can understand the nature of C, as follows. Given some candidate function w, we can solve for the corresponding u_1 and u_2 from the first two equations in (3.4.3):

$$A_{11}u_1 = b_1 - A_{13}w,$$
$$A_{22}u_2 = b_2 - A_{23}w.$$

The quantities u_1 and u_2 have the same values at the interface but their normal derivatives do not necessarily match. To obtain a solution of the original problem over the whole domain Ω, we should require that u_1 and u_2 have the same normal derivative at the interface. This is the equation, (3.4.4), that w solves. Therefore C can be regarded as a discretization of the difference of the Dirichlet-to-Neumann operator (DtN) on Ω_1 and Ω_2. Since DtN is a first-order operator the condition number of C should scale as $\mathcal{O}(h^{-1})$, where h is the grid spacing. This is an improvement over the condition number of the original problem (3.4.3), which is usually $\mathcal{O}(h^{-2})$.

3.4.2 Overlapping domain decomposition methods

Another option is to allow the subdomains to overlap, as also indicated in Figure 3.6. One can then set up an iterative procedure along the following lines: starting with some (arbitrary) boundary condition on $\partial\Omega_1^w$, one solves for u_1^1 (over Ω_1). From u_1^1, one can obtain some boundary condition on $\partial\Omega_2^w$ and use it to solve for u_2^1 (over Ω_2). From u_2^1, one can obtain some boundary condition on $\partial\Omega_1^w$ and use it to solve for u_1^2. From there one goes on to solve for u_2^2 as before. Continuing this procedure, one obtains $u_1^3, u_2^3, \ldots$ This is the well-known Schwarz iteration (see Figure 3.7).

The main issue is how to obtain the boundary conditions. They can be Dirichlet, Neumann, or some other type of boundary conditions; the original Schwarz iteration uses a simple Dirichlet boundary condition. If the boundary conditions are not chosen properly, such an iteration may not converge. The following is a simple example that illustrates this point.

Consider the problem $-u'' = 0$ with boundary conditions $u(0) = 1$ and $u(1) = 0$. The exact solution is $u(x) = 1 - x$. We define $\Omega_1 = [0, b]$ and $\Omega_2 = [a, 1]$ with overlap region $[a, b]$, where $0 < a < b < 1$. It is easy to see that if we use the Dirichlet boundary conditions at a and b, the algorithm converges. The convergence rate depends on the size of the overlap region (see Section 7.4):

$$r = \frac{a}{b}\left(\frac{1-b}{1-a}\right).$$

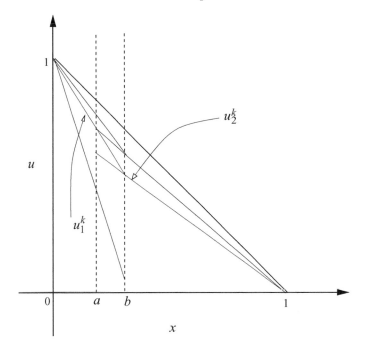

Figure 3.7. The overlapping domain decomposition for the one-dimensional problem $u'' = 0$. (courtesy of Xiang Zhou).

If, however, we use Neumann boundary conditions at a and b, it is easy to see that $\{u_1^k, u_2^k\}$ do not change and the procedure does not give the right solution. We will return to this example in Chapter 7.

3.5 Multiscale representation

To find the solution to any physical model, first we need a way of representing the solution. Consider the three examples of functions shown in Figure 3.8. The first example has a near discontinuity but it is isolated. The second example shows a repetitive pattern which in this case happens to be periodic.

It is intuitively quite clear that localized bases such as wavelet bases are suitable for the first example; periodic bases such as Fourier bases are suitable for the second example. It is not quite clear what to do about the third example. Clearly, if we restrict ourselves to the Fourier and wavelet representations then the relative efficiency of the two depends on the amplitude of the large gradients and the frequency of the oscillations.

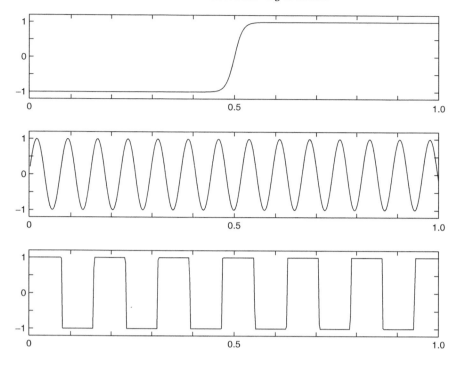

Figure 3.8. Three different kinds of multiscale function.

If we use a Fourier representation, the number of Fourier coefficients needed to achieve a particular accuracy depends strongly on the size of the large gradient but only weakly on the frequency. If we use a wavelet representation, the number of wavelet coefficients needed to achieve a particular accuracy depends linearly on the frequency but only weakly on the size of the gradient.

These simple examples illustrate the kind of multiscale behavior that we need to consider. In particular, the repetitive pattern displayed in the second and third examples is in some ways quite representative of the situation with scale separation when the macroscale of the system is much larger than the microscale. In this case the local microstructure is repetitive in some pointwise or statistical sense.

3.5.1 Hierarchical bases

Consider a partition of the unit interval $\Omega = [0, 1]$ into $N = 2^n$ subintervals of equal length and consider a piecewise-linear finite element space over this partition with Dirichlet boundary condition, as follows:

$$V_h = \{v_h : v_h \text{ is continuous and piecewise linear and } v_h(0) = v_h(1) = 0\}.$$
(3.5.1)

Denote the nodes in this partition by $\{x_j\}$, $j = 1, \ldots, N-1$, $x_j = jh$, $h = 1/N$. In standard finite element methods we use a nodal basis to represent functions in the space V_h. These are functions $\{\phi_k\}$ in V_h that satisfy

$$\phi_k(x_j) = \delta_{jk}, \quad j, k = 1, 2, \ldots, N-1.$$
(3.5.2)

This basis has the advantage that, for any function v_h in V_h, the coefficients in the expansion of v_h in terms of this nodal basis are simply the values of v_h at the nodes:

$$v_h(x) = \sum_{1}^{N-1} c_j \phi_j(x), \quad c_j = v_h(x_j).$$
(3.5.3)

However, it also has the disadvantage that the condition number of the stiffness matrix for the finite element method, defined as

$$A_h = (a_{ij}), \quad a_{ij} = \int_0^1 \phi_i(x) L \phi_j(x)\, dx$$

for an operator L, is usually very large. Recall that the condition number of a matrix A is defined as $\kappa(A) = \|A\|\|A^{-1}\|$ if the norm $\|\cdot\|$ is used to measure the vectors. For example, if L is a second-order elliptic differential operator L then typically $\kappa(A_h) \sim \mathcal{O}(h^{-2})$. This is an important issue since the condition number controls the accuracy and efficiency of the algorithms in numerical linear algebra [20] (see also Section 3.1).

This poor scaling behavior of $\kappa(A_h)$ is partly due to the locality of the nodal basis functions. An alternative basis set, the hierarchical basis, was suggested and analyzed by Yserentant [42] and is shown schematically in Figure 3.9. One advantage of this basis is that the condition number of the associated stiffness matrix is $\mathcal{O}(1)$.

When hierarchical bases are used to represent functions in the finite element space, the coefficients are no longer simply the values at the nodes or vertices. However, owing to the hierarchical nature of the basis, efficient hierarchical algorithms can be constructed to compute these coefficients [42].

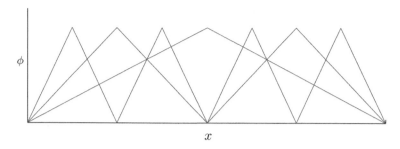

Figure 3.9. Illustration of hierarchical basis functions (courtesy of Xiang Zhou).

Hierarchical bases can be constructed for finite element spaces in high dimensions [42]. The condition number of the stiffness matrix in high dimensions is no longer $\mathcal{O}(1)$. For example, in two dimensions the condition number scales as $\mathcal{O}(|\log h|^2)$, which is still much better than the $\mathcal{O}(h^{-2})$ scaling for the nodal basis.

3.5.2 *Multi-resolution analysis and wavelet bases*

In multi-resolution analysis (MRA), we consider a sequence $\{V_j, j \in \mathbb{Z}\}$ of nested subspaces of the ambient space (say $L^2(\mathbb{R})$), where, roughly speaking, V_j contains functions with components up to the jth scale. The precise requirement for this ladder of subspaces was formalized by Mallat [30, 31]. A sequence of subspaces $\{V_j\}_{j \in \mathbb{Z}} \subset L^2(\mathbb{R})$ is an MRA if the following holds:

(1) $V_j \subset V_{j+1}, \quad j \in \mathbb{Z}$;
(2) V_j is the scaled version of V_0: $f(x) \in V_0$ if and only if $f(2^j x) \in V_j$;
(3) $\bigcap_{j \in \mathbb{Z}} V_j = \{0\}$, $\overline{\bigcup_{j \in \mathbb{Z}} V_j} = L^2(\mathbb{R})$;
(4) there exists a function ϕ such that the integer translations $\{\phi(\cdot - n), n \in \mathbb{Z}\}$ constitute an orthonormal basis for V_0.

Consequently, if we define $\phi_{k,j}$ by

$$\phi_{k,j}(x) = 2^{j/2}\phi(2^j x - k)$$

then, for each fixed j, the system $\{\phi_{k,j}, k \in \mathbb{Z}\}$ is an orthonormal basis for V_j.

Since $V_0 \subset V_1$, we can express ϕ in terms of the basis functions of V_1:

$$\phi(x) = \sum_k a_k \sqrt{2}\phi(2x - k). \qquad (3.5.4)$$

Functions satisfying (3.5.4) are called *scaling* or *refinable functions* and the sequence $\{a_k\}$ is called the *refinement mask*.

Let W_0 be the orthogonal complement of V_0 in V_1, i.e. $V_1 = V_0 \oplus W_0$, $V_0 \perp W_0$. Define a new function ψ by

$$\psi(x) = \sum_k b_k \sqrt{2}\phi(2x - k), \qquad (3.5.5)$$

where

$$b_k = (-1)^k a_{-k+1}, \quad k \in \mathbb{Z}.$$

The sequence $\{b_k\}$ is called the *wavelet mask*. It can be shown that the system $\{\psi(\cdot - k), k \in \mathbb{Z}\}$ forms an orthonormal basis of W_0 by using the fact that it forms an orthonormal basis of V_0. The function ψ is called the *wavelet function*.

Now let W_j be the jth dilation of W_0: $f(x) \in W_0$ if and only if $f(2^j x) \in W_j$. Then, by the definition of W_0 and MRA, W_j is also the orthogonal complement of V_j in V_{j+1}, i.e.

$$V_{j+1} = V_j \oplus W_j, \quad V_j \perp W_j.$$

Iterating this decomposition infinitely many times and letting $j \to \infty$, we obtain:

$$L^2(\mathbb{R}) = \overline{\oplus_{j \in \mathbb{Z}} W_j} = \oplus_{j \in \mathbb{Z}} W_j.$$

This states that $L^2(\mathbb{R})$ can be orthogonally decomposed into the sum of the spaces $\{W_j\}$. Furthermore, if we define $\psi_{k,j}$ by

$$\psi_{k,j}(x) = 2^{j/2}\psi(2^j x - k)$$

then, for each fixed j, the system $\{\psi_{k,j}, k \in \mathbb{Z}\}$ forms an orthonormal basis of W_j and the whole system $\{\psi_{k,j}, j, k \in \mathbb{Z}\}$ forms an orthonormal basis of $L^2(\mathbb{R})$. It is clear from the above discussion that the key point for constructing an orthonormal wavelet basis is the construction of the refinable function.

3.5.3 Examples

The first example is the Haar wavelet, obtained with the scaling function

$$\phi(x) = \begin{cases} 1, & 0 \le x < 1, \\ 0, & \text{otherwise.} \end{cases}$$

The function ϕ satisfies the following refinement relation:

$$\phi(x) = \phi(2x) + \phi(2x - 1) = \frac{1}{\sqrt{2}} \left(\sqrt{2}\phi(2x) + \sqrt{2}\phi(2x - 1) \right)$$

with refinement mask

$$a_0 = a_1 = \frac{1}{\sqrt{2}}$$

and $a_k = 0, k \ne 0, 1$. The Haar wavelet function is given by

$$\psi(x) = \begin{cases} 1, & 0 \le x < 1/2, \\ -1, & 1/2 \le x < 1, \\ 0, & \text{otherwise,} \end{cases}$$

with wavelet mask

$$b_0 = \frac{1}{\sqrt{2}}, \quad b_1 = -\frac{1}{\sqrt{2}}, \quad b_k = 0, \quad k \ne 0, 1.$$

It is clear that

$$\psi(x) = \phi(2x) - \phi(2x - 1).$$

The Haar wavelet is very localized, since it has the smallest support of all orthogonal wavelets. However, it is not very smooth, which means that it decays slowly in the Fourier domain and has rather poor approximation properties. One can construct compactly supported wavelets with better regularity and approximation properties; see the classic text by Daubechies [16]. A way of systematically constructing such compactly supported wavelets with better regularity and approximation properties has been developed by Daubechies herself. In her construction, the scaling functions are defined directly by their refinement masks instead of by an explicit analytical form.

Here, we give one example of a Daubechies' wavelet; interested readers will find more in [16]. The refinement mask of this example is defined

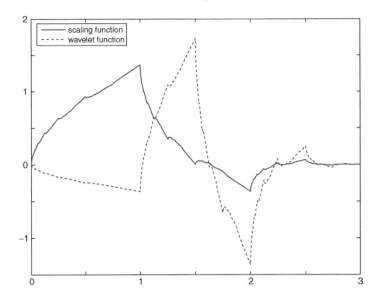

Figure 3.10. The scaling and wavelet functions of an example of Daubechies' wavelets.

as

$$a_0 = \frac{1 + \sqrt{3}}{4\sqrt{2}}, \qquad a_1 = \frac{3 + \sqrt{3}}{4\sqrt{2}}, \qquad a_2 = \frac{3 - \sqrt{3}}{4\sqrt{2}},$$

$$a_3 = \frac{1 - \sqrt{3}}{4\sqrt{2}}, \qquad a_k = 0, k \neq 0, 1, 2, 3.$$

The corresponding wavelet mask is

$$b_{-2} = \frac{1 - \sqrt{3}}{4\sqrt{2}}, \qquad b_{-1} = -\frac{3 - \sqrt{3}}{4\sqrt{2}}, \qquad b_0 = \frac{3 + \sqrt{3}}{4\sqrt{2}},$$

$$b_1 = -\frac{1 + \sqrt{3}}{4\sqrt{2}}, \qquad b_k = 0, \ k \neq 0, 1, 2, 3.$$

See Figure 3.10.

Given a function $f \in L^2(\mathbb{R})$, it can be approximated by its orthogonal projections into V_j for different j:

$$f_j = \sum_{k \in \mathbb{Z}} \langle f, \phi_{k,j} \rangle \phi_{k,j}. \qquad (3.5.6)$$

The accuracy of this approximation depends on the choice of j, which in practice is usually determined by the available data. For a given value of j, the accuracy then depends on the smoothness of f and the approximation properties of ϕ, and especially on the order of the Strang–Fix (SF) condition satisfied by ϕ [36]. Recall that a function ϕ satisfies the SF condition of order m if

$$\widehat{\phi}(0) \neq 0, \quad \widehat{\phi}^{(j)}(2\pi k) = 0, \quad j = 0, 1, 2, \ldots, m-1, \quad k \in \mathbb{Z} \backslash \{0\},$$

where $\widehat{\phi}$ is the Fourier transform of ϕ. Assume that the scaling function satisfies the SF condition of order m; then the spaces $\{V_j\}$ provide an order-m approximation, i.e. if $f \in W_2^m(\mathbb{R})$ then

$$\|f - f_j\| = O(2^{-jm}).$$

Here the Sobolev spaces $W_2^m(\mathbb{R})$, $m \in \mathbb{Z}_+$, are defined by

$$W_2^m(\mathbb{R}) := \left\{ f \in L^2(\mathbb{R}) : \|f\|_{W_2^m(\mathbb{R})} := \sqrt{2\pi} \left\| (1 + |\cdot|)^m \widehat{f} \right\| < \infty \right\}.$$

Details about the approximation order can be found in [18].

The piecewise-constant refinable function ϕ that leads to the Haar wavelet has approximation order 1. The scaling function of the above 4-tap wavelet has approximation order 2. Scaling functions with higher approximation orders can be found in [16].

Decomposition and reconstruction algorithms Given a function v in V_J, where V_J, represents the highest available resolution, we would like to express v as

$$v = v_0 + \sum_{k=0}^{J-1} w_k,$$

where $v_0 \in V_0, w_j \in W_j, j = 0, \ldots, J-1$, Mallat developed a very efficient hierarchical algorithm for performing this decomposition [30]; the spirit of Mallat's algorithm and the hierarchical algorithm of Yserentant [42] were discussed earlier. To see how the composition is done, let us consider a simpler problem. Assume that $v_j \in V_j$. We can write $v_j = v_{j-1} + w_{j-1}$ where $v_{j-1} \in V_{j-1}$ and $w_{j-1} \in W_{j-1}$ and $v_j = \sum_k c_{k,j}\phi_{k,j}$, $v_{j-1} = \sum_k c_{k,j-1}\phi_{k,j-1}$, $w_{j-1} = \sum_k d_{k,j-1}\psi_{k,j-1}$. Note that from (3.5.4) and (3.5.5) we have

$$\phi_{k,j-1} = \sum_m a_{m-2k} \phi_{m,j},$$

$$\psi_{k,j-1} = \sum_m b_{m-2k} \phi_{m,j}.$$

It is easy to see then that

$$c_{k,j-1} = \langle v_j, \phi_{k,j-1} \rangle$$

$$= \left\langle v_j, \sum_m a_{m-2k} \phi_{m,j} \right\rangle$$

$$= \sum_m a_{m-2k} \langle v_j, \phi_{m,j} \rangle$$

$$= \sum_m a_{m-2k} c_{m,j}.$$

Similarly,

$$d_{k,j-1} = \sum_m b_{m-2k} c_{m,j}.$$

Knowing $\{c_{m,j}\}$ we can compute $\{c_{m,j-1}, d_{m,j-1}\}$ very easily from the formulas above. This procedure can then be repeated.

The decomposition algorithm described above can also be used in the reverse direction to reconstruct the fine-scale coefficients in terms of the coefficients at the coarser scale, since we have

$$c_{k,j} = \langle v_j, \phi_{k,j} \rangle$$

$$= \sum_m c_{m,j-1} \langle \phi_{m,j-1}, \phi_{k,j} \rangle + \sum_m d_{m,j-1} \langle \psi_{m,j-1}, \phi_{k,j} \rangle$$

$$= \sum_m a_{m-2k} c_{m,j-1} + \sum_m b_{m-2k} b_{m,j-1}.$$

This gives a perfect reconstruction, in the sense that decomposition followed by reconstruction leads to the identity transform. It is important to note that, while scaling and wavelet functions are used in the analysis, only refinement and wavelet masks are used in the decomposition and reconstruction algorithms.

What makes wavelets useful is that for many classes of signals (or functions), their wavelet coefficients are approximately sparse, i.e. most

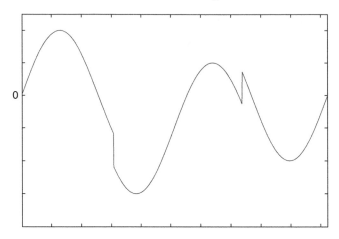

Figure 3.11. An example of a piecewise smooth function.

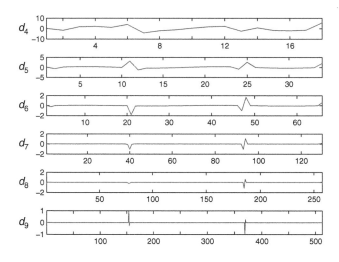

Figure 3.12. The wavelet coefficients d_k at various levels for the function shown in the previous figure.

wavelet coefficients are very small. In addition, the set of sizable coefficients is localized in the wavelet transform space. This is illustrated by Figure 3.11 and the sequences of Figure 3.12. The former shows a piecewise smooth function; the latter is a plot of the corresponding wavelet coefficients at various levels. As shown clearly in Figure 3.12, the wavelet coefficients are sparse.

In addition to sparsity and orthonormal wavelets, redundancy and the associated concept of *tight frames* have also been studied (see e.g. [35]).

It has been discovered that redundancy can often provide the flexibility needed to tune to the specific features of the particular application with which one is dealing.

3.6 Notes

This chapter is admittedly very concise, for a good reason: much has been written on the topics discussed here. Thorough discussions of them can be found in various textbooks and the review articles listed at the end of the chapter. We have found it useful to include at least a brief discussion of the main ideas in these classical multiscale algorithms for several reasons:

(1) these ideas are the basic building blocks for multiscale algorithms;
(2) this discussion makes it easier for us to see the difference between the philosophy behind these classical algorithms and the new ones to be presented later.

There are many linear-scaling multiscale algorithms that we have not covered in this chapter. Obvious examples include H-matrix techniques [10], other analysis-based fast transforms and algorithms based on discrete symbol calculus [19].

References for Chapter 3

[1] M. Ainsworth and J.T. Oden, *A Posteriori Error Estimation in Finite Element Analysis*, John Wiley, 2000.

[2] A.W. Appel, "An efficient program for many-body simulation," *SIAM J. Sci. Stat. Comput.*, vol. 6, pp. 85–103, 1985.

[3] I. Babuska and W.C. Rheinboldt, "Error estimates for adaptive finite element computations," *SIAM J. Numer. Anal.*, vol. 18, pp. 736–754, 1978.

[4] I. Babuska and W.C. Rheinboldt, "A posteriori error estimates for the finite element method," *Intl. J. Numer. Methods Engrg*, vol. 12, pp. 1597–1615, 1978.

[5] R.E. Bank and A. Weiser, "Some a posteriori error estimators for elliptic partial differential equations," *Math. Comp.*, vol. 44, 283–301, 1985.

[6] J. Barnes and P. Hut, "A hierarchical $O(N \log N)$ force-calculation algorithm," *Nature*, vol. 324, pp. 446–449, 1986.

[7] R. Beatson and L. Greengard, "A short course on fast multipole methods," in *Wavelets, Multilevel Methods and Elliptic PDEs*, M. Ainsworth, J. Levesley, W. Light and M. Marletta, eds., pp. 1–37, Oxford University Press, 1997.

[8] M.J. Berger, and P. Collela, "Local adaptive mesh refinement for shock hydrodynamics," *J. Comput. Phys.*, vol. 82, no. 1, pp. 64–84, 1989.

[9] G. Beylkin, R. Coifman and V. Rokhlin, "Fast wavelet transforms and numerical analysis," *Comm. Pure Appl. Math.*, vol. 44, no. 2, pp. 141–183, 1991.

[10] S. Börm, L. Grasedyck and W. Hackbusch, *Hierarchical Matrices*, Max-Planck Institute Lecture Notes, 2006.

[11] A. Brandt, "Multi-level adaptive solutions to boundary value problems," *Math. Comp.*, vol. 31, no. 138, pp. 333–390, 1977.

[12] A. Brandt, "Multiscale scientific computation: review 2001," in *Multiscale and Multiresolution Methods: Theory and Applications, Lecture Notes in Computer Science and Engineering*, T.J. Barth *et al.*, eds., vol. 20, pp. 3–96, Springer-Verlag, 2002.

[13] W.L. Briggs, V.E. Henson and S.F. McCormick, *A Multigrid Tutorial*, 2nd edn, SIAM, 2000.

[14] D.A. Caughey and A. Jameson, "Fast preconditioned multigrid solution of the Euler and Navier–Stokes equations for steady compressible flows," *Intl. J. Numer. Meth. Fluids*, vol. 43, pp. 537–553, 2003.

[15] S.D. Conte and C. de Boor, *Elementary Numerical Analysis*, McGraw-Hill, 1980.

[16] I. Daubechies, *Ten Lectures on Wavelets*, SIAM, 1992.

[17] S.F. Davis and J.E. Flaherty, "An adaptive finite element method for initial boundary value problems for partial differential equations," *SIAM J. Sci. Stat. Comp.*, vl. 3, pp. 6–27, 1982.

[18] C. de Boor, R.A. DeVore and A. Ron, "Approximation from shift-invariant subspaces of $L_2(\mathbb{R}^d)$," *Trans. Amer. Math. Soc.*, vol. 341, no. 2, pp. 787–806, 1994.

[19] L. Demanet and L. Ying, "Discrete symbol calculus," preprint.

[20] P. Deuflhard and A. Hohmann, *Numerical Analysis in Modern Scientific Computing: An Introduction*, Springer, 2000.

[21] W. E, M. Mu and H. Huang, "A posteriori error estimates for finite element methods," *Chinese Quart. J. Math.*, vol. 3, pp. 97–106, 1988.

[22] R.P. Fedorenko, "Iterative methods for elliptic difference equations," *Russian Math. Surveys*, vol. 28, pp. 129–195, 1973.

[23] A.L. Garcia, J.B. Bell, W.Y. Crutchfield and B.J. Alder "Adaptive mesh and algorithm refinement using direct simulation Monte Carlo," *J. Comput. Phys.*, vol. 154, pp. 134–155, 1999.

[24] J. Goodman and A.D. Sokal, "Multigrid Monte Carlo methods," *Phys. Rev. D*, vol. 40, no. 6, pp. 2035–2071, 1989.

[25] L. Greengard and V. Rokhlin, "A fast algorithm for particle simulations," *J. Comput. Phys.*, vol. 73, pp. 325, 1987.

[26] L. Greengard and J. Strain, "The fast Gauss transform," *SIAM J. Sci. Stat. Comp.* vol. 12, pp. 79–94, 1991.

[27] W. Hackbusch, "Convergence of multigrid iterations applied to difference equations," *Math. Comp.*, vol. 34, no. 150, pp. 425–440, 1980.

[28] W. Hackbusch, "A sparse matrix arithmetic based on $\mathcal{H}$-matrices. Part I: Introduction to $\mathcal{H}$-matrices," *Computing*, vol. 62, pp. 89–108, 1999.

[29] R. Li, T. Tang and P. Zhang, "Moving mesh methods in multiple dimensions based harmonic maps," *J. Comput. Phys.*, vol. 170, pp. 562–588, 2001.

[30] S. Mallat, "Multiresolution approximation and wavelets," *Trans. Amer. Math. Soc.*, vol. 315, pp. 69–88, 1989.

[31] S. Mallat, "A theory for multiresolution signal decomposition: the wavelet representation," *IEEE Trans. PAMI*, vol. 11, pp. 674–693, 1989.

[32] K. Miller and R.N. Miller, "Moving finite element. I," *SIAM J. Numer. Anal.*, vol. 18, pp. 1019–1032, 1981.

[33] Y. Ren and R.D. Russell, "Moving mesh techniques based upon equidistribution and their stability," *SIAM J. Sci. Stat. Comp.*, vol. 13, pp. 1265–1286, 1992.

[34] W. Ren and X. Wang, "An iterative grid redistribution method for singular problems in multiple dimensions," *J. Comput. Phys.*, vol. 159, pp. 246–273, 2000.

[35] Z. Shen, "Wavelet frames and image restorations", in *Proc. Int. Congress of Mathematicians*, Hyderabad, Hindustan Publishing Company, 2010.

[36] G. Strang and G. Fix, "A Fourier analysis of the finite element variational method", in *Constructive Aspects of Functional Analysis*, G. Geymonat, ed., pp. 793–840, 1973.

[37] A. Toselli and O. Widlund, *Domain Decomposition Methods*, Springer-Verlag, 2004.

[38] R. Verfürth, "A posteriori error estimators for the Stokes equations," *Numer. Math.*, vol. 55, pp. 309–325, 1989.

[39] A.M. Winslow, "Numerical solution of the quasilinear Poisson equation in a nonuniform triangle mesh," *J. Comput. Phys.*, vol. 1, pp. 149–172, 1967.

[40] W. Yang and T.-S. Lee, "A density-matrix divide-and-conquer approach for electronic structure calculations of large molecules," *J. Chem. Phys.*, vol. 103, no. 13, pp. 5674–5678, 1995.

[41] D. Young and L.A. Hageman, *Applied Iterative Methods*, Dover Publications, 2004.

[42] H. Yserentant, "On the multi-level splitting of finite element spaces," *Numer. Math.*, vol. 49, pp. 379–412, 1986.

[43] O.C. Zienkiewicz and J.Z. Zhu, "A simple error estimator and adaptive procedure for practical engineering analysis," *Int. J. Numer. Meth. Engrg.*, vol. 24, pp. 337–357, 1987.

4

The hierarchy of physical models

Given a physical system, we can model it at different levels of detail. This results in a hierarchy of models, each of which is a refinement of the models at the higher levels of the hierarchy (see Table 1.1). For example, if we want to model the dynamics of gases, we have the following options from which to choose:

(1) *continuum mechanics*, the Euler and Navier–Stokes equations;
(2) *kinetic theory*, the Boltzmann equation;
(3) *molecular dynamics*, the Newton's equation;
(4) *quantum mechanics*, the Schrödinger equation.

Continuum mechanics models, such as Euler's equation and the Navier–Stokes equations, are the crudest in this hierarchy. They model only the macrosopic density, velocity and temperature fields of the gas. Nevertheless, they are already quite sufficient in many engineering applications. Quantum mechanics models are the most detailed. They are required if we are interested in the details of the collision processes between gas particles. Kinetic theory and molecular dynamics models are of intermediate complexity, capturing respectively the phase-space probability distribution and the phase-space dynamics of the gas particles.

Of course, within each level of this hierarchy there are subhierarchies, i.e. models of different complexity and detail. For example, at the level of the quantum mechanical models for electronic structure analysis, we have:

(1) orbital-free density functional theories, such as the Thomas–Fermi–von Weizsäcker (TFW) model;
(2) Kohn–Sham density functional theory;
(3) The full quantum many-body problem.

The TFW model uses only the electron density to describe the electronic structure of the system. The Kohn–Sham density functional theory uses a set of (fictitious) one-body wave functions. The full quantum many-body theory starts from first principles. It describes an electronic system at the level of the many-body wave function.

Since they all describe the same physical system, the different models have to produce consistent results. Understanding this analytically is a main task in multiscale modeling. For example, one should be able to derive Euler's equation of gas dynamics from molecular dynamics models in the regime where Euler's equation is presumed to be valid. Among other things, this derivation should provide a microscopic expression for the constitutive relations in Euler's equation.

In this chapter, we will discuss the essential components of these models. We will start at the top of the hierarchy, the continuum mechanics models. This will be followed by the classical atomistic models in the form of molecular dynamics. We then introduce the basic ingredients of kinetic theory, which at least for gases serves as the bridge between the many-body molecular dynamics models and the continuum mechanics models. We will end this chapter at the bottom of the hierarchy with the quantum mechanics models. This chapter is certainly not going to be an exhaustive treatment of the different physical laws used in modeling. But we hope to convey the basic ideas.

4.1 Continuum mechanics

In continuum mechanics the system of interest is treated as a continuous medium. The basic objects that we study are fields such as velocity and density fields. In this section, we will discuss the ideas needed to formulate continuum models. We begin by introducing the concepts of stress and strain and the rate of strain. We then discuss nonlinear elasticity models of solids. Finally we discuss dynamic models of solids and fluids.

To describe the fields in continuum mechanics, we need a coordinate system. The most commonly used coordinates are *Lagrangian coordinates* and *Eulerian coordinates*. In the former, the description of a particle in the system is based on its initial position or its position in a fixed reference frame, which in solids is usually the undeformed configuration. In Eulerian coordinates, the description of the particle is based on its current

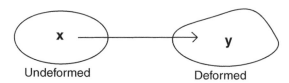

Figure 4.1. The Lagrangian and Eulerian coordinates of a continuous medium: **x** represents the Lagrangian coordinates and **y** represents the Eulerian coordinates.

position, after the system has changed as a result of deformation or flow. In Figure 4.1, **x** denotes the Lagrangian coordinates, and **y** denotes the Eulerian coordinates. The mapping from **x** to **y** is called the Lagrangian-to-Eulerian map; its inverse is the Eulerian-to-Lagrangian map. One main difference between solid and fluid mechanics is that models in solid mechanics, such as elasticity theory, typically use Lagrangian coordinates whereas in fluid mechanics we typically use Eulerian coordinates. One reason is that in fluids the deformation can be so large that it does not make much sense to refer back to the initial position of the material point. This difference creates some problems when modeling systems whose behavior is intermediate between fluids and solids, such as viscoelastic fluids or plastically deformed solids.

Continuum theory is often divided into two related parts: statics and dynamics. Models for statics are generally in the form of variational principles, asserting, for example, that the static configuration of a material minimizes the total free energy of the system taking into account the work done by external forces. Dynamic models are generally in the form of dynamic conservation laws of mass, momentum and energy.

A central concept in treating continuous media is the stress. It describes the forces acting on (virtual) surfaces inside the medium. Imagine an infinitesimally small piece of surface placed inside a continuous medium. As a consequence of the short-range molecular interactions there is a force between the material on one side of the surface and the material on the other side of the surface. This force is called the *stress*. It is the macroscopic manifestation of the microscopic short-range interactions between the atoms on the two sides of the surface. The long-range interactions, such as the Coulomb interaction, are usually represented as a body force. It is easy to see that the stress depends linearly on the orientation of the surface. At each point of the medium it is a mapping between the orientation of the surface and the force that we just described. Therefore it is a tensor of rank 2.

To obtain a closed model, we need to express the stress in terms of the displacement field and its derivatives. This is the role played by the *constitutive relation*. The following are some examples of constitutive relations:

(1) the equation of state for gases;
(2) the stored energy density function for solids;
(3) the stress–strain relation for solids;
(4) the stress as a function of the rate of strain for fluids.

Constitutive relations are macroscopic representations of the nature of the microscopic constituents of the system, i.e. the atoms that make up the system. This is where most empirical modeling in continuum theory comes in. For simple systems, linear constitutive relations have been remarkably successful. In fact, in many cases, constitutive modeling proceeds by [12]:

(1) writing down linear relations between the generalized forces and the generalized fluxes;
(2) using symmetry properties to reduce the number of independent coefficients in the linear relations.

These coefficients, for example the elastic moduli or the viscosity coefficients, can then be measured experimentally. From an abstract viewpoint, the only constraints on the coefficients are the second law of thermodynamics and the symmetry of the system. In fact, nonequilibrium thermodynamics, a subject that has had a great deal of success after the work of Onsager, Prigogine and others, was created to treat general systems using these principles [12].

A main purpose of multiscale, multi-physics modeling is to establish a microscopic foundation for the constitutive relations. This is not a new idea. There is a long history of work along this line (see for example [4, 10]). Back in the 1950s, Tsien initiated the subject of *physical mechanics* with the objective of providing the microscopic foundation for continuum mechanics [39]. The practical interest was to calculate material properties and transport coefficients using the principles of statistical and quantum mechanics, particularly at extreme conditions. For various reasons the subject did not become very popular after its initial success. The interest in multiscale, multi-physics modeling has given it a new impetus, and now would seem to be the right time to try to revive the subject.

4.1.1 Stress and strain in solids

Given a solid material, denote by Ω the region it occupies in the absence of any external loading. Our objective is to find its new configuration after some external load is applied to the material. We will refer to the configuration of the material before and after the loading as the reference and deformed configurations respectively.

Let $\mathbf{x}$ be the position of a material point in the reference configuration and let $\mathbf{y}$ be the position of the same material point in the deformed configuration. Obviously $\mathbf{y}$ is a function of $\mathbf{x}$: $\mathbf{y} = \mathbf{y}(\mathbf{x})$. We call $\mathbf{u}(\mathbf{x}) = \mathbf{y}(\mathbf{x}) - \mathbf{x}$ the *displacement field* and refer to

$$\mathbf{F} = \nabla \mathbf{u}$$

as the *deformation gradient tensor*.

Let $\delta \mathbf{x}$ be a small change in $\mathbf{x}$ in the reference configuration and $\delta \mathbf{y}$ be the corresponding change in the deformed configuration. To leading order, we have

$$\delta \mathbf{y} = (\mathbf{I} + \mathbf{F})\delta \mathbf{x},$$

where $\mathbf{I}$ is the identity tensor, and

$$\|\delta \mathbf{y}\|^2 - \|\delta \mathbf{x}\|^2 = (\delta \mathbf{x})^{\mathrm{T}} \left((\mathbf{I} + \mathbf{F})^{\mathrm{T}}(\mathbf{I} + \mathbf{F}) - \mathbf{I} \right) \delta \mathbf{x}$$
$$= 2(\delta \mathbf{x})^{\mathrm{T}} \mathbf{E} \delta \mathbf{x}.$$

The *strain tensor* $\mathbf{E}$ is defined as

$$\mathbf{E} = \frac{1}{2} \left((\mathbf{I} + \mathbf{F})^{\mathrm{T}}(\mathbf{I} + \mathbf{F}) - \mathbf{I} \right)$$

$$= \frac{1}{2}(\mathbf{F}^{\mathrm{T}} + \mathbf{F}) + \frac{1}{2}\mathbf{F}^{\mathrm{T}}\mathbf{F}.$$

In component form, we have

$$e_{ij} = \frac{1}{2}\left(\frac{\partial u_i}{\partial x_j} + \frac{\partial u_j}{\partial x_i} \right) + \frac{1}{2}\sum_l \frac{\partial u_l}{\partial x_i} \frac{\partial u_l}{\partial x_j}. \tag{4.1.1}$$

The tensor $\mathbf{E} = (e_{ij})$ characterizes the geometric change as a result of the deformation. It is most relevant for solids, since for fluids the deformation is usually quite large and therefore for fluids it is more convenient to discuss the *rate of strain*, which is defined as the symmetric part of the gradient of the velocity field:

$$\mathbf{D} = \frac{1}{2}(\nabla \mathbf{v} + (\nabla \mathbf{v})^{\mathrm{T}}),$$

where $\mathbf{v}$ is the velocity field.

Imagine an infinitesimal surface with area ΔS at a point inside the material, with normal vector $\mathbf{n}$. Let $\mathbf{g}\Delta S$ be the force acting on one side of the surface (the side opposite to $\mathbf{n}$) due to the other side (i.e. the side in the direction of $\mathbf{n}$). It is easy to see that there is a linear relation between $\mathbf{n}$ and $\mathbf{g}$, which can be expressed as

$$\mathbf{g} = \boldsymbol{\sigma}\mathbf{n};$$

$\boldsymbol{\sigma}$ is called the *stress tensor*.

There are two ways of defining the normal vector and the area of the infinitesimal surface, depending on whether we use the Lagrangian or the Eulerian coordinate system. When the latter is used, the resulting stress tensor is called the *Cauchy stress tensor*. When the Lagrangian coordinate system is used, the resulting stress tensor is called the *Piola–Kirkhoff stress tensor*. This distinction is very important for solids, since in this case the Lagrangian coordinate system is typically more convenient but the Cauchy stress is easier to use and to think about. Fortunately, there is a simple relation between the two notions.

Lemma *Let* $\mathbf{y} = \mathbf{y}(\mathbf{x})$ *be a nondegenerate map on* $\mathbb{R}^3$, *i.e.* $J(\mathbf{x}) = \det(\nabla_{\mathbf{x}}\mathbf{y}(\mathbf{x})) \neq 0$. *Denote by* dS_0 *an infinitesimal surface element at* $\mathbf{x}$ *and by* dS *its image under the mapping* $\mathbf{y}(\cdot)$. *Let* $\mathbf{n}_0$ *and* $\mathbf{n}$ *be the normals to* dS_0 *and* dS *respectively. Then*

$$\mathbf{n}\, dS = J(\mathbf{x})(\nabla_{\mathbf{x}}\mathbf{y}(\mathbf{x}))^{-T}\mathbf{n}_0\, dS_0. \tag{4.1.2}$$

For a proof of this lemma, see [29]. We will use $\boldsymbol{\sigma}$ to denote the Cauchy stress tensor and $\boldsymbol{\sigma}_0$ to denote the Piola–Kirkhoff stress tensor.

Let S_0 be an arbitrary closed surface in the reference configuration and let S be its image under the mapping $\mathbf{y}(\cdot)$. From the above lemma, we have

$$\int_S \boldsymbol{\sigma}(\mathbf{y})\mathbf{n}(\mathbf{y})\, dS(\mathbf{y}) = \int_{S_0} J(\mathbf{x})\boldsymbol{\sigma}(\mathbf{y}(\mathbf{x}))(\nabla_{\mathbf{x}}\mathbf{y}(\mathbf{x}))^{-T}\mathbf{n}_0(\mathbf{x})\, dS_0(\mathbf{x}).$$

Hence

$$\boldsymbol{\sigma}_0(\mathbf{x}) = J(\mathbf{x})\boldsymbol{\sigma}(\mathbf{y}(\mathbf{x}))(\nabla_{\mathbf{x}}\mathbf{y}(\mathbf{x}))^{-T}.$$

Note that, since $\boldsymbol{\sigma}_0\mathbf{n}_0 dS_0 = \boldsymbol{\sigma}\mathbf{n}dS$, the force vectors $\sigma(\mathbf{y})\mathbf{n}(\mathbf{y})$ and $\sigma_0\mathbf{n}_0(\mathbf{x})$ are collinear. Their magnitudes can be different as a result of the difference between dS_0 and dS. They are the same physical stress, i.e. the stress (force per unit area) in the deformed body at $\mathbf{y} = \mathbf{y}(\mathbf{x})$. Their difference is the result of the different ways in which area is measured.

In fluids, owing to the large deformations one normally uses the Cauchy stress tensor.

4.1.2 Variational principles in elasticity theory

The static configuration of a deformed elastic body is often characterized by a variational principle of the type:

$$\min \int_{\Omega} (W(\{\mathbf{u}\}) - \mathbf{f} \cdot \mathbf{u})\, d\mathbf{x}, \qquad (4.1.3)$$

where $\mathbf{f}$ is the external force density and $W(\{\mathbf{u}\})$ is called the *stored energy density* or *elastic energy density*. It is the potential energy or free energy of the material. The dependence on $\{\mathbf{u}\}$ means that W can depend on $\mathbf{u}, \nabla\mathbf{u}, \ldots$ This function plays the role of the constitutive relation in this type of model.

One of the simplest forms of the constitutive relation is obtained when W depends only on $\mathbf{E}$, i.e.

$$W(\{\mathbf{u}\}) = W_0(\mathbf{E}).$$

This assumption is often sufficient for the analysis of the elastic deformation of bulk materials.

In the absence of any additional information, we can study first the case when the deformation is small enough that we can further assume that the material response is linear, i.e. that $W_0(\mathbf{E})$ is a quadratic function of the strain tensor:

$$W_0(\mathbf{E}) = \tfrac{1}{2} \sum_{i,j,k,l} C_{ijkl} e_{ij} e_{kl}, \qquad (4.1.4)$$

where $\{e_{ij}\}$ is defined by (4.1.1). The constant term is neglected and the linear term vanishes since the undeformed state is at equilibrium (and therefore has no residual stress). The coefficients $\{C_{\alpha\beta\gamma\delta}\}$ are called the *elastic constants*. For a three-dimensional problem, C has $3^4 = 81$ components. We can always require C to be symmetric, i.e.

$$C_{ijkl} = C_{klij}.$$

In addition, since $\mathbf{E}$ is a symmetric tensor we have

$$C_{ijkl} = C_{ijlk}.$$

These relations reduce the number of independent elastic constants to 21 [27].

In most cases the material under consideration has additional symmetries which allow us to reduce the number of independent parameters even further. Here we will discuss two cases, those of isotropic materials and materials with cubic symmetry.

If a material has cubic symmetry, its elastic properties are invariant under the exchange of any two coordinates. Let T be such a transformation:

- $T(1) = 2$, $T(2) = 1$, $T(3) = 3$;
- $T(1) = 3$, $T(2) = 2$, $T(3) = 1$;
- $T(1) = 1$, $T(2) = 3$, $T(3) = 2$;

then

$$C_{T(i)T(j)T(k)T(l)} = C_{ijkl}.$$

In addition, W also has inversion symmetry. Let I be any of the following transformations:

- $I(x) = -x$, $I(y) = y$, $I(z) = z$;
- $I(x) = x$, $I(y) = -y$, $I(z) = z$;
- $I(x) = x$, $I(y) = y$, $I(z) = -z$;

then

$$e_{I(i)I(j)} = \begin{cases} e_{ij}, & \text{if } i = j; \\ -e_{ij}, & \text{if } i \neq j. \end{cases} \tag{4.1.5}$$

This implies that all elastic constants with any subscript appearing an odd number of times must vanish. For instance, C_{1112} must be 0.

These symmetry considerations allow us to reduce the number of independent elastic constants for a material with cubic symmetry to only three. They are:

$$C_{1111} = C_{2222} = C_{3333},$$
$$C_{1212} = C_{2323} = C_{3131},$$
$$C_{1122} = C_{1133} = C_{2233}.$$

Using this information, we can write $W_0(\mathbf{E})$ as follows:

$$W_0(\mathbf{E}) = \tfrac{1}{2}C_{1111}\left(e_{11}^2 + e_{22}^2 + e_{33}^2\right) + C_{1122}\left(e_{11}e_{22} + e_{11}e_{33} + e_{22}e_{33}\right)$$
$$+ 2C_{1212}\left(e_{12}^2 + e_{23}^2 + e_{13}^2\right).$$

In solid state physics, we often adopt the *Voigt notation*:

$$e_1 = e_{11}, \qquad e_2 = e_{22}, \qquad e_3 = e_{33},$$
$$e_4 = 2e_{23}, \qquad e_5 = 2e_{31}, \qquad e_6 = 2e_{12},$$
$$C_{11} = C_{1111}, \quad C_{12} = C_{1122}, \quad C_{44} = C_{1212}.$$

For isotropic materials we have the additional property that W is invariant under a 45° degree rotation around any coordinate axis. It is easy to verify that this implies the following relation:

$$C_{11} = C_{12} + 2C_{44}. \tag{4.1.6}$$

This reduces the number of independent elastic constants to two:

$$\lambda = C_{12},$$
$$\mu = C_{44}.$$

They are called the *Lamé constants*. Now $W_0(\mathbf{E})$ can be written as

$$W_0(\mathbf{E}) = \tfrac{1}{2}\lambda\left(\sum_j e_{jj}\right)^2 + \mu\sum_{jk} e_{jk}^2.$$

Let us now consider briefly nonlinear isotropic materials. We will limit ourselves to the case of incompressible materials, i.e. materials for which volume is invariant during the deformation. Since W is invariant under rotation, it is only a function of the principal values of the matrix $\mathbf{C} = (\mathbf{I} + \mathbf{F})^{\mathrm{T}}(\mathbf{I} + \mathbf{F})$ (recall that $\mathbf{F} = \nabla\mathbf{u}$):

$$W = W(\lambda_1, \lambda_2, \lambda_3),$$

where $\lambda_1, \lambda_2, \lambda_3$ are the eigenvalues of $\mathbf{C}$. Examples of empirical models of W include:

(1) *neo-Hookean models,*

$$W = a(\lambda_1 + \lambda_2 + \lambda_3 - 3),$$

where a is a constant;

(2) *Mooney–Rivlin models,*

$$W = a(\lambda_1 + \lambda_2 + \lambda_3 - 3) + b(\lambda_2\lambda_3 + \lambda_3\lambda_1 + \lambda_1\lambda_2 - 3),$$

where a and b are constants.

More examples can be found in [29].

So far we have considered the case when W depends only on $\nabla \mathbf{u}$. This is often inadequate, for example for the cases of

- small systems, for which the dependence on the strain gradient should be included;
- plates, sheets and rods, for which the dependence on the curvature should be included.

Including the dependence on the strain gradient in W increases the complexity of the models substantially. In principle, however, the same kinds of consideration still applies.

4.1.3 Conservation laws

We now turn to dynamics. The first step toward studying dynamics is to write down the three universal conservation laws, of mass, momentum and energy.

Conservation of mass We will first discuss mass conservation in Lagrangian coordinates. Let D_0 be an arbitrary subdomain in the reference configuration and let D be its image in the deformed configuration. The mass inside D must be the same as the mass inside D_0. If we denote by ρ_0 the density function in the reference configuration and ρ the density function in the deformed configuration, then we have

$$\int_{D_0} \rho_0(\mathbf{x})\,d\mathbf{x} = \int_D \rho(\mathbf{y})\,d\mathbf{y}.$$

Since

$$d\mathbf{y} = J(\mathbf{x})\,d\mathbf{x}$$

we obtain

$$\rho_0(\mathbf{x}) = \rho(\mathbf{y}(\mathbf{x}))J(\mathbf{x}).$$

In a time-dependent setting this can be written as

$$\rho_0(\mathbf{x}) = \rho(\mathbf{y}(\mathbf{x},t))J(\mathbf{x},t).$$

We now turn to the Eulerian form of mass conservation, which is more convenient, for fluids. Let D be an arbitrary subdomain in the deformed configuration that is fixed in time. The instantaneous change of the mass inside D is equal to the mass flux through the boundary of D:

$$\frac{d}{dt}\int_D \rho(\mathbf{y},t)\,d\mathbf{y} = -\int_{\partial D} \mathbf{J}_\rho \cdot \mathbf{n}\,dS(\mathbf{y}).$$

Here $\mathbf{n}$ is the unit outward normal of ∂D and $\mathbf{J}_\rho$ is the mass flux. It is easy to see that

$$\mathbf{J}_\rho = \rho\mathbf{v},$$

where $\mathbf{v}$ is the velocity field. Using the divergence theorem we get

$$\frac{d}{dt}\int_D (\partial_t\rho(\mathbf{y},t) + \nabla \cdot (\rho\mathbf{v}))\,d\mathbf{y} = 0.$$

Since this is true for any domain D, we have

$$\partial_t\rho + \nabla_\mathbf{y} \cdot (\rho\mathbf{v}) = 0.$$

For the next two conservation laws, we will first discuss the Eulerian formulation. The Lagrangian formulation will be discussed in the next subsection.

Conservation of momentum The rate of change of the total momentum in D is the result of: (1) materials moving in and out of D due to convection; and (2) the force acting on D. In general there are two kinds of force acting on D: body forces such as gravity and the surface force acting on the boundary of D, i.e. the stress $\boldsymbol{\sigma}$. We will neglect the body forces since they can always be added back later. Putting (1) and (2) together, we have

$$\frac{d}{dt}\int_D \rho(\mathbf{y},t)\mathbf{v}(\mathbf{y},t)\,d\mathbf{y} = -\int_{\partial D} \rho\mathbf{v}(\mathbf{v} \cdot \mathbf{n})\,dS(\mathbf{y}) + \int_{\partial D} \boldsymbol{\sigma} \cdot \mathbf{n}\,dS(\mathbf{y}).$$

In differential form, this gives

$$\partial_t(\rho\mathbf{v}) + \nabla_{\mathbf{y}} \cdot (\rho\mathbf{v} \otimes \mathbf{v} - \boldsymbol{\sigma}) = 0.$$

We can also write this as

$$\partial_t(\rho\mathbf{v}) + \nabla_{\mathbf{y}} \cdot \mathbf{J}_m = 0,$$

where $\mathbf{J}_m$ is the total momentum flux, given by

$$\mathbf{J}_m = \rho\mathbf{v} \otimes \mathbf{v} - \boldsymbol{\sigma}.$$

Conservation of angular momentum Similarly, the rate of change of the total angular momentum in D is given by

$$\frac{d}{dt} \int_D \rho(\mathbf{y}, t)\mathbf{y} \times \mathbf{v}(\mathbf{y}, t) \, d\mathbf{y} = -\int_{\partial D} \rho\mathbf{y} \times \mathbf{v}(\mathbf{v} \cdot \mathbf{n}) \, dS(\mathbf{y})$$
$$+ \int_{\partial D} \mathbf{y} \times \boldsymbol{\sigma} \cdot \mathbf{n} \, dS(\mathbf{y}).$$

This states that the rate of change of the angular momentum inside D is due to the angular momentum flux through ∂D and the torque generated by the stress on ∂D. The main consequence of this is that

$$\boldsymbol{\sigma} = \boldsymbol{\sigma}^{\mathrm{T}}$$

i.e. the Cauchy stress tensor is symmetric. For a proof, see [29].

Conservation of energy In the same way, we can write down the energy conservation law. The total energy density E is the sum of the kinetic energy density and the potential energy density:

$$E = \tfrac{1}{2}\rho|\mathbf{v}|^2 + \rho e,$$

where e is the internal energy per unit mass. Given a volume D, the rate of change of energy inside D is the result of energy been convected in and out of D, the work done by the external force acting on D and the *heat flux*:

$$\frac{d}{dt} \int_D E \, d\mathbf{y} = -\int_{\partial D} E(\mathbf{v} \cdot \mathbf{n}) \, dS(\mathbf{y}) + \int_{\partial D} \mathbf{v} \cdot (\boldsymbol{\sigma} \cdot \mathbf{n}) \, dS(\mathbf{y}) + \int_{\partial D} \mathbf{q} \cdot \mathbf{n} \, dS(\mathbf{y}).$$

Here $\mathbf{q}$ is the heat flux. In a differential form, this gives

$$\partial_t E + \nabla_{\mathbf{y}} \cdot (\mathbf{v}E - \boldsymbol{\sigma}\mathbf{v} + \mathbf{q}) = 0$$

or

$$\partial_t E + \nabla_{\mathbf{y}} \cdot \mathbf{J}_E = 0$$

where

$$\mathbf{J}_E = \mathbf{v}E - \boldsymbol{\sigma}\mathbf{v} + \mathbf{q}.$$

4.1.4 Dynamic theory of solids and thermoelasticity

Dynamic models of continuous media are derived from the following principles:

(1) the laws of conservation of mass, momentum and energy;
(2) the second law of thermodynamics, i.e. the Clausius–Duhem inequality;
(3) linearization and symmetry; in the absence of additional information, a linear constitutive relation is assumed and symmetry is used to reduce the number of independent parameters.

In this and the next subsection, we will demonstrate how such a program can be carried out for simple fluids and solids.

As we remarked earlier, for solids it is more convenient to use the Lagrangian coordinate system. Therefore our first task is to express the conservation laws in Lagrangian coordinates.

Let us begin with mass conservation: we derived earlier that

$$\rho(\mathbf{y}(\mathbf{x},t),t) = \frac{1}{\det(\mathbf{I}+\mathbf{F})}\rho_0(\mathbf{x}). \qquad (4.1.7)$$

As for mass conservation, momentum conservation requires, that for any subdomain in the reference configuration D_0,

$$\frac{d}{dt}\int_{D_0}\rho_0(\mathbf{x})\partial_t\mathbf{y}(\mathbf{x},t)\,d\mathbf{x} = \int_{\partial D_t}\boldsymbol{\sigma}(\mathbf{y},t)\hat{\mathbf{n}}(\mathbf{y},t)\,dS(\mathbf{y})$$

$$= \int_{\partial D_0}\boldsymbol{\sigma}_0(\mathbf{x},t)\hat{\mathbf{n}}_0(\mathbf{x},t)\,dS(\mathbf{x}). \qquad (4.1.8)$$

Here D_t is the image of D_0 under the mapping $\mathbf{x} \to \mathbf{y}(\mathbf{x},t)$ at time t, $\hat{\mathbf{n}}(\mathbf{y},t)$ is the unit outward normal vector of ∂D_t at $\mathbf{y}$, $\hat{\mathbf{n}}_0(\mathbf{x})$ is the normal vector of ∂D_0 at $\mathbf{x}$ and $\boldsymbol{\sigma}$ and $\boldsymbol{\sigma}_0$ are respectively the Cauchy and the Piola–Kirchhoff stress tensors. We have made use of the relation (4.1.2). Applying Green's formula to the right-hand side of (4.1.8) we get

$$\rho_0(\mathbf{x})\partial_t^2\mathbf{y}(\mathbf{x},t) = \nabla_{\mathbf{x}} \cdot \sigma_0(\mathbf{x},t). \qquad (4.1.9)$$

Similarly, energy conservation implies that

$$\frac{d}{dt} \int_{D_0} \left(\tfrac{1}{2}\rho_0(\mathbf{x})|\partial_t \mathbf{y}(\mathbf{x},t)|^2 + \rho_0(\mathbf{x})e(\mathbf{x},t) \right) d\mathbf{x} \tag{4.1.10}$$

$$= \int_{\partial D_0} \partial_t \mathbf{y}(\mathbf{x},t) \cdot \boldsymbol{\sigma}_0 \cdot \hat{\mathbf{n}}_0(\mathbf{x},t) \, dS(\mathbf{x}) - \int_{\partial D_0} \mathbf{q}(\mathbf{x},t) \cdot \hat{\mathbf{n}}_0(\mathbf{x},t) \, dS(\mathbf{x}).$$

Applying Green's formula again, we get

$$\frac{\partial}{\partial t} \left(\tfrac{1}{2}\rho_0(\mathbf{x})|\partial_t \mathbf{y}(\mathbf{x},t)|^2 + \rho_0(\mathbf{x})e(\mathbf{x},t) \right)$$
$$= \nabla_\mathbf{x} \cdot (\partial_t \mathbf{y}(\mathbf{x},t) \cdot \boldsymbol{\sigma}_0(\mathbf{x},t)) - \nabla_\mathbf{x} \cdot \mathbf{q}(\mathbf{x},t).$$

Taking (4.1.9) into consideration, this equation can be rewritten as

$$\rho_0 \partial_t e = (\nabla_\mathbf{x} \partial_t \mathbf{y}) : \boldsymbol{\sigma}_0 - \nabla_\mathbf{x} \cdot \mathbf{q}. \tag{4.1.11}$$

Here we have used the notation $\mathbf{A} : \mathbf{B}$ to denote the scalar product of two matrices:

$$\mathbf{A} : \mathbf{B} = \sum_{i,j} a_{ij} b_{ji},$$

where $\mathbf{A} = (a_{ij})$, $\mathbf{B} = (b_{ij})$.

Additional information comes from the second law of thermodynamics. Denoting by $s(\mathbf{x},t)$ the entropy per unit mass and by $T(\mathbf{x},t)$ the temperature of the system, the second law states that the rate at which entropy increases cannot be slower than the contribution due to the exchange of heat with the environment:

$$\frac{d}{dt} \int_{D_0} \rho_0(\mathbf{x}) s(\mathbf{x},t) \, d\mathbf{x} \geq - \int_{\partial D_0} \frac{\mathbf{q}(\mathbf{x},t)}{T(\mathbf{x},t)} \cdot \hat{\mathbf{n}}_0(\mathbf{x}) \, dS(\mathbf{x}). \tag{4.1.12}$$

In a differential form, this gives

$$\rho_0 \partial_t s \geq -\nabla_\mathbf{x} \cdot \left(\frac{\mathbf{q}}{T} \right).$$

This is referred to as the *Clausius–Duhem inequality*.

Define the *Helmholtz free energy* per unit mass as

$$h(\mathbf{x},t) = e(\mathbf{x},t) - T(\mathbf{x},t)s(\mathbf{x},t). \tag{4.1.13}$$

Assume that h is a function of the form

$$h(\mathbf{x},t) = h_0(\mathbf{F},T,\mathbf{x});$$

then

$$\partial_t h = \nabla_{\mathbf{F}} h_0 : \partial_t \mathbf{F} + \frac{\partial h_0}{\partial T} \partial_t T.$$

From (4.1.13), we also have

$$\partial_t h = \partial_t e - s \partial_t T - T \partial_t s.$$

In terms of $\mathbf{F}$, the energy conservation law (4.1.11) can be expressed as

$$\rho_0 \partial_t e = \partial_t \mathbf{F} : \boldsymbol{\sigma}_0 - \nabla_{\mathbf{x}} \cdot \mathbf{q}.$$

Using the Clausius–Duhem inequality

$$\rho_0 \partial_t s \geq -\frac{1}{T} \nabla_{\mathbf{x}} \cdot \mathbf{q} + \frac{1}{T^2} \mathbf{q} \cdot \nabla_{\mathbf{x}} T,$$

we get

$$\rho_0 \left(\nabla_{\mathbf{F}} h_0 : \partial_t \mathbf{F} + \frac{\partial h_0}{\partial T} \partial_t T \right) = \rho_0 \partial_t h \leq -\rho_0 s \partial_t T - \mathbf{q} \cdot \frac{\nabla_{\mathbf{x}} T}{T} + \partial_t \mathbf{F} : \boldsymbol{\sigma}_0,$$

or, equivalently,

$$\partial_t \mathbf{F} : (\rho_0 \nabla_{\mathbf{F}} h_0 - \boldsymbol{\sigma}_0) + \partial_t T \left(\rho_0 \frac{\partial h_0}{\partial T} + \rho_0 s \right) + \mathbf{q} \cdot \frac{\nabla_{\mathbf{x}} T}{T} \leq 0. \quad (4.1.14)$$

Since $\mathbf{F}, T$ can take arbitrary values, the inequality (4.1.14) implies that [23]

$$\frac{\partial W}{\partial \mathbf{F}} = \boldsymbol{\sigma}_0,$$

$$s = -\frac{\partial h}{\partial T},$$

$$\mathbf{q} \cdot \frac{\nabla_{\mathbf{x}} T}{T} \leq 0,$$

where $W = \rho_0 h_0$. These are constraints on the possible constitutive relations.

In the absence of additional information, it is simplest to assume a linear relation between the remaining flux $\mathbf{q}$ and the corresponding driving force $\nabla_{\mathbf{x}} T / T$:

$$\mathbf{q} = -k \frac{\nabla_{\mathbf{x}} T}{T}.$$

Here k, the *thermal conductivity*, must be positive. The other constitutive information needed is an expression for W, the stored energy density, which was discussed earlier.

4.1.5 Dynamics of fluids

Next we turn to fluids. As we said earlier, it is much more convenient to use Eulerian coordinates to describe fluids. We will focus on dynamic models of fluids. For convenience, we now use $\mathbf{u}$ to denote the velocity field (instead of the displacement field as earlier). In Eulerian coordinates, the basic principles take the following form:

- the conservation of mass,

$$\partial_t \rho + \nabla \cdot \mathbf{J}_\rho = 0; \tag{4.1.15}$$

- the conservation of momentum,

$$\partial_t (\rho \mathbf{u}) + \nabla \cdot \mathbf{J}_m = 0; \tag{4.1.16}$$

- the conservation of energy,

$$\partial_t E + \nabla \cdot \mathbf{J}_E = 0. \tag{4.1.17}$$

- the Clausius–Duhem principle, in other words, the second law of thermodynamics:

$$\rho \left(\partial_t s + \mathbf{u} \cdot \nabla s \right) + \nabla \cdot \left(\frac{\mathbf{q}}{T} \right) \geq 0, \tag{4.1.18}$$

where $\mathbf{J}_\rho, \mathbf{J}_m, \mathbf{J}_E$ are the mass, momentum and energy fluxes respectively and are given by

$$\mathbf{J}_\rho = \rho \mathbf{u},$$
$$\mathbf{J}_m = \rho \mathbf{u} \otimes \mathbf{u} - \boldsymbol{\sigma},$$
$$\mathbf{J}_E = E\mathbf{u} - \boldsymbol{\sigma} \cdot \mathbf{u} + \mathbf{q}.$$

It is convenient to introduce the *material derivative*. For any function φ, its material derivative is defined as

$$\dot{\varphi} = D_t \varphi = \frac{\partial}{\partial t} \varphi + \mathbf{u} \cdot \nabla \varphi.$$

Using the material derivative, the conservation laws can be written as

$$\dot{\rho} + \rho \nabla \cdot \mathbf{u} = 0, \tag{4.1.19}$$

$$\rho \dot{\mathbf{u}} = \nabla \cdot \boldsymbol{\sigma}, \tag{4.1.20}$$

$$\rho \dot{e} + \nabla \cdot \mathbf{q} = \mathrm{Tr}\left(\boldsymbol{\sigma}^{\mathrm{T}} \nabla \mathbf{u} \right), \tag{4.1.21}$$

$$\rho \dot{s} + \nabla \cdot \frac{\mathbf{q}}{T} \geq 0, \tag{4.1.22}$$

where Tr $\mathbf{A}$ denotes the trace of a matrix $\mathbf{A}$. To obtain (4.1.21), multiply both sides of (4.1.20) by $\mathbf{u}$ to get

$$\partial_t \left(\frac{\rho}{2} |\mathbf{u}|^2 \right) + \nabla \cdot \left(\frac{\rho}{2} |\mathbf{u}|^2 \mathbf{u} - \mathbf{u} \cdot \boldsymbol{\sigma} \right) = -\text{Tr} \left(\boldsymbol{\sigma}^T \nabla \mathbf{u} \right)$$

Combining this equation and (4.1.17) gives (4.1.21).

The Helmholtz free energy per unit mass is, as before,

$$h = e - Ts,$$

where T is the temperature of the system. We assume that h is only a function of T and ρ, $h = h_0(T, \rho)$; then

$$\rho \dot{h} = \rho \left(\dot{e} - \dot{T}s - T\dot{s} \right). \tag{4.1.23}$$

Using (4.1.21), we obtain on the one hand

$$\rho \dot{h} = \text{Tr} \left(\boldsymbol{\sigma}^T \nabla \mathbf{u} \right) - \nabla \cdot \mathbf{q} - \rho s \dot{T} - \rho T \dot{s}.$$

On the other hand, we also have

$$\rho \dot{h} = \rho \frac{\partial h_0}{\partial T} \dot{T} + \rho \frac{\partial h_0}{\partial \rho} \dot{\rho} = \rho \frac{\partial h_0}{\partial T} \dot{T} - \rho^2 \frac{\partial h_0}{\partial \rho} \left(\nabla \cdot \mathbf{u} \right). \tag{4.1.24}$$

Combining (4.1.23), (4.1.24) and (4.1.18), we get

$$\text{Tr} \left(\left(\boldsymbol{\sigma} + \frac{\partial h_0}{\partial \rho} \rho^2 \mathbf{I} \right)^T \nabla \mathbf{u} \right) - \mathbf{q} \cdot \frac{\nabla T}{T} - \rho \dot{T} \left(s + \frac{\partial h_0}{\partial T} \right) \geq 0. \tag{4.1.25}$$

Since (4.1.25) must hold for any value of $\dot{T}$, we have

$$s = -\frac{\partial h_0}{\partial T},$$

which is a well-known result in thermodynamics. Now (4.1.25) becomes

$$\text{Tr} \left(\left(\boldsymbol{\sigma} + \frac{\partial h_0}{\partial \rho} \rho^2 \mathbf{I} \right)^T \nabla \mathbf{u} \right) - \mathbf{q} \cdot \frac{\nabla T}{T} \geq 0. \tag{4.1.26}$$

Let $(p = (\partial h_0/\partial \rho)\rho^2$ be the thermodynamic pressure and write

$$\boldsymbol{\sigma} = -\tilde{p}\mathbf{I} + \boldsymbol{\sigma}_{\text{d}}, \tag{4.1.27}$$

where $\check{p} = \frac{1}{3}\operatorname{Tr}\boldsymbol{\sigma}$, and $\operatorname{Tr}\boldsymbol{\sigma}_d = 0$. Then (4.1.26) becomes

$$(p - \tilde{p})(\nabla \cdot \mathbf{u}) + \operatorname{Tr}\left((\boldsymbol{\sigma}_d)^{\mathrm{T}}\nabla\mathbf{u}\right) - \mathbf{q} \cdot \frac{\nabla T}{T} \geq 0. \qquad (4.1.28)$$

This is the constraint on the constitutive relation from thermodynamic considerations.

In the absence of additional information, we again resort to a linear relation between the fluxes and the driving forces. This means that $p - \tilde{p}$, $\boldsymbol{\sigma}_d$ and $\mathbf{q}$ should all be linear functions of $\nabla \cdot \mathbf{u}$, $\nabla\mathbf{u}$ and $\nabla T/T$. It is easy to see that for isotropic fluids the constitutive relation must be of the form

$$p - \tilde{p} = \kappa \nabla \cdot \mathbf{u},$$

$$\boldsymbol{\sigma}_d = \mu\left(\frac{\nabla\mathbf{u} + \nabla\mathbf{u}^{\mathrm{T}}}{2} - (\nabla \cdot \mathbf{u})\mathbf{I}\right),$$

$$\mathbf{q} = -k\frac{\nabla T}{T}.$$

The presence of other cross terms on the right-hand side would violate isotropy or invariance under reflection. Here κ, μ, k are constants, representing respectively the *bulk viscosity, shear viscosity* and *heat conductivity*. Together with the conservation laws (4.1.19)–(4.1.21), we obtain the familiar compressible Navier–Stokes equation in fluid dynamics.

So far we have only discussed the simplest example of a fluid system. For more complicated systems the principles are the same as those described above. As an example, let us consider a two-component fluid system. Let C be the mass concentration of one of the two components; then the conservation of mass reads

$$\rho\dot{C} + \nabla \cdot \mathbf{J}_C = 0,$$

where $\mathbf{J}_C$ is the diffusive mass current for the corresponding component. Now besides T and ρ, the free energy also depends on C, i.e. $h = h_0(T, \rho, C)$. The Clausius–Duhem principle gives

$$(p - \tilde{p})\nabla \cdot \mathbf{u} + \operatorname{Tr}\left((\boldsymbol{\sigma}_d)^{\mathrm{T}}\nabla\mathbf{u}\right) - \mathbf{q} \cdot \frac{\nabla T}{T} - \mathbf{J}_C \cdot \nabla\tilde{\mu} \geq 0, \quad (4.1.29)$$

where $\tilde{\mu}$ is the chemical potential and $\tilde{\mu} = \partial h / \partial C$. In the linear response regime the constitutive relations are

$$p - \tilde{p} = \kappa \nabla \cdot \mathbf{u},$$

$$\boldsymbol{\sigma}_d = \mu \left(\frac{\nabla \mathbf{u} + \nabla \mathbf{u}^T}{2} - (\nabla \cdot \mathbf{u})\mathbf{I} \right),$$

$$\mathbf{q} = -A_1 \nabla \tilde{\mu} - A_2 \frac{\nabla T}{T},$$

$$\mathbf{J}_C = -B_1 \nabla \tilde{\mu} - B_2 \frac{\nabla T}{T}.$$

This suggests that the temperature gradient may drive mass diffusion, a phenomenon that is commonly referred to as thermodiffusion or the *Soret effect* [12]. The coefficients A_1, B_1, A_2, B_2 have to satisfy the constraint that the left-hand side of (4.1.29), as a quadratic form of $(\nabla \mathbf{u}, \nabla \tilde{\mu}, \nabla T/T)$, is semi-positive-definite. Further constraints may come from Onsager's reciprocity relation [12], which in this case states that $A_2 = B_1$.

To summarize, the basic steps for developing (dynamic) continuum models are as follows.

(1) Write down the conservation laws.
(2) Find the constraints on the constitutive relations from the second law of thermodynamics.
(3) Find the constraints on the constitutive relations from symmetry considerations. In most cases, as a first approximation, one can postulate linear constitutive relations.

Such a general approach has its merits: it gives a framework that allows us to explore all the possible leading-order physical effects, particularly cross effects such as thermodiffusion.

4.2 Molecular dynamics

Next we turn to atomistic models in the form of classical molecular dynamics. In this section, we discuss the basic ideas of molecular dynamics, including the empirical interatomic potentials, the different ensembles, the continuum limits and the linear response theory for computing transport coefficients using atomistic modeling.

Consider a system of N atoms and denote by m_j and $\mathbf{y}_j$ the mass and position of the jth atom, respectively. Molecular dynamics models the evolution of this system using Newton's second law:

$$m_j \ddot{\mathbf{y}}_j = \mathbf{F}_j = -\nabla_{\mathbf{y}_j} V(\mathbf{y}_1, \ldots, \mathbf{y}_N), \quad j = 1, \ldots, N, \qquad (4.2.1)$$

where V is the potential energy function of the system. This system of equations can also be written in Hamilton's form ($\mathbf{p}_j = m_j \dot{\mathbf{y}}_j$):

$$\frac{d\mathbf{y}_j}{dt} = \nabla_{\mathbf{p}_j} H, \quad \frac{d\mathbf{p}_j}{dt} = -\nabla_{\mathbf{y}_j} H, \quad j = 1, \ldots, N, \qquad (4.2.2)$$

where the Hamiltonian H is given by

$$H = \sum_j \frac{1}{2m_j} |\mathbf{p}_j|^2 + V(\mathbf{y}_1, \ldots, \mathbf{y}_N).$$

4.2.1 Empirical potentials

The first issue we have to address is: what is the function V? As we will see later when discussing quantum mechanics, in principle V can be obtained from electronic structure models. In practice, however, this is a rather costly procedure, and most models of molecular dynamics still use empirical potentials that are carefully designed for specific purposes.

It should be emphasized that at this point in time there are no systematic ways of finding empirical interatomic potentials. Current practice is a combination of coming up with a good guess for the functional form of the potential and calibrating the parameters by using the experimental data as well as data from calculations based on first principles.

Empirical potentials for molecules Potentials for molecules are usually in the following form:

$$V = V_{\text{bonded}} + V_{\text{non-bonded}} = V_{\text{I}} + V_{\text{II}} \qquad (4.2.3)$$

where V_{I} is the contribution due to the covalently bonded interactions and V_{II} is the contribution due to the non-bonded interactions. The bonded interactions include contributions from the changes in *bond length, bond angle* and *torsion angle* (also called *dihedral angle*). In the example of propane ($CH_3--CH_2--CH_3$, see Figure 4.2), there are 10 bonds (two C–C bonds, eight C–H bonds), 18 angles (one C–C–C angle, 10 C–C–H angles and seven H–C–H angles), and 18 torsion angles (12 H–C–C–H angles and six H–C–C–C angles).

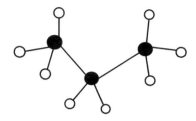

Figure 4.2. An illustration of a propane molecule.

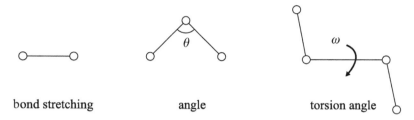

bond stretching   angle   torsion angle

Figure 4.3. An illustration of the basic variables associated with covalent bonds.

A typical form of V_I is

$$V_I = \sum_{\text{bonds}} \frac{k_i}{2}(l_i - l_{i,0})^2 + \sum_{\text{angles}} \frac{\xi_j}{2}(\theta_j - \theta_{j,0})^2$$

$$+ \sum_{\text{torsion angles}} \frac{V_n}{2}(\cos\omega_n - \cos\omega_{n,0})^2. \tag{4.2.4}$$

The first term is the energy due to bond stretching; l is the instantaneous bond length and l_0 is the equilibrium bond length. The second term is due to changes in the bond angles, which corresponds to bending; θ_j and $\theta_{j,0}$ are respectively the jth bond angle and its equilibrium value. The third term is the contribution from changes in the torsion angles, which corresponds to twisting. Finally, ω_n is the torsion angle between three consecutive bonds; $\omega_{n,0}$ is its equilibrium value. See Figure 4.3. Note that the functional forms of the terms are rather simplistic, being either quadratic or simple cosines. The motivation is the same as before, when we considered constitutive relations for solids and fluids: we always choose the simplest form that is consistent with what we already know. Additionally, the parameters in (4.2.4) can be computed from quantum mechanical calculations. This has become a standard practice when developing force fields [5].

In principle, one might also consider mixed terms such as those due to bond angle and torsion angle interactions, but these are only included when it is absolutely necessary to do so.

Next we discuss the non-bonded interaction. There are two main contributions: the electrostatic interaction and the *van der Waals interaction*,

$$V_{\mathrm{II}} = \frac{1}{2} \sum_{i \neq j} 4\varepsilon_{ij} \left(\left(\frac{\sigma_{ij}}{r_{ij}} \right)^{12} - \left(\frac{\sigma_{ij}}{r_{ij}} \right)^{6} \right) + \frac{1}{2} \sum_{i \neq j} \frac{q_i q_j}{4\pi\varepsilon_0 r_{ij}}. \qquad (4.2.5)$$

Here $r_{i,j} = |\mathbf{y}_i - \mathbf{y}_j|$, q_i is the partial charge of the ith atom and ε_0 is the permittivity of a vacuum. Their values need to be changed when solvent effects, which are often important when one is studying biomolecules, are taken into account.

The van der Waals interaction accounts for the effect of quantum fluctuations in the dipole moments. Consider two atoms a distance r apart, where r is much larger than the size of the atoms. Even though the atoms may not have permanent dipoles, quantum and thermal fluctuations will cause them to have temporary dipole moments. If one atom acquires some temporary dipole moment, it induces a polarization on the other atom. This induced polarization is of the order $\mathcal{O}(r^{-3})$ (see for example problem 3 in Chapter 11 of [32]). The energy of the dipole–dipole interaction is of the form

$$V(r) = -\frac{\alpha_1 \alpha_2}{r^6}, \qquad (4.2.6)$$

where α_1 and α_2 are the polarizability constants for the two atoms.

The potential in (4.2.6) is always negative, implying that the van der Waals interaction is attractive. At short distances, it is compensated by a repulsive force that prevents the two atoms from collapsing into each other. The repulsive force comes from either or both of from the charge–charge interaction and Pauli's exclusion principle, which prevents two electrons from occupying the same quantum state. Unfortunately, there is no derivation analogous to that for the van der Waals interaction for the form of the repulsive force. As a matter of convenience, it is often modeled by a r^{-12} term. This gives the Lennard-Jones form of the van der Waals interaction:

$$V(r) = 4\varepsilon \left(\left(\frac{\sigma}{r} \right)^{12} - \left(\frac{\sigma}{r} \right)^{6} \right). \qquad (4.2.7)$$

The two parameters in the *Lennard-Jones potential* are ε, which is scales the interaction energy and σ, which scales the length. When multiple species of atoms are encountered, a mixing rule is often used for the parameter values in the Lennard-Jones potential, e.g. for two atoms, A and B,

$$\sigma_{AB} = \tfrac{1}{2}(\sigma_{AA} + \sigma_{BB}), \quad \varepsilon_{AB} = \sqrt{\varepsilon_{AA}\varepsilon_{BB}}. \tag{4.2.8}$$

Empirical potentials for solids Solids can be classified into several different types according to the nature of the bonding between the atoms in the solid. These are: molecular, metallic, covalently bonded, ionic and hydrogen-bonded [4]. The last type occurs mostly in biological systems; we will not discuss it here. The first four types are outlined below.

(1) *Molecular crystals.* These occur mainly for inert gas elements in the last column of the periodic table, for which the atoms interact mainly through the van der Waals effect, modeled by the Lennard-Jones potential.

(2) *Metallic systems.* An important ingredient for modeling metallic systems is the charge density, which is often defined through

$$\rho_i = \sum_j \rho^*(r_{ij}),$$

where ρ^* is a fitting function. The most popular interatomic potential for metals is the *embedded atom model* (EAM) [11]

$$V = \tfrac{1}{2}\sum_{i\neq j} \Phi(r_{ij}) + \sum_i F(\rho_i). \tag{4.2.9}$$

Here the functions Φ, F are also fitting functions. One special case is the *Finnis–Sinclair potential*, where $F(\rho) = A\sqrt{\rho}$.

(3) *Covalently bonded systems.* These are somewhat similar to the molecular case. Examples include the diamond structure of silicon and the graphene structure of carbon. Two most popular potentials are the *Stillinger–Weber potential* and the *Tersoff potential*. The Stillinger–Weber potential takes the form

$$V = \tfrac{1}{2}\sum f(r_{ij}) + \sum (h(r_{ij}, r_{ik}, \theta_{jik}) + \text{permutations}), \tag{4.2.10}$$

where

$$h(r, r', \theta) = \lambda \exp\left(\frac{\gamma}{r-a} + \frac{\gamma}{r'-a}\right)\left(\cos\theta + \tfrac{1}{3}\right)^2 \tag{4.2.11}$$

This potential is similar to the potentials for molecules discussed above. The first term is the pairwise interaction. The second, three-body, term is the contribution due to variations in the bond angles.

The Tersoff potential, however, is quite different. It takes the form

$$V = \sum_{i \neq j} A e^{-\alpha r_{ij}} - b_{ij} B e^{-\beta r_{ij}}. \tag{4.2.12}$$

The interesting parameter here is b_{ij}, the local bond order, which depends on the coordination of the atoms and the bond strength [35].

(4) Ionic crystals, such as rock salt. In these systems the atoms interact mainly through the ion–ion electrostatic forces and the short-range repulsion. Born proposed the following interatomic potential [4]

$$V = \tfrac{1}{2} \sum_{i \neq j} \frac{q_i q_j}{4\pi\varepsilon_0 r_{ij}} + \sum \frac{A}{r_{ij}^n},$$

where q_i is the partial charge of the ith ion and A and n are empirical parameters. The difficulty with this potential and with modeling ionic systems in general is the long-range nature of the electrostatic interaction.

Empirical potentials for fluids Fluids are rarely modeled at the level of atomic detail. The models are often coarse-grained. The choice of empirical potential depends on how we coarse-grain the system. If the coarse-grained particles are treated as spherical point particles then the Lennard-Jones potential is often used. If they are treated as ellipsoids or rods then one may use an anisotropic form of the Lennard-Jones potential such as the Gay–Berne potential [1]. If the coarse-grained particles are chains, i.e. beads connected by springs, then bead–spring models are used. Such a model typically contains two components:

(1) the beads interact through a Lennard-Jones potential;
(2) neighboring beads on the same chain interact via a spring force.

An example of this type of potential is treated in Chapter 6. Coarse-grained models of fluids are discussed more systematically in [40].

4.2.2 Equilibrium states and ensembles

Consider the solutions of Hamilton's equation (4.2.2) with a random initial condition. Denote by $\rho(\mathbf{z}_1, \ldots, \mathbf{z}_N, t)$ the probability density of the particles at time t, where $\mathbf{z}_i = (\mathbf{y}_i, \mathbf{p}_i)$. It is easy to see that ρ satisfies *Liouville's equation*:

$$\partial_t \rho + \sum_{i=1}^{N} \nabla_{\mathbf{z}_i} \cdot (\mathbf{V}_i \rho) = 0, \qquad (4.2.13)$$

where $\mathbf{V}_i = (\nabla_{\mathbf{p}_i} H, -\nabla_{\mathbf{y}_i} H)$, $\nabla_{\mathbf{z}_i} = (\nabla_{\mathbf{y}_i}, \nabla_{\mathbf{p}_i})$.

Given a system of interacting particles (atoms or molecules, etc.), the first question we should think about is their equilibrium states. These are stationary solutions of the Liouville equation (4.2.13) that satisfy the condition of *detailed balance*:

$$\nabla_{\mathbf{z}_i} \cdot (\mathbf{V}_i \, \rho) = 0$$

for each $i = 1, \ldots, N$. It is obvious that probability densities of the form

$$\rho_0 = \frac{1}{Z_N} f(H)$$

are equilibrium states. Here f is a nonnegative function, and Z_N is the normalization constant, also called the partition function:

$$Z_N = \int_{\mathbb{R}^{6N}} f(H) \, d\mathbf{z}_1 \cdots d\mathbf{z}_N.$$

Of particular importance are the following special cases.

(1) The *microcanonical ensemble*: $f(H) = \delta(H - H_0)$ for some constant H_0.
(2) The *canonical ensemble*: $f(H) = e^{-\beta H}$ where $\beta = 1/(k_B T)$, k_B is the Boltzmann constant and T is the temperature.

The particular physical setting determines the particular equilibrium distribution that is realized. In the context of molecular dynamics, the microcanonical ensemble is realized through the *NVE ensemble*, in which the number of particles N, the volume V accessible to the particles and the total energy E are fixed. This models a closed system with no exchange of mass, momentum or energy with the environment. Similarly, the canonical ensemble is realized through the *NVT ensemble*, in which, instead of the total energy, the temperature of the system is fixed. Recall that temperature is defined as the average kinetic energy per degree of freedom due to the fluctuating part of the particle velocity, assuming that the average velocity is zero,

$$\frac{1}{N}\left\langle \sum_{j=1}^{N} \mathbf{v}_j \right\rangle = 0$$

where $\mathbf{v}_j = \mathbf{p}_j/m_j$; then

$$\frac{1}{N}\left\langle \sum_{j=1}^{N} \tfrac{1}{2}m_j|\mathbf{v}_j|^2 \right\rangle = \tfrac{3}{2}k_B T.$$

The NVE ensemble is obtained when one solves directly Hamilton's equations (4.2.2) in a box with, e.g. a reflection boundary condition. The value of the Hamiltonian is conserved in this case. However, to obtain the NVT ensemble, in which the temperature rather than the total energy of the system is fixed, we need to modify Hamilton's equations. There are several empirical ways of doing this. The simplest idea is the isokinetic energy trick (also called velocity rescaling). The idea is very straightforward for a discretized form of (4.2.2): at each time step, one rescales the velocities $\mathbf{v}_j$ simultaneously by a factor such that, after rescaling,

$$\frac{1}{N} \sum_{j=1}^{N} \tfrac{1}{2}m_j|\mathbf{v}_j|^2 = \tfrac{3}{2}k_B T$$

holds, where T is the desired temperature. This gives the right average temperature but it does not capture the temperature fluctuations correctly. Indeed in this case the quantity

$$\tilde{T} = \frac{1}{3k_B N} \sum_{j=1}^{N} \tfrac{1}{2}m_j|\mathbf{v}_j|^2$$

does not fluctuate. This is not the case for the true canonical ensemble. In fact, one can show that [18]

$$\frac{\langle (\tilde{T} - T)^2 \rangle}{\langle T \rangle^2} = \frac{2}{3N},$$

i.e. the fluctuation in $\tilde{T}$ is of order $N^{-1/2}$, which only vanishes in the limit $N \to \infty$.

Other popular ways of generating the NVT ensemble include the following.

(1) The Anderson thermostat. At random times, a random particle is picked and its velocity is changed to a random vector from the *Maxwell–Boltzmann distribution*:

$$\rho(\mathbf{v}_1, \ldots, \mathbf{v}_N) \sim \exp\left(-\beta \sum_j \frac{1}{2} m_j |\mathbf{v}_j|^2\right).$$

(2) Replacing Hamilton's equation by the Langevin equation

$$m_j \frac{d^2 \mathbf{y}_j}{dt^2} = -\frac{\partial V}{\partial \mathbf{y}_j} - \gamma_j \frac{d\mathbf{y}_j}{dt} + \sigma_j \dot{\mathbf{w}}_j, \quad j = 1, \ldots, N,$$

where γ_j is the friction coefficient for the jth particle, σ_j is the noise amplitude,

$$\sigma_j = \sqrt{2 k_B T \gamma_j},$$

and $\{\dot{\mathbf{w}}_1, \ldots, \dot{\mathbf{w}}_N\}$ are independent white noises.

(3) The Nóse–Hoover thermostat. This is a way of producing the canonical ensemble by adding an additional degree of freedom to the system. The effect of this additional degree of freedom is to supply or to take away energy from the system, according to its instantaneous thermal energy [18].

In general, these thermostats produce the equilibrium properties of the system correctly, but it is not clear that the dynamic quantities are produced correctly [18].

4.2.3 The elastic continuum limit; the Cauchy–Born rule

From the viewpoint of multiscale, multi-physics modeling, it is natural to ask how the atomistic models are related to the continuum models discussed in the last section. We will address this question from three different but related angles. In this subsection we will discuss the elastic continuum limit at zero temperature. In the next subsection we will discuss the relation between the continuum models and the atomistic models at the level of the conservation laws. Finally we will discuss how to compute macroscopic transport coefficients using atomistic models.

At zero temperature, $T = 0$, the canonical ensemble becomes a degenerate distribution concentrated at the minimizers of the total potential energy:

$$\min V(\mathbf{y}_1, \ldots, \mathbf{y}_N). \qquad (4.2.14)$$

This special case is helpful for understanding many issues in the theory of solids.

It is well known that at low temperatures, solids are crystals whose atoms reside on lattices. Common lattice structures include body-centered cubic (BCC), face-centered cubic (FCC), diamond, hexagonal closed packed (HCP) etc. [32]. For example, at room temperature and atmospheric pressure, iron (Fe) exists in a BCC lattice, aluminum (Al) exists in an FCC lattice and silicon (Si) exists in a diamond lattice. Lattices are divided into two types: simple and complex.

Simple lattices Simple lattices are also called Bravais lattices. Their lattice points take the form

$$L(\{\mathbf{e}_i\}, \mathbf{o}) = \left\{ \mathbf{x} : \mathbf{x} = \sum_{i=1}^{d} \nu^i \mathbf{e}_i + \mathbf{o}, \quad \nu^i \text{ integers} \right\}, \qquad (4.2.15)$$

where $\{\mathbf{e}_i\}_{i=1}^{d}$ is a set of basis vectors, d is the dimension and $\mathbf{o}$ is a particular lattice site. Note that the basis vectors are not unique.

Of the examples listed above, BCC and FCC are simple lattices. For the FCC lattice, one set of basis vectors is

$$\mathbf{e}_1 = \frac{a}{2}(0, 1, 1), \quad \mathbf{e}_2 = \frac{a}{2}(1, 0, 1), \quad \mathbf{e}_3 = \frac{a}{2}(1, 1, 0).$$

For the BCC lattice, we may choose

$$\mathbf{e}_1 = \frac{a}{2}(-1, 1, 1), \quad \mathbf{e}_2 = \frac{a}{2}(1, -1, 1), \quad \mathbf{e}_3 = \frac{a}{2}(1, 1, -1)$$

as the basis vectors. Here, and in what follows, we use a to denote the lattice constant.

Another example of a simple lattice is the two-dimensional triangular lattice. Its basis vectors can be chosen as

$$\mathbf{e}_1 = a(1, 0), \quad \mathbf{e}_2 = \frac{a}{2}(1, \sqrt{3}).$$

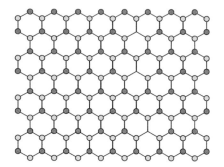 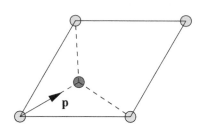

Figure 4.4. A graphene sheet as an example of a complex lattice. The right-hand panel illustrates a unit cell (courtesy of Pingbing Ming).

Complex lattices Crystal lattices that are not simple are called complex (see Figure 4.4). It can be shown that every complex lattice is a union of simple lattices, each of which is a shift of any other; i.e. the lattice points can be expressed in the form

$$L = L(\{\mathbf{e}_i\}, \mathbf{o}) \cup L(\{\mathbf{e}_i\}, \mathbf{o} + \mathbf{p}_1) \cup \cdots \cup L(\{\mathbf{e}_i\}, \mathbf{o} + \mathbf{p}_k).$$

Here $\mathbf{p}_1, \ldots, \mathbf{p}_k$ are the shift vectors. For example, a two-dimensional hexagonal lattice with lattice constant a can be regarded as the union of two triangular lattices with shift vector $\mathbf{p}_1 = a(-1/2, -\sqrt{3}/6)$ (see Figure 4.4). The diamond lattice is made up of two interpenetrating FCC lattices with shift vector $\mathbf{p}_1 = (a/4)(1, 1, 1)$. The HCP lattice is obtained by stacking two simple hexagonal lattices with shift vector $\mathbf{p}_1 = a(1/2, \sqrt{3}/6, \sqrt{6}/3)$. Some solids consist of more than one species of atom. Sodium chloride (NaCl), for example, has equal numbers of sodium and chloride ions placed at alternating sites of a simple cubic lattice. This can be viewed as the union of two FCC lattices, one for the sodium ions and one for the chloride ions.

The picture described above is only for single crystals and it neglects lattice defects such as stacking faults, vacancies and dislocations. Moreover, most crystalline solids are polycrystals, i.e. unions of single crystal grains with different orientations. Nevertheless, it is certainly true that the basic building blocks of crystalline solids are crystal lattices.

Turning now to the physical models, we have previously discussed two approaches for analyzing the deformation of solids: the continuum theory discussed in the last section and the atomistic model. It is natural to ask how these two approaches are related to each other. From the viewpoint

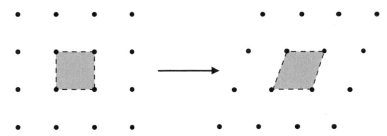

Figure 4.5. Illustration of the Cauchy–Born rule: $W(\mathbf{A})$ is equal to the energy density of a homogeneously deformed crystal obtained through the relation $\mathbf{y} = \mathbf{x} + \mathbf{A}\mathbf{x}$ (courtesy of Pingbing Ming).

of atomistic modeling, the relevant question is the macroscopic continuum limit of the atomistic models. From the viewpoint of the continuum theory, the relevant question is the microscopic foundation of the continuum models.

Recall that, in continuum elasticity theory, the potential energy of an elastic solid is given by

$$\int_\Omega W(\nabla \mathbf{u})\, d\mathbf{x}, \qquad (4.2.16)$$

where W is the stored energy density. In the subsection 4.2.2, we discussed ways of obtaining empirical expressions for W. We now ask: can we derive W from an atomistic model?

This question was considered by Cauchy, who derived atomistic expressions for the elastic moduli. Cauchy's work was extended by Born, who also considered complex lattices [4]. For crystalline solids the *Cauchy–Born rule* establishes an expression for W in terms of an atomistic model.

Let us consider first the case when the crystal is a simple lattice. In this case, consider a homogeneously deformed crystal with deformation gradient $\mathbf{A}$, i.e. the positions of the atoms in the deformed configuration are given by $\mathbf{y}_j = \mathbf{x}_j + \mathbf{A}\mathbf{x}_j$ for all j (see Figure 4.5). Knowing the positions of the atoms in the deformed crystal we can compute the energy density (the energy per unit cell) of the deformed crystal using the atomistic model. This gives the value of $W_{CB}(\mathbf{A})$.

Next consider the case when the crystal is a complex lattice. In this case, to define the configuration of a homogeneously deformed crystal we not only have to specify the deformation gradient $\mathbf{A}$, we also need to specify the shift vectors $\mathbf{p}_1, \ldots, \mathbf{p}_k$. Once they are specified, we can compute the energy density of the homogeneously deformed crystal using the atomistic model. We denote this value by $w(\mathbf{A}, \mathbf{p}_1, \ldots, \mathbf{p}_k)$. We then define

$$W_{CB}(\mathbf{A}) = \min_{\mathbf{p}_1, \ldots, \mathbf{p}_k} w(\mathbf{A}, \mathbf{p}_1, \ldots, \mathbf{p}_k). \qquad (4.2.17)$$

To see where the Cauchy–Born rule comes from, consider the case of a simple lattice with atoms interacting via a two-body potential

$$V = \frac{1}{2} \sum_{i \neq j} V_0 \left(|\mathbf{y}_i - \mathbf{y}_j| \right).$$

Cauchy's idea was to assume that the deformation of the atoms follows a smooth displacement vector field $\mathbf{u}$, i.e. [16]

$$\mathbf{y}_j = \mathbf{x}_j + \mathbf{u}(\mathbf{x}_j).$$

Then the total energy of the system is given by

$$V = \frac{1}{2} \sum_{i \neq j} V_0 \left(|(\mathbf{x}_j + \mathbf{u}(\mathbf{x}_j)) - (\mathbf{x}_i + \mathbf{u}(\mathbf{x}_i))| \right),$$

$$= \frac{1}{2} \sum_{i \neq j} V_0 \left(|\mathbf{x}_j - \mathbf{x}_i + \nabla \mathbf{u}(\mathbf{x}_i)(\mathbf{x}_j - \mathbf{x}_i) + \mathcal{O}(|\mathbf{x}_j - \mathbf{x}_i|^2)| \right).$$

To leading order we have

$$V \approx \frac{1}{2} \sum_{i \neq j} V_0 \left(|(\mathbf{I} + \nabla \mathbf{u}(\mathbf{x}_i))(\mathbf{x}_j - \mathbf{x}_i)| \right)$$

$$= \sum_j W(\nabla \mathbf{u}(\mathbf{x}_j)) V_a \approx \int W(\nabla \mathbf{u}(\mathbf{x})) \, d\mathbf{x}, \qquad (4.2.18)$$

where

$$W(\mathbf{A}) = \frac{1}{2V_a} \sum_{i \neq 0} V_0(|(\mathbf{I} + \mathbf{A})(\mathbf{x}_i)|)$$

and V_a is the volume of the unit cell of the undeformed lattice. This is nothing other than the Cauchy–Born rule. In one dimension, with

$x_k = ka$, equation (4.2.18) becomes

$$
\begin{aligned}
V &= \frac{1}{2} \sum_i \left(\sum_k V_0 \left(\left(1 + \frac{du}{dx}(x_i) \right) ka \right) \right) \\
&= \sum_i W \left(\frac{du}{dx}(x_i) \right) a \\
&\approx \int W \left(\frac{du}{dx}(x) \right) dx,
\end{aligned}
\tag{4.2.19}
$$

where

$$
W(A) = \frac{1}{2a} \sum_k V_0((1 + A)ka).
\tag{4.2.20}
$$

Calculations of this type have been carried out for very general cases in [3]. In essence, such a calculation demonstrates that, to leading order, the atomistic model can be regarded as a Riemann sum of the nonlinear elasticity model, where the stored energy density is given by the Cauchy–Born rule. The validity of the Cauchy–Born rule at zero temperature has been studied in great detail, in [14] for the static case and in [15] for the dynamic case. The essence of this analysis is to find out under what conditions Cauchy's assumption that the atoms follow a smooth displacement field is valid. These conditions are shown to be in the form of stability conditions for the phonons as well as the elastic stiffness tensor. These conclusions seem to be sharp and quite general.

We will refer to the nonlinear elasticity model in which $W = W_{\mathrm{CB}}$ as the *Cauchy–Born elasticity model*.

So far we have only discussed the highly idealized situation when the temperature is zero. The extension of the Cauchy–Born construction to finite temperatures is still something of an open problem.

4.2.4 Nonequilibrium theory

From a statistical viewpoint, the basic object for the nonequilibrium theory is Liouville's equation. It contains all the statistical information we need. However, Liouville's equation is a very complicated object since it describes the dynamics of a function of many variables, the probability density. In the following we will try to make a more direct link with nonequilibrium continuum theory. We will do so by deriving, from the equations of molecular dynamics, conservation laws similar to those in the continuum theory. We will restrict ourselves to the case when the interatomic potential has a two-body form:

$$V(\mathbf{y}_1,\ldots,\mathbf{y}_N) = \frac{1}{2}\sum_{i\neq j} V_0(|\mathbf{y}_i - \mathbf{y}_j|).$$

Given the output from molecular dynamics, i.e. $\{(\mathbf{y}_j(t), \mathbf{v}_j(t)),\ j = 1,\ldots,N\}$, we define the empirical mass, momentum and energy densities as follows:

$$\tilde{\rho}(\mathbf{y},t) = \sum_j m_j\, \delta\left(\mathbf{y} - \mathbf{y}_j(t)\right),$$

$$\tilde{\mathbf{m}}(\mathbf{y},t) = \sum_j m_j\mathbf{v}_j(t)\, \delta\left(\mathbf{y} - \mathbf{y}_j(t)\right),$$

$$\tilde{E}(\mathbf{y},t) = \sum_j \frac{1}{2}m_j|\mathbf{v}_j(t)|^2\, \delta\left(\mathbf{y} - \mathbf{y}_j(t)\right)$$

$$+ \frac{1}{2}\sum_j \left(\sum_{i\neq j} V_0(|\mathbf{y}_j(t) - \mathbf{y}_i(t)|)\right) \delta\left(\mathbf{y} - \mathbf{y}_j(t)\right).$$

Here $\delta(\cdot)$ is the delta function. Letting $\mathbf{f}_{ij} = -\nabla_{\mathbf{y}_i} V_0(|\mathbf{y}_i - \mathbf{y}_j|)$, we can define analogs of the stress tensor and energy flux:

$$\tilde{\boldsymbol{\sigma}} = \sum_i m_i\mathbf{v}_i(t) \otimes \mathbf{v}_i(t)\delta(\mathbf{y} - \mathbf{y}_i(t))$$

$$+ \frac{1}{2}\sum_i\sum_{j\neq i}(\mathbf{y}_i(t) - \mathbf{y}_j(t)) \otimes \mathbf{f}_{ij}(t)$$

$$\times \int_0^1 \delta\left(\mathbf{y} - [\mathbf{y}_j(t) + \lambda(\mathbf{y}_i(t) - \mathbf{y}_j(t))]\right) d\lambda, \qquad (4.2.21)$$

$$\tilde{\mathbf{J}} = \sum_i \mathbf{v}_i(t)\left(\frac{1}{2}m_i\mathbf{v}_i(t)^2 + \frac{1}{2}\sum_{j\neq i} V_0(|\mathbf{y}_i(t) - \mathbf{y}_j(t)|)\right)\delta\left(\mathbf{y} - \mathbf{y}_i(t)\right)$$

$$+ \frac{1}{4}\sum_{j\neq i}(\mathbf{v}_j(t) + \mathbf{v}_i(t))\cdot \mathbf{f}_{ij}(t) \otimes (\mathbf{y}_i(t) - \mathbf{y}_j(t))$$

$$\times \int_0^1 \delta\left(\mathbf{y} - [\mathbf{y}_j(t) + \lambda(\mathbf{y}_i(t) - \mathbf{y}_j(t))]\right) d\lambda. \qquad (4.2.22)$$

Equation (4.2.21) is the well-known *Irving–Kirkwood formula* [22]. It is a straightforward but tedious calculation to show that the following equations hold as a consequence of the molecular dynamics equation (4.2.1):

$$\partial_t \tilde{\rho} + \nabla_{\mathbf{y}} \cdot \tilde{\mathbf{m}} = 0, \tag{4.2.23}$$

$$\partial_t \tilde{\mathbf{m}} + \nabla_{\mathbf{y}} \cdot \tilde{\boldsymbol{\sigma}} = 0, \tag{4.2.24}$$

$$\partial_t \tilde{E} + \nabla_{\mathbf{y}} \cdot \tilde{\mathbf{J}} = 0. \tag{4.2.25}$$

These are the analogs of the conservation laws in continuum theory.

Even though molecular dynamics naturally uses Eulerian coordinates, these formulas also have their analogs in Lagrangian coordinates. Denote $\mathbf{x}_j^0 = \mathbf{x}_j(0)$, which can be thought of as being the Lagrangian coordinate of the jth atom, and let [30]

$$\tilde{\rho}(\mathbf{x}, t) = \sum_j m_j \delta(\mathbf{x} - \mathbf{x}_j^0),$$

$$\tilde{\mathbf{m}}(\mathbf{x}, t) = \sum_j m_j \mathbf{v}_j(t) \delta(\mathbf{x} - \mathbf{x}_j^0),$$

$$\tilde{E}(\mathbf{x}, t) = \sum_j \frac{1}{2} m_j |\mathbf{v}_j(t)|^2 \delta(\mathbf{x} - \mathbf{x}_j^0) + \frac{1}{2} \sum_j \left(\sum_{i \neq j} V_0(|\mathbf{x}_j^0 - \mathbf{x}_i^0|) \right) \delta\left(\mathbf{x} - \mathbf{x}_j^0\right),$$

$$\tilde{\boldsymbol{\sigma}} = \frac{1}{2} \sum_i \sum_{j \neq i} (\mathbf{x}_i^0 - \mathbf{x}_j^0) \otimes \mathbf{f}_{ij}(t) \int_0^1 \delta\left(\mathbf{x} - \left(\mathbf{x}_j^0 + \lambda(\mathbf{x}_i^0 - \mathbf{x}_j^0)\right)\right) d\lambda,$$

$$\tilde{\mathbf{J}} = \frac{1}{4} \sum_{j \neq i} (\mathbf{v}_j(t) + \mathbf{v}_i(t)) \cdot \mathbf{f}_{ij}(t) \otimes (\mathbf{x}_i^0 - \mathbf{x}_j^0)$$
$$\times \int_0^1 \delta\left(\mathbf{x} - \left(\mathbf{x}_j^0 + \lambda(\mathbf{x}_i^0 - \mathbf{x}_j^0)\right)\right) d\lambda.$$

We then have [30, 41]

$$\partial_t \tilde{\mathbf{m}} + \nabla_{\mathbf{x}} \cdot \tilde{\boldsymbol{\sigma}} = 0, \tag{4.2.26}$$

$$\partial_t \tilde{E} + \nabla_{\mathbf{x}} \cdot \tilde{\mathbf{J}} = 0. \tag{4.2.27}$$

In principle, one way of deriving the continuum models is to perform averaging on the conservation laws we have just discussed. In practice, we face the closure problem: in order to calculate the averaged fluxes, we need to know the many-particle probability density of the atoms. We get a closed form model if this probability density can be expressed in terms of the averaged conserved quantities (the mass density and averaged momentum and energy densities). However, this is usually not the case. Therefore approximations have to be made in order to solve the closure problem. This point will be made more clearly in the next subsection.

4.2.5 Linear response theory and the Green–Kubo formula

The connection between the continuum and atomistic models established above is quite general but it may not be explicit enough to offer significant physical insights. Here we consider the linear response regime and establish a connection between the transport coefficients, such as the diffusion constants and viscosity coefficients, in terms of microscopic quantities. We can always put a system into the linear response regime by limiting the size of the external driving force.

First, let us consider the diffusion of a tagged particle in a fluid. We assume that its probability distribution ρ satisfies the diffusion equation:

$$\partial_t \rho(\mathbf{x}, t) = D\Delta\rho(\mathbf{x}, t). \tag{4.2.28}$$

Assume that the particle is at the origin initially. Since

$$\langle |\mathbf{x}(t)|^2 \rangle = \int |\mathbf{x}|^2 \rho(\mathbf{x}, t)\, d\mathbf{x},$$

we have

$$\frac{\partial}{\partial t} \int |\mathbf{x}|^2 \rho(\mathbf{x}, t)\, d\mathbf{x} = D \int |\mathbf{x}|^2 \rho(\mathbf{x}, t)\, d\mathbf{x}$$
$$= D \int \nabla \cdot (|\mathbf{x}|^2 \nabla\rho)\, d\mathbf{x} - D \int (\nabla|\mathbf{x}|^2) \cdot \nabla\rho\, d\mathbf{x}$$
$$= -2D \int \mathbf{x} \cdot \nabla\rho\, d\mathbf{x}$$
$$= 2D \int (\nabla \cdot \mathbf{x})\rho\, d\mathbf{x}$$
$$= 6D$$

in three dimensions.

This far the discussion is limited to the macroscopic model (4.2.28). Now we consider the dynamics of a diffusing particle. Using $\mathbf{y}(t) = \int_0^t \mathbf{v}(t_1)\,dt_1$, we have

$$\langle |\mathbf{y}|^2(t) \rangle = \left\langle \int_0^t \int_0^t \mathbf{v}(t_1)\mathbf{v}(t_2)\,dt_1\,dt_2 \right\rangle$$

$$= \int_0^t \int_0^t \langle \mathbf{v}(t_1 - t_2)\mathbf{v}(0) \rangle\,dt_1\,dt_2,$$

$$\frac{d}{dt}\langle |\mathbf{y}|^2(t) \rangle \to \int_0^\infty \langle \mathbf{v}(t)\mathbf{v}(0) \rangle\,dt$$

as $t \to \infty$. Hence we have

$$D = \tfrac{1}{6} \int_0^\infty \langle \mathbf{v}(t)\mathbf{v}(0) \rangle\,dt.$$

This equation relates the transport coefficient D to the integral of an autocorrelation function for the microscopic system, here the velocity autocorrelation function $g(t) = \langle \mathbf{v}(t)\mathbf{v}(0) \rangle$ of the diffusing particle. The function $g(\cdot)$ can be calculated using microscopic simulations such as molecular dynamics. Similar expressions can be found for the viscosity, heat conductivity etc. Such relationships are known as the *Green–Kubo formulas* [9].

4.3 Kinetic theory

Kinetic theory serves as a convenient bridge between continuum and atomistic models. It can be viewed either as an approximation to the hierarchy of equations for the many-particle probability densities obtained from molecular dynamics or as a microscopic model for the phase-space probability distribution of a particle from which continuum models can be derived.

The basic object of interest in kinetic theory is the one-particle distribution function. In the classical kinetic theory of gases, a particle is described by its coordinates in phase space, i.e. its position and momentum. Therefore classical kinetic theory describes the behavior of the one-particle phase-space distribution function [8].

Kinetic theory has been extended in many different directions. One interesting extension is the kinetic models of complex fluids such as liquid crystals made up of rod-like molecules [13] in which additional variables are introduced to describe the conformation of the molecules. For example, for rod-like molecules it is important to include the orientation of the molecules as part of the description.

The main difficulty in formulating models of kinetic theory is in the modeling of the particle–particle interactions. Indeed, for this reason kinetic theory is mostly used when particle–particle interactions are weak, e.g. for gases and dilute plasmas. Another case where kinetic theory is used is when the particle–particle interaction can be modeled accurately by a mean field approximation.

In this section we will briefly summarize the main features of kinetic theory for simple systems, i.e. particles without internal degrees of freedom.

4.3.1 The BBGKY hierarchy

At a fundamental level, kinetic equation can be viewed as an approximation to the hierarchy of equations for the many-particle distribution functions for interacting particle systems. This hierarchy of equations is called the *BBGKY hierarchy*, named after Bogoliubov, Born, Green, Kirkwood and Yvon [36].

We will consider a system of N identical particles interacting with only a two-body potential. In this case, the Hamiltonian is given by

$$H = \sum_i \frac{1}{2m} |\mathbf{p}_i|^2 + \frac{1}{2} \sum_{j \neq i} V_0(|\mathbf{x}_i - \mathbf{x}_j|).$$

Let $\mathbf{z}_i = (\mathbf{p}_i, \mathbf{x}_i)$, and let $\rho_N(\mathbf{z}_1, \ldots, \mathbf{z}_N, t)$ be the probability density function of this interacting particle system in the $\mathbb{R}^{6N}$-dimensional phase space at time t. Liouville's equation (4.2.13) can be written as [36]

$$\frac{\partial \rho_N}{\partial t} + H_N \rho_N = 0, \tag{4.3.1}$$

where

$$H_N = \sum_{i=1}^{N} \frac{1}{m} (\mathbf{p}_i \cdot \nabla_{\mathbf{x}_i}) - \sum_{i<j} \hat{L}_{ij}$$

and

$$\hat{L}_{ij} = \frac{1}{2} \left(-(\mathbf{f}_{ij} \cdot \nabla_{\mathbf{p}_i}) - (\mathbf{f}_{ji} \cdot \nabla_{\mathbf{p}_j}) \right), \quad \mathbf{f}_{ij} = -\nabla_{\mathbf{x}_i} V_0(|\mathbf{x}_i - \mathbf{x}_j|).$$

Define the one-particle probability density as

$$\rho_1(\mathbf{z}_1, t) = \int \cdots \int \rho_N(\mathbf{z}_1, \ldots, \mathbf{z}_N, t)\, d\mathbf{z}_2 \cdots d\mathbf{z}_N.$$

We define the k-particle probability density similarly:

$$\rho_k(\mathbf{z}_1, \ldots, \mathbf{z}_k, t) = \int \cdots \int \rho_N(\mathbf{z}_1, \ldots, \mathbf{z}_N, t)\, d\mathbf{z}_{k+1} \cdots d\mathbf{z}_N.$$

Since the particles are identical, it makes sense to limit our attention to the case when the distribution functions $\{\rho_k\}$, $k = 1, \ldots, N$, are symmetric functions. Let

$$H_k = \sum_{i=1}^{k} \frac{1}{m}(\mathbf{p}_i \cdot \nabla_{\mathbf{x}_i}) - \sum_{i<j\leq k} \hat{L}_{ij}.$$

Integrating over the variables $\mathbf{z}_{k+1}, \ldots, \mathbf{z}_N$, we get from (4.3.1)

$$\frac{\partial \rho_k}{\partial t} + H_k \rho_k$$

$$= \int \cdots \int d\mathbf{z}_{k+1} \cdots d\mathbf{z}_N \left(-\sum_{i=k+1}^{N} \frac{1}{m}(\mathbf{p}_i \cdot \nabla_{\mathbf{x}_i}) \right.$$

$$\left. + \sum_{k+1\leq i<j} \hat{L}_{ij} + \sum_{i\leq k, j\geq k+1} \hat{L}_{ij} \right) \rho_N(\mathbf{z}_1, \ldots, \mathbf{z}_N, t).$$

Assuming that ρ_N decays to 0 sufficiently fast as $\|\mathbf{z}_i\|$ goes to ∞, both the first and second terms on the right-hand side vanish after the integration. For the third term, using the fact that the ρ_k are symmetric functions we have, for each $j \geq k + 1$,

$$\int \cdots \int d\mathbf{z}_{k+1} \cdots d\mathbf{z}_N\, \hat{L}_{ij} \rho_N(\mathbf{z}_1, \ldots, \mathbf{z}_N, t)$$

$$= \int \cdots \int d\mathbf{z}_{k+1} \cdots d\mathbf{z}_N\, \hat{L}_{i,k+1} \rho_N(\mathbf{z}_1, \ldots, \mathbf{z}_N, t)$$

$$= \int d\mathbf{z}_{k+1} \hat{L}_{i,k+1} \rho_{k+1}(\mathbf{z}_1, \ldots, \mathbf{z}_{k+1}).$$

Therefore,

$$\frac{\partial \rho_k}{\partial t} + H_k \rho_k = (N-k) \sum_{i\leq k} \int d\mathbf{z}_{k+1} \hat{L}_{i,k+1} \rho_{k+1}(\mathbf{z}_1, \ldots, \mathbf{z}_{k+1}).$$

In particular, the equation for ρ_1 reads

$$\frac{\partial \rho_1}{\partial t} + \frac{1}{m}(\mathbf{p}_1 \cdot \nabla_{\mathbf{x}_1})\rho_1 = (N-1)\int d\mathbf{z}_2 \hat{L}_{12}\rho_2(\mathbf{z}_1, \mathbf{z}_2, t).$$

The equations just derived form a hierarchy of equations for all the k-particle distribution functions. Note that the equation for ρ_1 depends on ρ_2, the equation for ρ_2 depends on ρ_3 and so on. This is a familiar situation, which we see repeatedly in modeling: to obtain a closed system with a small number of equations and unknowns we must place some assumptions on this hierarchy of distribution functions. In particular, to obtain a closed equation for ρ_1, we must make closure assumptions that express ρ_2 in terms of ρ_1 only. The resulting equation for ρ_1 is called the kinetic equation.

4.3.2 The Boltzmann equation

Consider now a dilute system of particles in the gas phase. We will model the particle–particle interaction by a collision, i.e. particles do not interact unless two particles collide, and we will model the collision process probablistically. This allows us to derive a closed model in terms of ρ_1 only. By doing so we neglect the details of the interaction process. We will present an intuitive way of deriving the kinetic equation, i.e. the Boltzmann equation, by studying the collision process directly. Our presentation is to a large extent influenced by the treatment in [29].

From this point on we will write the one-particle probability density $\rho_1(\mathbf{z}_1, t)$ as $f(\mathbf{x}, \mathbf{v}, t)$, where $\mathbf{x}$ and $\mathbf{v}$ are the position and velocity coordinates respectively. We will consider elastic collisions only, i.e. the total energy of the particles is conserved during the collision process.

Let P_1 and P_2 be two particles with velocities $\mathbf{v}_1$, $\mathbf{v}_2$ before a collision and $\mathbf{v}'_1$, $\mathbf{v}'_2$ after it. Since their total momentum and energy are conserved during the collision process, we must have

$$\mathbf{v}_1 + \mathbf{v}_2 = \mathbf{v}'_1 + \mathbf{v}'_2, \tag{4.3.2}$$

$$|\mathbf{v}_1|^2 + |\mathbf{v}_2|^2 = |\mathbf{v}'_1|^2 + |\mathbf{v}'_2|^2. \tag{4.3.3}$$

Define

$$\mathbf{v} = \tfrac{1}{2}(\mathbf{v}_1 + \mathbf{v}_2), \quad \mathbf{v}' = \tfrac{1}{2}(\mathbf{v}'_1 + \mathbf{v}'_2),$$
$$\mathbf{u} = \mathbf{v}_1 - \mathbf{v}_2, \quad \mathbf{u}' = \mathbf{v}'_1 - \mathbf{v}'_2;$$

the conservation laws (4.3.2) and (4.3.3) imply that

$$\mathbf{v} = \mathbf{v}',$$
$$|\mathbf{u}| = |\mathbf{u}'|.$$

The second of these identities implies that $\mathbf{u}'$ can be obtained by rotating $\mathbf{u}$ by some solid angle ω:

$$\mathbf{u}' = R_\omega \mathbf{u}.$$

Here $\omega \in \mathbb{S}^2$ (the unit sphere in $\mathbb{R}^3$) is called the scattering angle and R_ω is the rotation operator by the angle ω. Knowing ω, we can relate $(\mathbf{v}'_1, \mathbf{v}'_2)$ uniquely to $(\mathbf{v}_1, \mathbf{v}_2)$:

$$\mathbf{v}'_1 = \tfrac{1}{2}(\mathbf{v}' + \mathbf{u}') = \tfrac{1}{2}(\mathbf{v} + R_\omega \mathbf{u}),$$
$$\mathbf{v}'_2 = \tfrac{1}{2}(\mathbf{v}' - \mathbf{u}') = \tfrac{1}{2}(\mathbf{v} - R_\omega \mathbf{u})$$

and vice versa.

The outcome of the collision process is determined uniquely by the scattering angle ω. In kinetic theory ω is considered to be a random variable, whose (unnormalized) probability density $\sigma(\omega)$ is called the *scattering cross section*. The total scattering cross section,

$$\sigma_{\text{tot}} = \int_{\mathbb{S}^2} \sigma(\omega)\, d\omega,$$

which has the dimension of area, can be roughly thought of as the area (say, normal to $\mathbf{v}_1 - \mathbf{v}_2$) available for the particles to collide. Outside this area, particles pass by each other without collision. One should note that $\sigma(\omega)$ depends on $\mathbf{v}_1$ and $\mathbf{v}_2$. In principle, $\sigma(\omega) = \sigma((\mathbf{v}_1, \mathbf{v}_2) \to (\mathbf{v}'_1, \mathbf{v}'_2))$ can be obtained from a more detailed model for the collision process, such as a quantum mechanical model. In practice it is often modeled empirically.

However, independently of the details of $\sigma(\omega)$, it is intuitively clear that the following properties hold for the collision process for simple spherical particles:

(1) $\sigma(\omega)$ is invariant under rotation and reflection transformations of the velocity (or momentum) space;

(2) $\sigma(\omega)$ is reversible, i.e.

$$\sigma((\mathbf{v}_1, \mathbf{v}_2) \to (\mathbf{v}'_1, \mathbf{v}'_2)) = \sigma((\mathbf{v}'_1, \mathbf{v}'_2) \to (\mathbf{v}_1, \mathbf{v}_2)). \qquad (4.3.4)$$

More intuitively the rate of change of f is given by

$$\frac{\partial f}{\partial t} + (\mathbf{v} \cdot \nabla_{\mathbf{x}})f = J_+ - J_-,$$

where J_- is the loss term due to collisions of particles having velocity $\mathbf{v}$ with other particles and J_+ is a gain term due to the creation of particles with velocity $\mathbf{v}$ as a result of the collision process. To derive expressions for J_+ and J_-, let us fix $\mathbf{v}$ and the center of collision $\mathbf{x}$. Consider the collision of particles with velocity $\mathbf{v}$ and particles with velocity $\mathbf{w}$. Their relative velocity is $\mathbf{v} - \mathbf{w}$. The current of such colliding particles (which pass through the plane normal to $\mathbf{v} - \mathbf{w}$ at the center of collision $\mathbf{x}$) with velocity in the neighborhood of $\mathbf{w}$ with size $d\mathbf{w}$ is equal to $I = |\mathbf{v} - \mathbf{w}| f(\mathbf{x}, \mathbf{w}, t) \, d\mathbf{w}$. The rate at which particles scatter in a solid angle ω after the collision process is $I\sigma(\omega)$. Therefore the loss term is equal to

$$J_- = f(\mathbf{x}, \mathbf{v}, t) \int_{\mathbb{R}^3} \sigma_{\text{tot}} |\mathbf{v} - \mathbf{w}| f(\mathbf{x}, \mathbf{w}, t) \, d\mathbf{w}$$

$$= \int_{\mathbb{R}^3} \int_{\mathbb{S}^2} f(\mathbf{x}, \mathbf{v}, t) f(\mathbf{x}, \mathbf{w}, t) \sigma(\omega) |\mathbf{v} - \mathbf{w}| \, d\omega \, d\mathbf{w}$$

$$= f(\mathbf{x}, \mathbf{v}, t) \int_{\mathbb{S}^2} \sigma(\omega) \, d\omega \int_{\mathbb{R}^3} f(\mathbf{x}, \mathbf{w}, t) |\mathbf{v} - \mathbf{w}| \, d\mathbf{w}.$$

Similarly, to compute the gain term we consider collision processes of the type:

$$(\mathbf{v}', \mathbf{w}') \to (\mathbf{v}, \mathbf{w}).$$

Since $\mathbf{v}$ is fixed, we can parametrize such collision processes by $\mathbf{w}$ and ω. Therefore we have

$$J_+ = \int_{\mathbb{R}^3} \int_{\mathbb{S}^2} f(\mathbf{x}, \mathbf{v}', t) f(\mathbf{x}, \mathbf{w}', t) |\mathbf{v}' - \mathbf{w}'| \sigma((\mathbf{v}', \mathbf{w}') \to (\mathbf{v}, \mathbf{w})) \, d\omega \, d\mathbf{w}$$

$$= \int_{\mathbb{R}^3} \int_{\mathbb{S}^2} f(\mathbf{x}, \mathbf{v}', t) f(\mathbf{x}, \mathbf{w}', t) |\mathbf{v}' - \mathbf{w}'| \, d\omega \, d\mathbf{w}$$

$$= \int_{\mathbb{R}^3} \int_{\mathbb{S}^2} f(\mathbf{x}, \mathbf{v}', t) f(\mathbf{x}, \mathbf{w}', t) \sigma(\omega) |\mathbf{v} - \mathbf{w}| \, d\omega \, d\mathbf{w}.$$

Putting these together, we obtain the *Boltzmann equation*

$$\frac{\partial f}{\partial t} + (\mathbf{v} \cdot \nabla_{\mathbf{x}})f = B(f, f), \tag{4.3.5}$$

where

$$B(f, f)$$
$$= \int_{\mathbb{R}^3} \int_{\mathbb{S}^2} \sigma(\omega) |\mathbf{w} - \mathbf{v}| \left(f(\mathbf{x}, \mathbf{v}', t) f(\mathbf{x}, \mathbf{w}', t) - f(\mathbf{x}, \mathbf{v}, t) f(\mathbf{x}, \mathbf{w}, t) \right) d\omega \, d\mathbf{w};$$

A more detailed derivation can be found in [29].

4.3.3 The equilibrium states

Next we consider homogeneous equilibrium states such that $f(\mathbf{x}, \mathbf{v}, t) = f_0(\mathbf{v})$. From (4.3.5) we see that we must have

$$B(f_0, f_0) \equiv 0. \tag{4.3.6}$$

Equation (4.3.6) is an integral constraint stating that the net gain is equal to the net loss, summed over all collisions. The following lemma gives a much stronger statement: the gain and loss terms are equal for each individual pairwise collision.

Lemma *Assume that $\sigma(\omega) > 0$, for all $\omega \in \mathbb{S}^2$. Then, for any equilibrium states f_0, the following holds:*

$$f_0(\mathbf{v}') f_0(\mathbf{w}') = f_0(\mathbf{v}) f_0(\mathbf{w}), \tag{4.3.7}$$

where $\mathbf{v}', \mathbf{w}'$ and $\mathbf{v}, \mathbf{w}$ are related by the conservation laws of momentum and energy, i.e. $(\mathbf{v}', \mathbf{w}')$ is a possible outcome after the collision of a pair of particles with velocities $\mathbf{v}$ and $\mathbf{w}$ respectively.

Proof Let

$$H(f) = \iint f \ln f \, d\mathbf{x} \, d\mathbf{v}.$$

Then

$$\frac{dH}{dt} = \iint (\ln f + 1) \frac{\partial f}{\partial t} \, d\mathbf{x} \, d\mathbf{v}$$
$$= \iint (\ln f + 1) B(f, f) \, d\mathbf{x} \, d\mathbf{v}$$
$$= \iiint \sigma(\omega) |\mathbf{v} - \mathbf{w}| (f_1' f_2' - f_1 f_2)(1 + \ln f_1) \, d\omega \, d\mathbf{w} \, d\mathbf{x} \, d\mathbf{v}.$$

Here, and in what follows, we use the short-hand notation $f_1 = f_0(\mathbf{v})$, $f_2 = f_0(\mathbf{w})$, $f_1' = f_0(\mathbf{v}')$, $f_2' = f_0(\mathbf{w}')$. Now we will make use of the symmetry properties of our system. First, if we exchange $\mathbf{v}$ and $\mathbf{w}$, the result should not change, i.e.

$$\frac{dH}{dt} = \iiint \sigma(\omega)|\mathbf{v} - \mathbf{w}|(f_1'f_2' - f_1 f_2)(1 + \ln f_2) \, d\omega \, d\mathbf{w} \, d\mathbf{x} \, d\mathbf{v}.$$

Hence,

$$\frac{dH}{dt} = \frac{1}{2} \iiint \sigma(\omega)|\mathbf{v} - \mathbf{w}|(f_1'f_2' - f_1 f_2)(2 + \ln(f_1 f_2)) \, d\omega \, d\mathbf{w} \, d\mathbf{x} \, d\mathbf{v}.$$

Secondly, because of (4.3.4), if we exchange the pairs $(\mathbf{v}, \mathbf{w})$ and $(\mathbf{v}', \mathbf{w}')$, again the result should not change:

$$\frac{dH}{dt} = \frac{1}{2} \iiint \sigma(\omega)|\mathbf{v} - \mathbf{w}|(f_1 f_2 - f_1'f_2')(2 + \ln(f_1'f_2')) \, d\omega \, d\mathbf{w} \, d\mathbf{x} \, d\mathbf{v}.$$

Hence,

$$\frac{dH}{dt} = \frac{1}{4} \iiint \sigma(\omega)|\mathbf{v} - \mathbf{w}|(f_1'f_2' - f_1 f_2)\left(\ln(f_1 f_2) - \ln(f_1'f_2')\right) \, d\omega \, d\mathbf{w} \, d\mathbf{x} \, d\mathbf{v}.$$

Using the fact that, for any $a, b > 0$,

$$(b - a)(\ln b - \ln a) \geq 0, \tag{4.3.8}$$

we conclude that

$$\frac{dH}{dt} \leq 0. \tag{4.3.9}$$

In addition, if $dH/dt \equiv 0$, we must have

$$f_0(\mathbf{v}')f_0(\mathbf{w}') = f_0(\mathbf{v})f_0(\mathbf{w})$$

for all $\omega, \mathbf{v}$ and $\mathbf{w}$, since the equality in (4.3.8) holds only when $a = b$. $\qquad\square$

Equation (4.3.7) is referred to as the condition of *detailed balance.*

Equally important is the statement in (4.3.9), which holds for any solutions of the Boltzmann equation. This statement is referred to as *Boltzmann's H-theorem*. One consequence of the H-theorem is the irreversibility of the Boltzmann equation. This is an intriguing statement, since the collision process, upon which Boltzmann's equation was derived, is reversible.

Using the detailed balance condition, we can derive the general form of the equilibrium states $f_0(\mathbf{v})$. For this purpose, let us examine further the quantities that are conserved during the collision process. We know from

(4.3.2) and (4.3.3) that particle number, total momentum and kinetic energy are conserved during collision. We can show that if the collision cross section is nondegenerate, i.e. if $\sigma(\omega)$ is strictly positive, then there are no additional conserved quantities. In fact, assume that on the contrary there is such an conserved quantity, say η; then we would have

$$\eta(\mathbf{v}) + \eta(\mathbf{w}) = \eta(\mathbf{v}') + \eta(\mathbf{w}')$$

for all $\mathbf{v}, \mathbf{w}, \omega$. Fix $(\mathbf{v}, \mathbf{w})$; then $\mathbf{v}', \mathbf{w}'$ are uniquely determined by ω. An additional conserved quantity would also means an additional constraint on the possible values of ω, i.e. ω could only take values on a low-dimensional subset of $\mathbb{S}^2$. This would contradict the strict positivity of σ, which guarantees that ω can take any value on $\mathbb{S}^2$.

Taking logs, we now write the relation of detailed balance (4.3.7) in terms of the abbreviated variables as

$$\ln f_1' + \ln f_2' = \ln f_1 + \ln f_2.$$

This means that $\ln f$ is also a conserved quantity. Thus, $\ln f$ must be a linear combination of $1, \mathbf{v}, |\mathbf{v}|^2/2$, namely

$$f = \exp\left(C_1 + C_2 \cdot \mathbf{v} + C_3 \frac{|\mathbf{v}|^2}{2} \right),$$

or, equivalently,

$$f = A \exp(-C_3(\mathbf{v} - \mathbf{v}_0)^2).$$

To determine the constants A and C_3, define the number density

$$n \equiv \int f \, d\mathbf{v} = A \exp\left(-C_3(\mathbf{v} - \mathbf{v}_0)^2\right) = A \left(\frac{\pi}{C_3}\right)^{3/2}. \quad (4.3.10)$$

Then

$$\int \mathbf{v} f \, d\mathbf{v} = n\mathbf{v}_0.$$

In addition, it is easy to compute the internal energy of the system:

$$\int \frac{1}{2} m |\mathbf{v} - \mathbf{v}_0|^2 f \, d\mathbf{v} = \frac{3\rho}{4C_3},$$

where m is the mass of the particles and $\rho = nm$. Defining temperature as the thermal energy per degree of freedom, i.e.

$$\frac{\int \frac{1}{2}m|\mathbf{v} - \mathbf{v}_0|^2 f \, d\mathbf{v}}{\int f \, d\mathbf{v}} = \frac{3}{2}k_B T,$$

then we obtain

$$C_3 = \frac{m}{2k_B T}.$$

Using (4.3.10), we obtain the well-known *Maxwell–Boltzmann distribution*:

$$f_0(\mathbf{v}) = n \left(\frac{m}{2\pi k_B T} \right)^{3/2} \exp \left(-\frac{m|\mathbf{v} - \mathbf{v}_0|}{2k_B T} \right).$$

4.3.4 Macroscopic conservation laws

We are interested in conserved quantities of the form

$$I(\mathbf{x}, t) = \int \eta(\mathbf{v}) f(t, \mathbf{x}, \mathbf{v}) \, d\mathbf{v}.$$

Using the Boltzmann equation (4.3.5), we have

$$\int \eta(\mathbf{v}) \left(\frac{\partial f}{\partial t} + \mathbf{v} \cdot \nabla_{\mathbf{x}} f \right) d\mathbf{v} = \int \eta(\mathbf{v}) B(f, f) \, d\mathbf{v}.$$

If η satisfies

$$\int \eta(\mathbf{v}) B(f, f) \, d\mathbf{v} = 0 \qquad\qquad (4.3.11)$$

then I must be a conserved quantity. In addition, let $J(\mathbf{x}, t) = \int \eta(\mathbf{v})\mathbf{v} f \, d\mathbf{v}$; then we obtain a conservation law in local form:

$$\frac{\partial I}{\partial t} + \nabla_{\mathbf{x}} \cdot J = 0,$$

where J is the corresponding flux. In the following, we will write $\nabla_{\mathbf{x}} = \nabla$.

We have already seen that $\eta(\mathbf{v}) = 1$, $\mathbf{v}$, and $|\mathbf{v}|^2/2$ satisfy the condition (4.3.11). They give rise to the conservation laws of mass, momentum and energy respectively. We now examine these conservation laws in some detail.

Define the averaging operator

$$\langle \eta \rangle = \frac{\int \eta f \, d\mathbf{v}}{\int f \, d\mathbf{v}}.$$

First we let $\eta(\mathbf{v}) = 1$. Then $I = n = \rho/m$, $J = n\mathbf{u}$ where $\mathbf{u} = \langle\mathbf{v}\rangle$ is the mean velocity. Mass conservation is obtained:

$$\frac{\partial\rho}{\partial t} + \nabla\cdot(\rho\mathbf{u}) = 0. \tag{4.3.12}$$

In what follows, it is convenient to split the velocity into the sum of its mean and fluctuation:

$$\mathbf{v} = \mathbf{u} + \mathbf{v}',$$

where $\mathbf{v}'$ has mean 0. Let $\eta(\mathbf{v}) = \mathbf{v}$; then

$$I = \int \mathbf{v}f\,d\mathbf{v} = n\mathbf{u} = \frac{1}{m}\rho\mathbf{u},$$

and the momentum flux becomes

$$J = \int \mathbf{v}\otimes\mathbf{v}f\,d\mathbf{v} = \int (\mathbf{u}+\mathbf{v}')\otimes(\mathbf{u}+\mathbf{v}')f\,d\mathbf{v}'$$

$$= n\mathbf{u}\otimes\mathbf{u} + \int \mathbf{v}'\otimes\mathbf{v}'f\,d\mathbf{v}'.$$

Let us define the tensor $\boldsymbol{\tau}$ as follows:

$$\boldsymbol{\tau} = \frac{1}{m}\int \mathbf{v}'\otimes\mathbf{v}'f\,d\mathbf{v}';$$

then the conservation of momentum reads

$$\frac{\partial(\rho\mathbf{u})}{\partial t} + \nabla\cdot(\rho\mathbf{u}\otimes\mathbf{u}) + \nabla\cdot\boldsymbol{\tau} = 0. \tag{4.3.13}$$

Finally, let $\eta(\mathbf{v}) = |\mathbf{v}|^2/2$; then

$$I = \int \frac{|\mathbf{v}|^2}{2}f\,d\mathbf{v} = \int \frac{|\mathbf{u}+\mathbf{v}'|^2}{2}f\,d\mathbf{v}'$$

$$= \frac{|\mathbf{u}|^2}{2}n + \int \frac{|\mathbf{v}'|^2}{2}f\,d\mathbf{v}'.$$

Now define the internal energy density and temperature using the following relations

$$e = \frac{3}{2}k_B T = \left\langle \frac{|\mathbf{v}'|^2}{2}\right\rangle;$$

then we have

$$I = \frac{1}{m}\left(\rho\frac{|\mathbf{u}|^2}{2} + \rho e\right) = \frac{1}{m}E.$$

The energy flux is

$$J = \int \frac{|\mathbf{v}|^2}{2} \mathbf{v} f \, d\mathbf{v}$$

$$= \frac{1}{m} \left(\rho \frac{|\mathbf{u}|^2}{2} \mathbf{u} + \boldsymbol{\tau} \cdot \mathbf{u} + e\rho \mathbf{u} \right) + \int \frac{|\mathbf{v}'|^2}{2} \mathbf{v}' f \, d\mathbf{v}'.$$

Letting

$$\mathbf{q} = \frac{1}{m} \int \frac{|\mathbf{v}'|^2}{2} \mathbf{v}' f \, d\mathbf{v}',$$

we obtain

$$\partial_t E + \nabla \cdot (E\mathbf{u} + \boldsymbol{\tau} \cdot \mathbf{u}) + \nabla \cdot \mathbf{q} = 0. \qquad (4.3.14)$$

4.3.5 The hydrodynamic regime

The Knudsen number The most important parameter in kinetic theory is the *Knudsen number*, defined as the ratio of the mean free path of the particles λ and the system size L:

$$Kn = \frac{\lambda}{L}.$$

Here by the system size we mean the scale over which averaged quantities such as the density, momentum and energy change. In a typical dense gas, λ is of the order $10^{-7} - 10^{-8}$ m, while the system size is, say $\mathcal{O}(1\,\mathrm{m})$. Hence the Knudsen number is very small. As we will see below, in this case hydrodynamic approximations are very accurate. However, there are exceptions, for example:

(1) at a shock, the relevant system size becomes the width of the shock, which can be very small;
(2) at a boundary, the relevant system size becomes the boundary layer width;
(3) when modeling high-frequency waves, the time scale of interest is specified by the frequency of the waves and the Knudsen number is the ratio of the frequency of the waves and the collision frequency;
(4) when the gas is dilute, the mean free path becomes large.

In these cases the full Boltzmann equation is needed to model the dynamics of the gas.

If we nondimensionalize the Boltzmann equation, using $\mathbf{x} = L\mathbf{x}'$, $f = f'/(\lambda^3 v^3)$, $\mathbf{v} = v\mathbf{v}'$, $\mathbf{w} = v\mathbf{w}'$, $t = t'L/v$, $\sigma = \lambda^2 \sigma'$, then, omitting the primes, we obtain

$$\frac{\partial f}{\partial t} + \mathbf{v} \cdot \nabla_{\mathbf{x}} f = \frac{1}{\varepsilon} B(f, f), \qquad (4.3.15)$$

where $\varepsilon = \mathrm{Kn}$ is the Knudsen number.

The local equilibrium approximation We now turn to the asymptotic analysis of (4.3.15). When $\varepsilon \ll 1$, we have to leading order

$$B(f, f) = 0.$$

Therefore, as discussed above, f must be of the form

$$f(\mathbf{x}, \mathbf{v}, t) = \frac{\rho(\mathbf{x}, t)}{(2\pi T(\mathbf{x}, t))^{3/2}} \exp\left(-\frac{m|\mathbf{v} - \mathbf{u}(\mathbf{x}, t)|^2}{2k_B T(\mathbf{x}, t)}\right).$$

This is called the *local equilibrium approximation.*

It is straightforward to calculate that under the local equilibrium approximation we have

$$\boldsymbol{\tau} = p\mathbf{I}, \quad \mathbf{q} = 0,$$

where

$$p = \rho k_B T.$$

Together with the conservation laws, this gives Euler's equations for an ideal gas:

$$\partial_t \rho + \nabla \cdot (\rho \mathbf{u}) = 0,$$
$$\partial_t (\rho \mathbf{u}) + \nabla \cdot (\rho \mathbf{u} \otimes \mathbf{u}) + \nabla p = 0,$$
$$\partial_t E + \nabla \cdot ((E + p)\mathbf{u}) = 0.$$

The Chapman–Enskog expansion The leading-order result does not include any dissipative effects such as those due to viscosity or thermal conduction. To see these we have to go to the next order. The standard way of doing this is the Chapman–Enskog expansion. Here we give a brief summary of this procedure. More details can be found in, for example, [29].

We look for approximate solutions of the Boltzmann equation in the form:

$$f = f_0(1 + \varepsilon \varphi_1 + \varepsilon^2 \varphi_2 + \cdots).$$

To leading order, we have

$$B(f_0, f_0) = 0, \tag{4.3.16}$$

$$\frac{\partial f_0}{\partial t} + \mathbf{v} \cdot \nabla f_0 = B(f_0, f_0\varphi_1) + B(f_0\varphi_1, f_0) \equiv -\frac{\rho}{m} f_0 J(\varphi_1). \tag{4.3.17}$$

The last relation also serves as the definition of the operator J. Equation (4.3.16) shows that f_0 actually follows the Maxwell–Boltzmann distribution. For simplicity, we write $\varphi = \varphi_1$. The left-hand side of (4.3.17) can be calculated using the local equilibrium approximation and Euler's equation for an ideal gas, and this gives

$$J(\varphi) = -\frac{m}{\rho k_B \rho T^2} (\tfrac{1}{2} m |\mathbf{v} - \mathbf{u}|^2 - \tfrac{5}{2} k_B T)(\mathbf{v} - \mathbf{u}) \cdot \nabla T$$

$$- \frac{m}{\rho k_B T}((\mathbf{v} - \mathbf{u}) \otimes (\mathbf{v} - \mathbf{u}) - \tfrac{1}{3}|\mathbf{v} - \mathbf{u}|^2 I) : \mathbf{D},$$

where $\mathbf{D} \equiv \tfrac{1}{2}\left(\nabla\mathbf{u} + (\nabla\mathbf{u})^{\mathsf{T}}\right)$ is the strain rate.

Using linearity, we can write φ as a linear combination of $(\partial T/\partial x_i)$ and D_{ij}:

$$\varphi = -\frac{m}{\rho k_B T^2} \sum_i a_i \frac{\partial T}{\partial x_i} - \frac{m}{\rho k_B T} \sum_{ij} b_{ij} D_{ij},$$

where $\{a_i\}, \{b_{ij}\}$ satisfy

$$J(a_i) = \tilde{a}_i = (\tfrac{1}{2}|\mathbf{v} - \mathbf{u}|^2 - \tfrac{5}{2} k_B T)(v_i - u_i),$$

$$J(b_{ij}) = \tilde{b}_{ij} = (v_i - u_i)(v_j - u_j) - \tfrac{1}{3}|\mathbf{v} - \mathbf{u}|^2 \delta_{ij}.$$

It can be concluded by symmetry considerations that the solutions to these equations should be of the form

$$a_i = \alpha(T, |\mathbf{v} - \mathbf{u}|^2)\tilde{a}_i,$$

$$b_{ij} = \beta(T, |\mathbf{v} - \mathbf{u}|^2)\tilde{b}_{ij},$$

for some functions α and β. From this we get

$$\mathbf{q} = -\kappa \nabla T, \quad \tau = \rho k_B T \mathbf{I} + 2\mu \mathbf{D}$$

where $\mathbf{I}$ is the identity tensor,

$$\kappa = \frac{\varepsilon}{2\rho k_B T^2} \int f_0 |\mathbf{v}'|^4 \alpha(T, |\mathbf{v}'|^2)(\tfrac{1}{2}|\mathbf{v}'|^2 - \tfrac{5}{2} k_B T) \, d\mathbf{v}',$$

$$\mu = -\frac{\varepsilon}{2\rho k_B T} \int f_0 \beta(T, |\mathbf{v}'|^2)(\mathbf{v}' \otimes \mathbf{v}') : (\mathbf{v}' \otimes \mathbf{v}' - \tfrac{1}{3}|\mathbf{v}'|^2 I) \, d\mathbf{v}'.$$

Remark Note that the viscous and heat conduction terms are $\mathcal{O}(\varepsilon)$ compared with the terms obtained under the local equilibrium approximation.

There are two general approaches for obtaining more accurate approximations.

(1) In the first, higher-order moments are included. Grad's 13-moment model is a primary example of this kind [20, 38].
(2) In the second, higher-order derivatives are included. The Burnett equation is a typical example of this class of models [17].

4.3.6 Other kinetic models

A considerably simplified model in kinetic theory is the *BGK model*, which approximates the collision term by a simple relaxation form [2]:

$$\frac{\partial f}{\partial t} + (\mathbf{v} \cdot \nabla_{\mathbf{x}})f = M_f - f,$$

where M_f is the local Maxwell–Boltzmann distribution,

$$M_f(\mathbf{x}, \mathbf{v}, t) = n(\mathbf{x}, t) \left(\frac{m}{2\pi k_B T(\mathbf{x}, t)} \right)^{3/2} \exp\left(-\frac{m|\mathbf{v} - \mathbf{u}(\mathbf{x}, t)|^2}{2k_B T(\mathbf{x}, t)} \right),$$

and n, $\mathbf{u}$, T are obtained from f through

$$n(\mathbf{x}, t) = \int f \, d\mathbf{v}, \quad n(\mathbf{x}, t)\mathbf{u}(\mathbf{x}, t) = \int \mathbf{v} f(\mathbf{x}, \mathbf{v}, t) \, d\mathbf{v},$$

$$\frac{3}{2} k_B T(\mathbf{x}, t) n(\mathbf{x}, t) = \int \frac{m}{2} |\mathbf{v} - \mathbf{u}(\mathbf{x}, t)|^2 f(\mathbf{x}, \mathbf{v}, t) \, d\mathbf{v}.$$

Note that this model is still quite nonlocal and nonlinear, owing to the term M_f. However, it is certainly much simpler than the original Boltzmann equation. For some problems, this simple model gives fairly accurate results compared to the original Boltzmann equation.

4.4 Electronic structure models

Strictly speaking, none of the models we have discussed so far can be regarded as being a "real" description: they all contain aspects of empirical modeling. Models in molecular dynamics use empirical interatomic potentials. Kinetic theory models rely on the empirical scattering cross section σ. In continuum theory, the constitutive relations are empirical in nature. For a real first-principle model we need quantum mechanics; we should consider the quantum many-body problem. As we discuss below, this quantum many-body problem contains no empirical parameters. The difficulty, however, lies in the complexity of its mathematics. For this reason, approximate models have been developed which describe the electronic structure of matter at various levels of detail. These approximate models include:

(1) *The Hartree–Fock model*, in which the many-body electronic wave function is limited to the form of a single Slater determinant [34].

(2) *Kohn–Sham density functional theory*, a form of the density functional theory (DFT) that uses fictitious orbitals. This theory can in principle be made exact by using the right functionals. But, in practice, approximations have to be made for the exchange-correlation functional [34].

(3) *Orbital-free density functional theory*. The Thomas–Fermi model is the simplest example of this type. These models do not involve any wave functions. They use the electron density as the only variable. They are in general less accurate than Kohn–Sham DFT but are much easier to handle [7].

(4) *Tight-binding models*. These are much simplified electronic structure models in which the electronic wave functions are expressed as linear combinations of a minimal set of atomic orbitals. The electron–electron interaction is usually neglected, except that the Pauli exclusion principle is imposed.

In the following we will first discuss the quantum many-body problem. We will then discuss approximate models. We will focus on the zero-temperature case and neglect consideration of the spins.

4.4.1 The quantum many-body problem

Quantum mechanics is built upon two fundamental principles:

(1) the form of the Hamiltonian operator and the Schrödinger equation.;

(2) the Pauli exclusion principle.

Consider a system consists of N_{nuc} nuclei and N electrons. It is described by a Hamiltonian, which, in atomic units, is given by

$$H = \sum_{I=1}^{N_{\text{nuc}}} \frac{\mathbf{P}_I^2}{2M_I} + \sum_{i=1}^{N} \frac{1}{2}\mathbf{p}_i^2 + V(\mathbf{R}_1, \ldots, \mathbf{R}_{N_{\text{nuc}}}, \mathbf{x}_1, \ldots, \mathbf{x}_N).$$

Here M_I is the mass of the Ith nucleus; the mass of an electron is 1 in atomic units; $\mathbf{R}_I$ is the position coordinate of the Ith nucleus; $\mathbf{x}_1, \ldots, \mathbf{x}_N$ are the spatial coordinates for the N-electron system; $\mathbf{P}_I$, $\mathbf{p}_i$ are the corresponding momentum operators, $\mathbf{p}_i = -i\nabla_{\mathbf{x}_i}$, $\mathbf{P}_I = -i\nabla_{\mathbf{R}_I}$. Finally, V is the interaction potential of the nuclei as well as the electrons and is given by

$$V(\mathbf{R}_1, \ldots, \mathbf{R}_{N_{\text{nuc}}}, \mathbf{x}_1, \ldots, \mathbf{x}_N)$$
$$= \frac{1}{2}\sum_{I \neq J} \frac{Z_I Z_J}{|\mathbf{R}_I - \mathbf{R}_J|} + \frac{1}{2}\sum_{i \neq j} \frac{1}{|\mathbf{x}_i - \mathbf{x}_j|} - \sum_{i,I} \frac{Z_I}{|\mathbf{x}_i - \mathbf{R}_I|},$$

where Z_I is the charge on the Ith nucleus. We can write the Hamiltonian as

$$H = -\sum_{I=1}^{N_{\text{nuc}}} \frac{1}{2M_I}\Delta_{\mathbf{R}_I} - \sum_{i=1}^{N} \Delta_{\mathbf{x}_i} + V(\mathbf{R}_1, \ldots, \mathbf{R}_{N_{\text{nuc}}}, \mathbf{x}_1, \ldots, \mathbf{x}_N). \quad (4.4.1)$$

The ground state of a system of electrons and nuclei can be found by solving the Schrödinger equation

$$H\Psi = E\Psi, \quad \Psi = \Psi(\mathbf{R}_1, \ldots, \mathbf{R}_{N_{\text{nuc}}}; \mathbf{x}_1, \ldots, \mathbf{x}_N), \quad (4.4.2)$$

subject to the symmetry constraints determined by the statistics of the particles (nuclei or electrons). For example, since electrons are fermions, Ψ must be an antisymmetric function in $(\mathbf{x}_1, \ldots, \mathbf{x}_N)$, i.e. Ψ changes sign if any two coordinates $\mathbf{x}_i$ and $\mathbf{x}_j$ are interchanged:

$$\Psi(\mathbf{R}_1, \ldots, \mathbf{R}_{N_{\text{nuc}}}; \mathbf{x}_1, \ldots, \mathbf{x}_i, \ldots, \mathbf{x}_j, \ldots, \mathbf{x}_N)$$
$$= -\Psi(\mathbf{R}_1, \ldots, \mathbf{R}_{N_{\text{nuc}}}; \mathbf{x}_1, \ldots, \mathbf{x}_j, \ldots, \mathbf{x}_i, \ldots, \mathbf{x}_N), \quad 1 \leq i < j \leq N.$$
$$(4.4.3)$$

This quantum many-body problem contains no empirical parameters. One only has to enter the mass $\{M_I\}$ and charge $\{Z_I\}$ of the participating nuclei; then, the description is complete.

Born–Oppenheimer approximation The time scales of the nuclear and electronic degrees of freedom are determined by the respective masses. Since M_I is usually on the scale of 10^3 or larger, the electrons relax at a much faster pace than the nuclei. This suggests making the approximation that the electronic structure is always at the ground state corresponding to the state of the nuclei. Under this approximation the electronic state is slaved by the state of the nuclei. This is the *Born–Oppenheimer approximation*, also referred to as the adiabatic approximation. In addition, in many cases, treating the nuclei classically provides sufficient accuracy.

The Born–Oppenheimer approximation allows us to solve the coupled nuclei–electron problem in two steps.

(1) Given the positions of the nuclei, we can find the ground state of the electrons by solving

$$H_e \Psi_e = E_e \Psi_e, \tag{4.4.4}$$

where Ψ_e is the antisymmetric wave function for the electronic degrees of freedom and

$$H_e = -\sum_i \frac{1}{2} \Delta_{\mathbf{x}_i} + V(\mathbf{R}_1, \dots, \mathbf{R}_{N_{\mathrm{nuc}}}, \mathbf{x}_1, \dots, \mathbf{x}_N). \tag{4.4.5}$$

(2) We find the state of the nuclei using the classical Hamiltonian

$$H = \sum_I \frac{P_I^2}{2M_I} + E_e(\mathbf{R}_1, \dots, \mathbf{R}_{N_{\mathrm{nuc}}}), \tag{4.4.6}$$

where $E_e = \langle \Psi_e | H_e | \Psi_e \rangle$ is the effective potential due to the electrons.

We often write the Hamiltonian H_e as:

$$H_e = -\sum_i \frac{1}{2} \Delta_{\mathbf{x}_i} + \sum_i V_{\mathrm{ext}}(\mathbf{x}_i) + V_{ee}(\mathbf{x}_1, \dots, \mathbf{x}_N) + V_{nn}(\mathbf{R}_1, \dots, \mathbf{R}_{N_{\mathrm{nuc}}}),$$

$$\tag{4.4.7}$$

where

$$V_{\mathrm{ext}}(\mathbf{x}) = -\sum_I \frac{Z_I}{|\mathbf{x} - \mathbf{R}_I|}, \quad V_{ee}(\mathbf{x}_1, \dots, \mathbf{x}_N) = \frac{1}{2} \sum_{i \neq j} \frac{1}{|\mathbf{x}_i - \mathbf{x}_j|},$$

$$V_{nn}(\mathbf{R}_1, \dots, \mathbf{R}_{N_{\mathrm{nuc}}}) = \frac{1}{2} \sum_{I \neq J} \frac{Z_I Z_J}{|\mathbf{R}_I - \mathbf{R}_J|}. \tag{4.4.8}$$

The term V_{nn} makes an important contribution to E_e. However, in the Schrödinger equation (4.4.4) the positions of nuclei are fixed and V_{nn} is just a constant potential. Therefore, we often use

$$H_e = -\sum_i \frac{1}{2}\Delta_{\mathbf{x}_i} + \sum_i V_{\text{ext}}(\mathbf{x}_i) + V_{ee}(\mathbf{x}_1, \ldots, \mathbf{x}_N), \qquad (4.4.9)$$

with a slight abuse of notation. It is understood that the V_{nn} term should be added back into the Born–Oppenheimer surface energy E_e.

It is well known that the Schrödinger equation admits an equivalent variational formulation. For example, corresponding to (4.4.4) we can define the functional

$$I(\Psi_e) = \langle \Psi_e | H_e | \Psi_e \rangle = \int \sum_{i=1}^{N} \left(\frac{1}{2} |\nabla_{\mathbf{x}_i} \Psi_e|^2 + V |\Psi_e|^2 \right) \, d\mathbf{x}_1 \cdots \, d\mathbf{x}_N.$$
$$(4.4.10)$$

We then have

$$E_e = \inf_{\Psi_e} I(\Psi_e), \qquad (4.4.11)$$

subject to the normalization and antisymmetry constraints.

Interatomic potentials based on first principles Equation (4.4.6) includes the effective interatomic potential $E_e(\mathbf{R}_1, \ldots, \mathbf{R}_{N_{\text{nuc}}})$, which is derived from the first principles. It gives a rigorous starting point for thinking about interatomic potentials. However, it is very hard to use this in practice. Even for a nano-sized system, the wave function depends on so many variables that it is practically impossible to solve for such wave functions. Therefore further approximations have to be made. We now proceed to discuss some of these approximations.

4.4.2 Hartree and Hartree–Fock approximations

The simplest approximation is the *Hartree approximation*, which assumes that the N-electron wave function takes the form of a product of single-electron wave functions:

$$\Psi_e(\mathbf{x}_1, \ldots, \mathbf{x}_N) = \psi_1(\mathbf{x}_1) \cdots \psi_N(\mathbf{x}_N), \qquad (4.4.12)$$

with normalization constraints

$$\int |\psi_i(\mathbf{x})|^2 \, d\mathbf{x} = 1, \quad j = 1, \ldots, N.$$

Putting this ansatz into the variational principle, we get

$$I_{\mathrm{H}}(\{\psi_i\}) = \int_{\mathbb{R}^3} \left(\sum_{i=1}^{N} \tfrac{1}{2}|\nabla\psi_i|^2 + V_{\mathrm{ext}} \sum_{i=1}^{N} |\psi_i|^2 \right) d\mathbf{x}$$

$$+ \tfrac{1}{2} \iint_{\mathbb{R}^3 \times \mathbb{R}^3} \sum_{i \neq j} \frac{|\psi_i(\mathbf{x})|^2 |\psi_j(\mathbf{y})|^2}{|\mathbf{x} - \mathbf{y}|} \, d\mathbf{x}\,d\mathbf{y}.$$

The corresponding Euler–Lagrange equations become

$$\left(-\tfrac{1}{2}\Delta + V_{\mathrm{ext}}(\mathbf{x}) + \sum_{j \neq i} \int_{\mathbb{R}^3} \frac{|\psi_j(\mathbf{y})|^2}{|\mathbf{x} - \mathbf{y}|} d\mathbf{y} \right) \psi_i(\mathbf{x}) = \varepsilon_i \psi_i(\mathbf{x}), \quad i = 1, \ldots, N,$$

where ε_i is the Lagrange multiplier corresponding to the normalization constraint.

Hartree's approximation has a simple interpretation, namely that each electron feels a mean field potential due to the Coulomb interaction with the other electrons in the system. However, it violates the Pauli exclusion principle. This problem is addressed in the *Hartree–Fock approximation*, in which the wave function Ψ_e is assumed to be in the form of a single *Slater determinant*:

$$\Psi_e(\mathbf{x}_1, \ldots, \mathbf{x}_N) = \frac{1}{\sqrt{N!}} \det \begin{pmatrix} \psi_1(\mathbf{x}_1) & \cdots & \psi_1(\mathbf{x}_N) \\ \vdots & \ddots & \vdots \\ \psi_N(\mathbf{x}_1) & \cdots & \psi_N(\mathbf{x}_N) \end{pmatrix}. \quad (4.4.13)$$

Here $\{\psi_1, \ldots, \psi_N\}$ is a set of orthonormal (one-particle) wave functions:

$$\int_{\mathbb{R}^3} \psi_i^*(\mathbf{x}) \psi_j(\mathbf{x}) \, d\mathbf{x} = \delta_{ij}. \quad (4.4.14)$$

The Hartree–Fock functional is obtained from the functional (4.4.10) using the Hartree–Fock approximation (4.4.13):

$$I_{\mathrm{HF}}(\{\psi_i\}) = \sum_{i=1}^{N} \int_{\mathbb{R}^3} \tfrac{1}{2}|\nabla\psi_i(\mathbf{x})|^2 + V_{\mathrm{ext}}(\mathbf{x})|\psi_i(\mathbf{x})|^2 \, d\mathbf{x}$$

$$+ \tfrac{1}{2} \sum_{i,j=1}^{N} \iint \frac{\psi_i^*(\mathbf{x})\psi_i(\mathbf{x})\psi_j^*(\mathbf{y})\psi_j(\mathbf{y})}{|\mathbf{x} - \mathbf{y}|} \, d\mathbf{x}\,d\mathbf{y}$$

$$- \tfrac{1}{2} \sum_{i,j=1}^{N} \iint \frac{\psi_i^*(\mathbf{x})\psi_i(\mathbf{y})\psi_j^*(\mathbf{y})\psi_j(\mathbf{x})}{|\mathbf{x} - \mathbf{y}|} \, d\mathbf{x}\,d\mathbf{y}.$$

The Hartree–Fock approximation is the solution to the variational problem

$$\inf_{\{\psi_i\}} I_{\mathrm{HF}}(\{\psi_i\})$$

subject to the orthonormality constraint (4.4.14). Obviously the Hartree–Fock approximation gives an upper bound to the energy of the many-body system. The Euler–Lagrange equations for this variational problem can be written as

$$H\psi_i(\mathbf{x}) - \sum_j \int_{\mathbb{R}^3} \frac{\psi_j(\mathbf{x})\psi_i(\mathbf{y})\psi_j^*(\mathbf{y})}{|\mathbf{x} - \mathbf{y}|}\, \mathrm{d}\mathbf{y} = \varepsilon_i \psi_i(\mathbf{x}) \ , \quad i = 1, \ldots, N,$$

(4.4.15)

where

$$H = -\tfrac{1}{2}\Delta + V_{\mathrm{ext}}(\mathbf{x}) + \int_{\mathbb{R}^3} \frac{\rho(\mathbf{y})}{|\mathbf{x} - \mathbf{y}|} d\mathbf{y},$$

with density ρ defined by

$$\rho(\mathbf{x}) = \sum_{i=1}^{N} |\psi_i(\mathbf{x})|^2.$$

The term

$$\sum_j \int_{\mathbb{R}^3} \frac{\psi_j(\mathbf{x})\psi_i(\mathbf{y})\psi_j^*(y)}{|\mathbf{x} - \mathbf{y}|} d\mathbf{y}$$

is called the *exchange term*, since it comes from the antisymmetry of the wave function. For a mathematical introduction to the Hartree–Fock model, we refer readers to [31].

4.4.3 Density functional theory

Density functional theory is an alternative approach for describing the electronic structure of matter [21, 26, 34]. Its original motivation was the use of only the density of the electrons, rather than the wave functions, as the basic object for describing the electronic structure of matter (in its ground state). Given the N-body wave function Ψ_e, the electron density is defined by

$$\rho(\mathbf{x}) = N \int |\Psi_e(\mathbf{x}, \mathbf{x}_2, \ldots, \mathbf{x}_N)|^2 d\mathbf{x}_2 \cdots d\mathbf{x}_N,$$

(4.4.16)

which is the probability of finding an electron at point **x**. At a first sight, it would seem quite impossible to describe the electronic structure of an arbitrary system just by the electron density, since ρ seems to contain much less information than the many-body wave function. However, we should note that both the wave function and the density are very much constrained by the fact that they are associated with the ground state of some Hamiltonian operator. In fact, Hohenberg and Kohn showed that there is a one-to-one correspondence between the external potential of an electronic system and the electron density of its ground state [21], provided that the ground state is nondegenerate. Since the external potential determines the entire system, the ground state wave function and hence the ground state energy are uniquely determined by the ground state electron density. Indeed, not only the total energy but all the energy terms, such as the kinetic energy and Coulomb interaction energy, are uniquely determined by the ground state electron density. They can be expressed as universal functionals of the ground state electron density.

Instead of discussing the details of the Hohenberg–Kohn theorem, it is more convenient for us to introduce the formulation of the Levy–Lieb constrained variational problem. Imagine solving the variational problem (4.4.11) in two steps: first we fix a density function ρ and minimize the functional I over all wave functions whose density is ρ and then we minimize over all ρ. More precisely, define the functional

$$F(\rho) = \inf_{\Psi \mapsto \rho} \langle \Psi | H_e | \Psi \rangle = \left\langle \Psi \left| -\sum_i \tfrac{1}{2}\Delta_{\mathbf{x}_i} + V_{ee} \right| \Psi \right\rangle, \qquad (4.4.17)$$

where the constraint $\Psi \mapsto \rho$ means that ρ is equal to the density given by Ψ using (4.4.16). We also assume Ψ to be antisymmetric.

The ground energy of the system can now be expressed as

$$E = \inf_{\rho} \left(F(\rho) + \int V_{\text{ext}}\, \rho \, d\mathbf{x} \right).$$

The functional F contains the kinetic energy of the electrons and the Coulomb interaction between the electrons. The quantity F is a universal functional, in the sense that it does not depend on the external potential V_{ext}. However, the explicit form of F is not known, and further approximations are necessary in order to make this formulation useful in practice.

Kohn–Sham density functional theory The first step, made by Kohn and Sham [26], is to restrict Ψ in (4.4.17) to the form of a Slater determinant (4.4.13), where the the one-body wave functions $\{\psi_1, \ldots, \psi_N\}$ satisfy the orthonormality constraints (4.4.14). This is similar in appearance to the Hartree–Fock approximation, but the underlying philosophy is quite different. In the Hartree–Fock approximation we aim at obtaining an upper bound of the energy by restricting the form of the many-body wave function. Here we are mapping the many-body system to a virtual system of noninteracting electrons (i.e. whose wave function is given by a Slater determinant), with an effective potential carefully chosen so that the density of the electrons is the same as the density of the system in which we are interested. The question reduces to the choice of this effective potential. To see this, we first compute the energy of a system of non-interacting electrons. Since

$$
\left\langle \Psi \left| -\sum_i \frac{1}{2}\Delta_{\mathbf{x}_i} + V_{ee} \right| \Psi \right\rangle
$$

$$
= \sum_{i=1}^{N} \int_{\mathbb{R}^3} \frac{1}{2}|\nabla\psi_i(\mathbf{x})|^2 + \frac{1}{2}\iint_{\mathbb{R}^3\times\mathbb{R}^3} \frac{\rho(\mathbf{x})\rho(\mathbf{y})}{|\mathbf{x}-\mathbf{y}|}\,d\mathbf{x}\,d\mathbf{y}
$$

$$
- \frac{1}{2}\sum_{i,j=1}^{N} \iint_{\mathbb{R}^3\times\mathbb{R}^3} \frac{\psi_i^*(\mathbf{x})\psi_i(\mathbf{y})\psi_j^*(\mathbf{y})\psi_j(\mathbf{x})}{|\mathbf{x}-\mathbf{y}|}\,d\mathbf{x}\,d\mathbf{y},
$$

the error made by neglecting electron–electron interactions is given by the *correlation energy*

$$
E_{\mathrm{c}}(\rho)
$$

$$
= F(\rho) - \inf_{\{\psi_i\}\mapsto\rho} \left(\sum_{i=1}^{N} \int_{\mathbb{R}^3} \frac{1}{2}|\nabla\psi_i(\mathbf{x})|^2\,d\mathbf{x} + \frac{1}{2}\iint_{\mathbb{R}^3\times\mathbb{R}^3} \frac{\rho(\mathbf{x})\rho(\mathbf{y})}{|\mathbf{x}-\mathbf{y}|}\,d\mathbf{x}\,d\mathbf{y} \right.
$$

$$
\left. - \frac{1}{2}\sum_{i,j=1}^{N} \iint_{\mathbb{R}^3\times\mathbb{R}^3} \frac{\psi_i^*(\mathbf{x})\psi_i(\mathbf{y})\psi_j^*(\mathbf{y})\psi_j(\mathbf{x})}{|\mathbf{x}-\mathbf{y}|}\,d\mathbf{x}\,d\mathbf{y} \right).
$$

$$
(4.4.18)
$$

here $\{\psi_i\} \mapsto \rho$ means that $\rho(x) = \sum_i |\psi_i(x)|^2$. Note that the correlation energy is a functional of ρ.

The last term on the right-hand side of (4.4.18) will give a nonlocal potential in the Euler–Lagrange equation, as we discussed in the context of the Hartree–Fock approximation. This is sometimes unpleasant to deal with, and one replaces it by an explicit functional of ρ called the *exchange energy*:

$$E_{\mathrm{x}}(\rho) = F(\rho) - \inf_{\{\psi_i\} \mapsto \rho} \sum_{i=1}^{N} \int_{\mathbb{R}^3} \tfrac{1}{2} |\nabla \psi_i(\mathbf{x})|^2 d\mathbf{x}$$

$$- E_c(\rho) - \tfrac{1}{2} \iint_{\mathbb{R}^3 \times \mathbb{R}^3} \frac{\rho(\mathbf{x})\rho(\mathbf{y})}{|\mathbf{x} - \mathbf{y}|} d\mathbf{x}d\mathbf{y}. \qquad (4.4.19)$$

Note that E_x is also a functional of ρ. Defining the *exchange-correlation energy*

$$E_{\mathrm{xc}}(\rho) = E_{\mathrm{x}}(\rho) + E_{\mathrm{c}}(\rho);$$

we can write

$$F(\rho) = T(\rho) + \tfrac{1}{2} \iint_{\mathbb{R}^3 \times \mathbb{R}^3} \frac{\rho(\mathbf{x})\rho(\mathbf{y})}{|\mathbf{x} - \mathbf{y}|} d\mathbf{x}\, d\mathbf{y} + E_{\mathrm{xc}}(\rho).$$

Here $T(\rho)$ is the new kinetic energy functional:

$$T(\rho) = \inf_{\{\psi_i\} \mapsto \rho} \sum_{i=1}^{N} \int_{\mathbb{R}^3} \tfrac{1}{2} |\nabla \psi_i(\mathbf{x})|^2 \, d\mathbf{x}.$$

Notice that, up to this point, what we have done is just to reformulate the original many-body problem – no approximations have been made. We still do not know the form of $E_{\mathrm{x}}(\rho)$ and $E_{\mathrm{c}}(\rho)$, which account for the error made by approximating the many-body wave function using a single Slater determinant and approximating the electron–electron Coulomb interaction by the *Hartree energy*,

$$E_{\mathrm{H}}(\rho) = \tfrac{1}{2} \iint_{\mathbb{R}^3 \times \mathbb{R}^3} \frac{\rho(\mathbf{x})\rho(\mathbf{y})}{|\mathbf{x} - \mathbf{y}|} d\mathbf{x}\, d\mathbf{y}.$$

The main insight of Kohn and Sham [26] was that by approximating the kinetic energy using the one-body wave functions, one takes care of a large portion of the kinetic energy. The remaining contribution is small and can be approximated with acceptable accuracy using simple ideas such as the *local density approximation* (LDA):

$$E_{\mathrm{xc}}(\rho) = \int \varepsilon_{\mathrm{xc}}(\rho(\mathbf{x}))d\mathbf{x},$$

where $\varepsilon_{\mathrm{xc}}(\rho)$ is a local function (rather than a functional) of ρ that remains to be modeled. For example, for a homogeneous free electron gas, its exchange energy can be easily calculated and is given by [34]:

$$E_{\mathrm{x}} = -C_{\mathrm{D}} V \rho^{4/3},$$

where V is the volume and C_{D} is some universal constant (the subscript D refers to Dirac, who was the first to obtain this relation). On this basis it is assumed that for systems close to a homogeneous electron gas the exchange energy can be approximated as

$$E_{\mathrm{x}}(\rho) = -C_D \int \rho(\mathbf{x})^{4/3} \, d\mathbf{x},$$

in the spirit of the local density approximation. One can also include local density gradients to get better accuracy,

$$E_{\mathrm{xc}}(\rho) = \int \varepsilon_{\mathrm{xc}}(\rho(\mathbf{x}), \nabla\rho(\mathbf{x})) \, d\mathbf{x};$$

this is known as the *generalized gradient approximation* (GGA) [33].

These approximate exchange–correlation functionals can often give satisfactory results in practical computations. However, in general it is difficult to make systematic corrections to improve the accuracy.

Under the approximations for the exchange-correlation energy, the Kohn–Sham energy functional is given by

$$I_{\mathrm{KS}}(\rho) = T(\rho) + \int V_{\mathrm{ext}}\rho \, d\mathbf{x} + E_{\mathrm{xc}}(\rho) + \tfrac{1}{2} \iint \frac{\rho(\mathbf{x})\rho(\mathbf{y})}{|\mathbf{x} - \mathbf{y}|} \, d\mathbf{x} \, d\mathbf{y}, \quad (4.4.20)$$

where:

(1) $T(\rho)$ is the universal kinetic energy functional for a noninteracting electron gas;

(2) V_{ext} is the external potential, which is, for a nuclei–electron system, given by

$$V_{\mathrm{ext}}(\mathbf{x}) = -\sum_I \frac{Z_I}{|\mathbf{x} - \mathbf{R}_I|};$$

(3) E_{xc} is the exchange–correlation energy, which, assuming LDA or GGA approximations, is an explicit functional of ρ;

(4) the last term on the right-hand side is the Hartree energy.

Notice that we may also consider an energy functional defined in terms of the one-body wave functions:

$$I_{\mathrm{KS}}(\{\psi_i\})$$
$$= \sum_i \int \tfrac{1}{2}|\nabla\psi_i|^2 \, d\mathbf{x} + \int V_{\mathrm{ext}} \, \rho \, d\mathbf{x} + E_{\mathrm{xc}}(\rho) + \tfrac{1}{2} \iint \frac{\rho(\mathbf{x})\rho(\mathbf{y})}{|\mathbf{x}-\mathbf{y}|} \, d\mathbf{x} \, d\mathbf{y}.$$

The formulation presented above treats every electron in the system on an equal footing. In practice, one often associates the core electrons with the nucleus and treats only the valence electrons. This has a tremendous practical advantage, since the wave functions associated with the core electrons typically have rapid variations and are therefore difficult to compute numerically; in addition, they usually do not participate in the binding between atoms or chemical reactions. However, in doing so we also have to represent the effect of the nuclei differently, by introducing some effective potential, often called the *pseudo-potential* [33]. We can then rewrite the Kohn–Sham functional as

$$I_{\mathrm{KS}}(\rho) = T(\rho) + J(\rho) + E_{\mathrm{xc}}(\rho), \qquad (4.4.21)$$

where $T(\rho)$ is the kinetic energy term and

$$J(\rho) = \frac{1}{2} \iint \frac{(\rho(\mathbf{x})-m(\mathbf{x}))(\rho(\mathbf{y})-m(\mathbf{y}))}{|\mathbf{x}-\mathbf{y}|} \, d\mathbf{x} \, d\mathbf{y}.$$

Here $m(\mathbf{x})$ is the background charge distribution due to the nuclei and the core electrons. It usually takes the form

$$m(\mathbf{x}) = \sum_I m_{\mathrm{a},I}(\mathbf{x}-\mathbf{R}_I),$$

where $m_{\mathrm{a},I}$ is the contribution from the Ith nucleus with its core electrons and $\mathbf{R}_I$ is its position. Note that, in the new form (4.4.21), the effect of the nuclei–nuclei Coulomb interaction is included in the functional.

Thomas–Fermi and orbital-free density functional theory In *Thomas–Fermi theory*, we model $T(\rho)$ by considering the kinetic energy of a homogeneous free (i.e. noninteracting) electron gas. This gives rise to [34]

$$T_{\mathrm{TF}}(\rho) = C_F \int \rho^{5/3}(\mathbf{x}) \, d\mathbf{x}, \qquad (4.4.22)$$

where C_F is a universal constant. This result is the consequence of a simple scaling argument. Since we are considering a homogeneous free electron gas, the end result can only depend on the pointwise value of ρ (not on, for example, the gradient of ρ). Obviously, on dimensional grounds we have $[\Psi]^2 = [\rho]$, $[\rho] = 1/L^3$, where L denotes the dimension of length. Therefore we have

$$[|\nabla\Psi|^2] = [\rho]/L^2 = [\rho]^{5/3}$$

and this gives (4.4.22).

The simplicity of the Thomas–Fermi theory comes at a large price: the Thomas–Fermi model does not allow binding between atoms. Indeed, the well-known Teller theorem states that if we divide a neutral electronic system (i.e. one in which the total charge vanishes) into two disjoint neutral subsystems then, within Thomas–Fermi theory, the total energy is a decreasing function of the distance between the two subsystems [34].

Many ideas have been proposed to remedy this problem. One well-known example is the *Thomas–Fermi–von Weizsäcker (TFW) model*:

$$T_{\text{TFW}}(\rho) = T_{\text{TF}}(\rho) + \lambda \int \frac{|\nabla\rho|^2}{\rho}\, d\mathbf{x},$$

where λ is some parameter. The TFW model does allow binding between atoms.

The second idea is to add integral terms. The best-known example of such a functional is the *Wang–Teter functional* [7]

$$T_{\text{WT}}(\rho) = T_{\text{TFW}}(\rho) + \iint \rho^{5/6}(\mathbf{x}) K_{\text{WT}}(\mathbf{x},\mathbf{y}) \rho^{5/6}(\mathbf{y})\, d\mathbf{x}\, d\mathbf{y}.$$

Here K_{WT} is a kernel function, constructed in order to reproduce accurately the linear response of a homogeneous free electron gas; it is given by the Lindhard function [7].

4.4.4 Tight-binding models

Tight-binding models (TBMs) are the simplest electronic structure models with explicit wave functions. They are a minimalist type of model and are based on:

(1) representing the wave functions using a minimum basis set, often in the form of atomic orbitals which are the ground and excited states for the electron in the hydrogen atom; and

Figure 4.6. Setup of the tight-binding model for the analysis of the binding of two atoms (solid circles).

(2) neglecting electron–electron interactions except that the Pauli exclusion principle is still imposed.

Under these approximations the state space becomes finite and the Hamiltonian operator becomes a matrix. This makes it possible to carry out explicit analytical calculations as well as efficient numerical computations. Even though the TBM approach is quite crude, it is very useful as both a theoretical and practical tool. Its accuracy can be improved to some extent by calibrating the matrix elements of the Hamiltonian against results from more accurate models.

Let us start with the example of a hydrogen ion H_2^+. We will set up coordinates with the two atoms on the real axis; see Figure 4.6. The wave function Ψ is expressed as a linear combination of a minimal set of atomic orbitals:

$$\Psi = c_1\varphi_1 + c_2\varphi_2, \tag{4.4.23}$$

where φ_1 is the $1s$ orbital of atom 1, and φ_2 is the $1s$ orbital of atom 2 (see [24]). The Hamiltonian operator is of the form

$$H = T + V_1 + V_2, \tag{4.4.24}$$

where T is the kinetic energy operator and V_1 and V_2 are the Coulomb potentials associated with atoms 1 and 2 respectively. We will discuss the explicit form of H later. Our problem is then reduced to the eigenvalue problem

$$H\Psi = \varepsilon\Psi \tag{4.4.25}$$

for finding c_1, c_2 and ε.

Taking the inner product of (4.4.25) with respect to φ_1 and φ_2 respectively, we get

$$(\varphi_1, H\Psi) = \varepsilon(\varphi_1, \Psi),$$
$$(\varphi_2, H\Psi) = \varepsilon(\varphi_2, \Psi).$$

We can express these equations in a matrix form:

$$\mathbf{H}\begin{pmatrix} c_1 \\ c_2 \end{pmatrix} = \varepsilon\mathbf{S}\begin{pmatrix} c_1 \\ c_2 \end{pmatrix}, \tag{4.4.26}$$

where

$$\mathbf{H} = \begin{pmatrix} H_{11} & H_{12} \\ H_{21} & H_{22} \end{pmatrix} = \begin{pmatrix} (\varphi_1, H\varphi_1), & (\varphi_1, H\varphi_2) \\ (\varphi_2, H\varphi_1), & (\varphi_2, H\varphi_2) \end{pmatrix} \tag{4.4.27}$$

$$\mathbf{S} = \begin{pmatrix} S_{11} & S_{12} \\ S_{21} & S_{22} \end{pmatrix} = \begin{pmatrix} (\varphi_1, \varphi_1), & (\varphi_1, \varphi_2) \\ (\varphi_2, \varphi_1), & (\varphi_2, \varphi_2) \end{pmatrix}$$

are the Hamiltonian and mass matrices respectively.

Next we discuss how the elements of these two matrices can be approximated. In *orthogonal TBM*, the overlap integral between the different atomic orbitals is neglected, $(\varphi_1, \varphi_2) \sim 0$, and $\mathbf{S}$ is approximated by the identity matrix. For the Hamiltonian matrix,

$$\begin{aligned} H_{11} &= \langle \varphi_1 | T + V_1 + V_2 | \varphi_1 \rangle \\ &= \langle \varphi_1 | T + V_1 | \varphi_1 \rangle + \langle \varphi_1 | V_2 | \varphi_1 \rangle \\ &= \varepsilon_{1s}^0 + \Delta \\ &= \varepsilon_s. \end{aligned} \tag{4.4.28}$$
$$\tag{4.4.29}$$

We call ε_s the on-site energy; $\varepsilon_{1s}^0 = \langle \varphi_1 | T + V_1 | \varphi_1 \rangle$ is the energy of the $1s$ state of the hydrogen atom. Next,

$$\begin{aligned} H_{12} &= \langle \varphi_1 | T + V_1 + V_2 | \varphi_2 \rangle \\ &= \langle \varphi_1 | T + V_1 | \varphi_2 \rangle + \langle \varphi_1 | V_2 | \varphi_2 \rangle \\ &= \varepsilon_{1s}^0 (\varphi_1, \varphi_2) + \langle \varphi_1 | V_2 | \varphi_2 \rangle \\ &\approx \langle \varphi_1 | V_2 | \varphi_2 \rangle = -t < 0. \end{aligned} \tag{4.4.30}$$

The last equation defines the parameter t. Since φ_1 and φ_2 are positive, V_2 is negative and t is positive. The values of ε_s and t depend on the distance between the two ions. Expressions for them can be obtained empirically with input from the results of more accurate models.

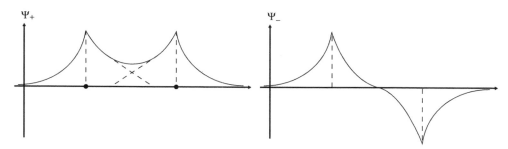

Figure 4.7. Bonding (left) and antibonding (right) states

To summarize, we can write **H** as

$$\mathbf{H} = \begin{pmatrix} \varepsilon_{\mathrm{s}} & -t \\ -t & \varepsilon_{\mathrm{s}} \end{pmatrix}. \tag{4.4.31}$$

The eigenvalues for this matrix are

$$\varepsilon_+ = \varepsilon_{\mathrm{s}} - t < \varepsilon_{\mathrm{s}}, \tag{4.4.32}$$
$$\varepsilon_- = \varepsilon_{\mathrm{s}} + t,$$

with corresponding eigenvectors

$$\Psi_+ = \frac{1}{\sqrt{2}}(\varphi_1 + \varphi_2), \tag{4.4.33}$$

$$\Psi_- = \frac{1}{\sqrt{2}}(\varphi_1 - \varphi_2).$$

Here Ψ_+ is the bonding state, since an electron in this state has a higher probability of being found between the two nuclei, and Ψ_- is the antibonding state; see Figure 4.7.

If we take into account the overlap then Ψ_+ and Ψ_- are still the eigenvectors but, using standard chemical notation, the eigenvalues change to

$$\varepsilon_+ = \frac{\langle \Psi_+|H|\Psi_+\rangle}{\langle \Psi_+|\Psi_+\rangle} = \varepsilon_{\mathrm{s}} + V_{ss\sigma} - \langle \varphi_1|\varphi_2\rangle V_{ss\sigma}, \tag{4.4.34}$$

$$\varepsilon_- = \frac{\langle \Psi_-|H|\Psi_-\rangle}{\langle \Psi_-|\Psi_-\rangle} = \varepsilon_{\mathrm{s}} - V_{ss\sigma} - \langle \varphi_1|\varphi_2\rangle V_{ss\sigma}, \tag{4.4.35}$$

where

$$V_{ss\sigma} = \frac{H_{12} - \langle \varphi_1|\varphi_2\rangle \varepsilon_{\mathrm{s}}}{1 - \langle \varphi_1|\varphi_2\rangle^2} = \frac{\langle \varphi_1|V_2|\varphi_2\rangle}{1 - \langle \varphi_1|\varphi_2\rangle^2} < 0. \tag{4.4.36}$$

The first term on the right-hand side of (4.4.34) is the on-site energy. The second term is negative, corresponding to an attractive interaction. The third term is only important when there is significant overlap, i.e. at short distance. This is a repulsive term.

Next we consider the problem of a one-dimensional chain of N lithium atoms. We will use the periodic boundary condition. We denote by x_j the position of the jth atom: $x_j = ja$, where a is the distance between neighboring atoms. The wave function is represented as a linear combination of the $1s$ orbitals of the atoms:

$$\Psi = \sum_{j=1}^{N} c_j \varphi_j, \qquad (4.4.37)$$

where φ_j is the $1s$ orbital of the jth atom. Again for simplicity, we will use the orthogonal TBM. For the Hamiltonian matrix we will use the nearest neighbor approximation. In this case the Hamiltonian matrix has only two parameters, ε_s, the on-site energy and t, the hopping element:

$$H = \begin{pmatrix} \varepsilon_s & -t & & & & -t \\ -t & \varepsilon_s & -t & & 0 & \\ & -t & \varepsilon_s & \cdot & & \\ & & \cdot & \cdot & \cdot & \\ 0 & & & \cdot & \cdot & \cdot \\ & & & & \cdot & \cdot & -t \\ -t & & & & -t & \varepsilon_s \end{pmatrix}. \qquad (4.4.38)$$

The eigenvalues and eigenvectors for this matrix can be found by a discrete Fourier transform:

$$\varepsilon_k = \varepsilon_s - 2t \, \cos ka, \qquad (4.4.39)$$

$$\Psi_k = \frac{1}{\sqrt{N}} \sum_{l=-N/2}^{N/2} e^{ikx_l} \varphi_l, \qquad (4.4.40)$$

where l runs over the integers between $-N/2$ and $N/2$; l and k are related by

$$k = \frac{2\pi l}{Na} \qquad (4.4.41)$$

and k runs over the so-called first Brillouin zone,

$$-\frac{\pi}{a} < k \leq \frac{\pi}{a}. \qquad (4.4.42)$$

The ranges of l and k can be shifted by periods of N and $2\pi/a$ respectively.

Since each state can accommodate two electrons, one with an up-spin and one with a down-spin, only half the states are filled in the ground state of the system. In addition, when N is large the energy gap between the occupied states and the unoccupied states is very small. Therefore the system behaves like a metal.

4.5 Notes

The main objective of this chapter was to give an overview of physical models at different levels of detail and how they are related to each other. These physical models are the main ingredients used in multiscale, multi-physics modeling.

From a mathematical viewpoint, a huge amount of work has been done on continuum models such as those in elasticity theory and fluid dynamics. Some work has been done on kinetic theory. Much less has been done on atomistic and electronic structure models. As we will see below, this relative lack of understanding of atomistic and electronic structure models is a basic obstacle in building the foundations of multiscale, multi-physics modeling.

Understanding the connection between the different models is also a very important component of multiscale modeling. Here we have illustrated how continuum models of nonlinear elasticity can be linked to atomistic models through the Cauchy–Born rule, and how hydrodynamic equations of gas dynamics can be derived from the Boltzmann equation. An important topic that we have not discussed is the hydrodynamic limit of interacting particle systems. We refer the interested reader to [25, 37].

References for Chapter 4

[1] M.P. Allen and D.J. Tildesley, *Computer Simulation of Liquids*, Clarendon Press, 2001.

[2] P.L. Bhatnagar, E.P. Gross and M. Krook, "A model for collision processes in gases. I. Small amplitude processes in charged and neutral one-component systems," *Phys. Rev.*, vol. 94, pp. 511–525, 1954.

[3] X. Blanc, C. Le Bris and P.-L. Lions, "From molecular models to continuum mechanics," *Arch. Rat. Mech. Anal.*, vol. 164, pp. 341–381, 2002.

[4] M. Born and K. Huang, *Dynamical Theory of Crystal Lattices*, Oxford University Press, 1954.

[5] B.R. Brooks, R.E. Bruccoleri, B.D. Olafson, D.J. States, S. Swaminathan and M. Karplus, "CHARMM: a program for macromolecular energy, minimization, and dynamics calculations," *J. Comp. Chem.*, vol. 4, pp. 187–217, 1983.

[6] R. Car, unpublished lecture notes.

[7] E.A. Carter, "Orbital-free density functional theory dynamics: evolution of thousands of atoms with quantum forces," in *Advances in Computational Engineering and Sciences*, S.N. Atluri and F.W. Brust, eds., Tech Science Press, 2000.

[8] C. Cercignani, *The Boltzmann Equation and its Applications*, Springer-Verlag, 1988.

[9] D. Chandler, *Introduction to Modern Statistical Mechanics*, Oxford University Press, 1987.

[10] S. Chapman and T.G. Cowling, *The Mathematical Theory of Non-Uniform Gases: An Account of the Kinetic Theory of Viscosity, Thermal Conduction, and Diffusion in Gases*, Cambridge University Press, 1939.

[11] M.S. Daw and M.I. Baskes, "Semiempirical, quantum mechanical calculation of hydrogen embrittlement in metals," *Phys. Rev. Lett.*, vol. 50, pp. 1285–1288, 1983.

[12] S.R. de Groot and P. Mazur, *Non-Equilibrium Thermodynamics*, Dover Publications, 1984.

[13] M. Doi and S.F. Edwards, *The Theory of Polymer Dynamics*, Oxford University Press, 1986.

[14] W. E and P.B. Ming, "Cauchy–Born rule and the stability of the crystalline solids: static problems," *Arch. Rat. Mech. Anal.*, vol. 183, pp. 241–297, 2007.

[15] W. E and P.B. Ming, "Cauchy–Born rule and the stability of the crystalline solids: dynamic problems," *Acta Math. Appl. Sin. Engl. Ser.* vol. 23, pp. 529–550, 2007.

[16] J.L. Ericksen, "On the Cauchy–Born rule," *Math. Mech. Solids*, vol. 13, pp. 199–220, 2008.

[17] J.H. Ferziger and H.G. Kaper, *Mathematical Theory of Transport in Gases*, North-Holland, 1972.

[18] D. Frenkel and B. Smit, *Understanding Molecular Simulation: From Algorithms to Applications*, 2nd edition, Academic Press, 2001.

[19] N. Goldenfeld, *Lectures on Phase Transitions and the Renormalization Group*, Perseus Books, 1992.

[20] H. Grad, "On the kinetic theory of rarefied gases," *Comm. Pure Appl. Math.* vol. 2, pp. 331–407, 1949.

[21] P. Hohenberg and W. Kohn, "Inhomogeneous electron gas," *Phys. Rev.*, vol. 136, pp. B864–B871, 1964.

[22] J.H. Irving and J.G. Kirkwood, "The statistical mechanical theory of transport processes IV," *J. Chem. Phys.*, vol. 18, pp. 817–829, 1950.

[23] R.D. James, Unpublished lecture notes.

[24] E. Kaxiras, *Atomic and Electronic Stucture of Solids*, Cambridge University Press, 2003.

[25] C. Kipnis and C. Landim, *Scaling Limits of Interacting Particle Systems*, Springer-Verlag, 1999.

[26] W. Kohn and L. Sham, "Self-consistent equations including exchange and correlation effects," *Phys. Rev.*, vol. 140, pp. A1133–A1138, 1965.

[27] L. Landau and E.M. Lifshitz, *Theory of Elasticity*, 3rd edition, Pergamon Press, 1986.

[28] C.D. Levermore, "Moment closure hierarchies for kinetic theories," *J. Stat. Phys.*, vol. 83, nos. 5–6, pp. 1021–1065, 1996.

[29] T. Li and T. Qin, *Physics and Partial Differential Equations* (in Chinese), The Higher Education Press, 2000.

[30] X. Li and W. E, "Multiscale modeling of the dynamics of solids at finite temperature," submitted to *J. Mech. Phys. Solids*, vol. 53, pp. 1650–1685, 2005.

[31] E.H. Lieb and B. Simon, "The Hartree–Fock theory for Coulomb systems," *Commun. Math. Phys.*, vol. 53, pp. 185–194, 1977.

[32] M. Marder, *Condensed Matter Physics*, Wiley-Interscience, 2000.

[33] R.M. Martin, *Electronic Structure: Basic Theory and Practical Methods*, Cambridge University Press, 2004.

[34] R.G. Parr and W. Yang, *Density Functional Theory of Atoms and Molecules*, Oxford University Press, 1989.

[35] D.G. Pettifor, *Bonding and Structure of Molecules and Solids*, Clarendon Press, 1995.

[36] L.E. Reichl, *A Modern Course in Statistical Physics*, University of Texas Press, 1980.

[37] H. Spohn, *Large Scale Dynamics of Interacting Particles*, Springer-Verlag, 1991.

[38] H. Struchtrup, *Macroscopic Transport Equations for Rarefied Gas Flows*, Springer-Verlag, 2005.

[39] S. Tsien, *Lectures on Physical Mechanics* (in Chinese), Science Press, 1962.

[40] G.A. Voth, ed., *Coarse-graining of Condensed Phase and Biomolecular Systems*, CRC Press, 2009.

[41] M. Zhou, "A new look at the atomic level virial stress: on continuum–molecular system equivalence," *Proc. Roy. Soc. London A*, vol. 459, pp. 2347–2392, 2003.

5

Examples of multi-physics models

After this chapter, the remaining chapters will be devoted to numerical algorithms for multiscale problems. However, before launching into a discussion about numerics, we would like to make the point that developing better physical models based on multiscale, multi-physics considerations is at least as important. After all, the ultimate goal of modeling is to get a better understanding of the physical problem and obtaining better models is a very important way of doing that. In addition, there is a very close relation between multiscale algorithms and multiscale models.

Speaking in broad terms, there has been a long history of multiscale, multi-physics models. One class of such models is associated with the resolution of singularities (shocks, interfaces and vortices) by the introduction of additional regularizing terms that often represent more detailed physics. In Ginzburg–Landau theory, for example, we add gradient terms to the Landau expansion of the free energy to account for the contribution due to the spatial variation of the order parameter. This allows us to resolve the internal structure of the interface between different phases or the internal structure of the vortices, both of which are at the microscale. Adding viscous and heat conduction terms to the equations of gas dynamics has a similar effect, namely, it allows us to resolve the internal structure of the shocks.

Another class of examples is those that explicitly involve the coupling of models at different levels of physics. Car–Parrinello or first-principle-based molecular dynamics is one such example (see the next chapter). Other examples include coupled Brownian dynamics and hydrodynamics models for polymer fluids, coupled atomistic and continuum models for solids.

We will discuss three examples in this chapter. In the first example, we describe how one may use kinetic theory or Brownian dynamics to supply the constitutive relation for the stress in polymer fluids; the ultimate result is a coupled Brownian dynamics–hydrodynamics model. In the second example we describe how one may derive continuum nonlinear elasticity models for nanostructures using atomistic models.

The final example is the moving contact line problem, where better boundary conditions are needed at the contact line in order to resolve the singularity that arises in classical continuum theory. When water is poured in or drained out of a cup the three different phases, air, water and solid cup, intersect at a line called the *contact line*. The dynamics of the contact line is often regarded as being one of the few remaining unsolved problems in classical fluid dynamics. It is well known that if we use the classical Navier–Stokes equation with a no-slip boundary condition to describe the problem then we obtain a singularity at the moving contact line with an infinite rate of energy dissipation [22]. Even though this singularity is unavoidable at the macroscopic scale, many proposals have been made to remove it at smaller scales, by changing either the governing PDE or the boundary condition. While these proposals do succeed in removing the singularity, it is far from being clear whether they reflect the actual physics going on near the contact line. One can also use molecular dynamics or coupled atomistic–continuum models such as those discussed in Chapter 7. These approaches are very useful for discovering new microscopic physical phenomena, such as the existence of the slip region, but they are inconvenient as analytical tools. We will describe the approach advocated in [36], in which thermodynamic considerations are used to suggest the form of the correct continuum model, e.g. the form of the boundary condition, and microscopic simulations are then used to measure the detailed functional dependence of the constitutive relations. This kind of multiscale modeling should be useful for many other problems of this type, such as crack propagation and triple junction dynamics.

5.1 Brownian dynamics models of polymer fluids

As we saw earlier, the hydrodynamic equations for incompressible fluids can always be written in the form

$$\rho_0(\partial_t \mathbf{u} + (\mathbf{u} \cdot \nabla)\mathbf{u}) + \nabla p = \nabla \cdot \boldsymbol{\tau},$$
$$\nabla \cdot \mathbf{u} = 0, \tag{5.1.1}$$

where $\mathbf{u}$ is the velocity field, p is the pressure and $\boldsymbol{\tau}$ is the viscous stress. Consider a polymer in a solvent. Then $\boldsymbol{\tau}$ can be written as $\boldsymbol{\tau} = \boldsymbol{\tau}_{\mathrm{s}} + \boldsymbol{\tau}_{\mathrm{p}}$ where $\boldsymbol{\tau}_{\mathrm{s}}$ is the stress due to the solvent and $\boldsymbol{\tau}_{\mathrm{p}}$ is the polymer contribution to the stress. In general we can model $\boldsymbol{\tau}_{\mathrm{s}}$ by the simple linear constitutive relation $\boldsymbol{\tau}_{\mathrm{s}} = \eta_{\mathrm{s}}\mathbf{D}$, $\mathbf{D} = \frac{1}{2}(\nabla\mathbf{u} + \nabla\mathbf{u}^{\mathrm{T}})$, where η_{s} is the viscosity of the solvent. The difficulty lies in the modeling of $\boldsymbol{\tau}_{\mathrm{p}}$.

The traditional approach is to model $\boldsymbol{\tau}_{\mathrm{p}}$ via some empirical constitutive relations [3]. These constitutive relations are extensions of the constitutive relation for Newtonian fluids discussed in the previous chapter. Examples of such empirical constitutive relations include the following.

(1) *Generalized Newtonian models*:

$$\boldsymbol{\tau}_{\mathrm{p}} = \eta(\mathbf{D})\,\mathbf{D},$$

   i.e. the viscosity coefficient may depend on the strain rate.
(2) *Maxwell models* (for viscoelastic fluids):

$$\boldsymbol{\tau}_{\mathrm{p}}(\mathbf{x}, t) = \int_{-\infty}^{t} \frac{\eta_{\mathrm{s}}}{\lambda} \exp\left(-\frac{t - s}{\lambda}\right) \mathbf{D}(\mathbf{x}, s)\,ds, \qquad (5.1.2)$$

   where $(\eta_{\mathrm{s}}/\lambda)\exp(-(t-s)/\lambda)$ is the relaxation modulus representing the memory effect in the viscoelastic fluid.
(3) *Oldroyd models*:

$$\lambda \overset{\nabla}{\boldsymbol{\tau}}_{\mathrm{p}} + \boldsymbol{\tau}_{\mathrm{p}} + f(\boldsymbol{\tau}_{\mathrm{p}}, \mathbf{D}) = \eta\,\mathbf{D}, \qquad (5.1.3)$$

   where

$$\overset{\nabla}{\boldsymbol{\tau}}_{\mathrm{p}} = \frac{\partial \boldsymbol{\tau}_{\mathrm{p}}}{\partial t} + (\mathbf{u} \cdot \nabla)\boldsymbol{\tau}_{\mathrm{p}} - \nabla\mathbf{u} \cdot \boldsymbol{\tau}_{\mathrm{p}} - \boldsymbol{\tau}_{\mathrm{p}} \cdot (\nabla\mathbf{u})^{\mathrm{T}} \qquad (5.1.4)$$

   is the upper convective derivative [3] and λ is a relaxation time parameter. Well-known examples include the *Oldroyd-B model*, for which $f(\boldsymbol{\tau}_{\mathrm{p}}, \mathbf{D}) = 0$, and the *Johnson–Segalman model*, for which $f(\boldsymbol{\tau}_{\mathrm{p}}, \mathbf{D}) = \alpha\lambda(\boldsymbol{\tau}_{\mathrm{p}} \cdot \mathbf{D} + \mathbf{D} \cdot \boldsymbol{\tau}_{\mathrm{p}})$.

The other extreme is of course the full atomistic model. This is possible but at the present time it is not very productive if one wants to study rheological behavior. A more efficient approach is to coarse-grain the atomistic model, for example, by putting atoms into groups to form beads. Typical coarse-grained models include:

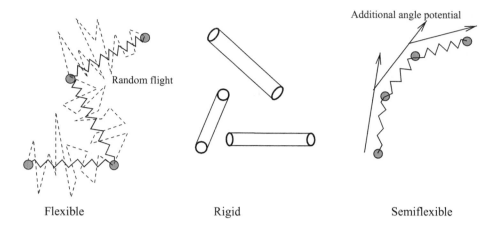

Figure 5.1. Illustration of flexible, semiflexible and rigid polymers (courtesy of Tiejun Li).

(1) bead and spring models (flexible and semiflexible);
(2) dumbbell models;
(3) rigid-rod models.

Such coarse-grained models have been very useful [4, 42]. Types (1) and (3) and illustrated in Figure 5.1.

In general, the inertial force for the polymer is much smaller than the friction force against the solvent. Therefore, it is common practice to neglect the inertial effects and consider so-called *Brownian dynamics* models.

There are two important length scales in these systems. One is the bending persistence length of a polymer, l_p, which is defined to be the correlation length for the tangent vectors at different positions along the polymer. The other is the length of the polymer, l. According to the relative size of l and l_p, polymers are classified as

flexible	semiflexible	rigid
$l \gg l_p$	$l \sim \mathcal{O}(l_p)$	$l \ll l_p$

We will consider a dumbbell model in the dilute limit of polymer solutions and will neglect the polymer–polymer interaction. A dumbbell consists of two beads connected by a spring. For simplicity, we will assume that the two beads have the same mass and friction coefficient, denoted by m and ζ respectively. The forces acting on the dumbbells are as follows.

(1) The inertial force $m\ddot{\mathbf{r}}_i$, $i = 1, 2$, where the dot denotes the time derivative. As mentioned above, in general this is neglected.
(2) A frictional force $-\zeta(\dot{\mathbf{r}}_i - \mathbf{u}(\mathbf{r}_i))$, $i = 1, 2$.
(3) A spring force $\mathbf{F}_i = -\nabla_{\mathbf{r}_i}\Psi(\mathbf{r}_1, \mathbf{r}_2)$, $i = 1, 2$, where Ψ is the potential energy of the spring. The choice of the potential Ψ will be discussed later. In general, we should have $\mathbf{F}_1 + \mathbf{F}_2 = 0$.
(4) The Brownian force $\eta_i = \sigma\dot{\mathbf{w}}_i$, $i = 1, 2$, where $\dot{\mathbf{w}}_1$ and $\dot{\mathbf{w}}_2$ are independent white noises; $\sigma = \sqrt{2k_BT\zeta}$ by the fluctuation–dissipation theorem.

A force balance gives

$$\zeta(\dot{\mathbf{r}}_1 - \mathbf{u}(\mathbf{r}_1)) = \mathbf{F}_1 + \sigma\dot{\mathbf{w}}_1, \qquad (5.1.5)$$

$$\zeta(\dot{\mathbf{r}}_2 - \mathbf{u}(\mathbf{r}_2)) = \mathbf{F}_2 + \sigma\dot{\mathbf{w}}_2, \qquad (5.1.6)$$

where $\sigma = \sqrt{2k_BT\zeta}$. If we add and subtract the two equations, we obtain equations for the center of resistance $\mathbf{r}_c = \frac{1}{2}(\mathbf{r}_1 + \mathbf{r}_2)$ and the conformation vector $\mathbf{Q} = \mathbf{r}_2 - \mathbf{r}_1$:

$$\dot{\mathbf{r}}_c = \mathbf{u}(\mathbf{r}_c) + \sqrt{\frac{2k_BT}{\zeta}}\frac{\dot{\mathbf{w}}_1 + \dot{\mathbf{w}}_2}{2} \qquad (5.1.7)$$

$$\dot{\mathbf{Q}} = \nabla\mathbf{u}(\mathbf{r}_c)\mathbf{Q} - \frac{2}{\zeta}\mathbf{F}_1 + \sqrt{\frac{2k_BT}{\zeta}}(\dot{\mathbf{w}}_2 - \dot{\mathbf{w}}_1).$$

Here we have made the following approximations: $\frac{1}{2}(\mathbf{u}(\mathbf{r}_1) + \mathbf{u}(\mathbf{r}_2)) \sim \mathbf{u}(\mathbf{r}_c)$ and $\mathbf{u}(\mathbf{r}_2) - \mathbf{u}(\mathbf{r}_1) \sim \nabla\mathbf{u}(\mathbf{r}_c)\mathbf{Q}$. Since $\frac{1}{\sqrt{2}}(\mathbf{w}_2 - \mathbf{w}_1)$ and $\frac{1}{\sqrt{2}}(\mathbf{w}_2 + \mathbf{w}_1)$ are independent standard Wiener processes, we can rewrite the two equations above as

$$\dot{\mathbf{r}}_c = \mathbf{u}(\mathbf{r}_c) + \sqrt{\frac{k_BT}{\zeta}}\dot{\mathbf{w}}_c, \qquad (5.1.8)$$

$$\dot{\mathbf{Q}} = \nabla\mathbf{u}(\mathbf{r}_c)\mathbf{Q} - \frac{2}{\zeta}\mathbf{F}(\mathbf{Q}) + \sqrt{\frac{4k_BT}{\zeta}}\dot{\mathbf{w}}_a. \qquad (5.1.9)$$

where $\dot{\mathbf{w}}_c$ and $\dot{\mathbf{w}}_a$ are standard independent white noises.

We now turn to the potential of the spring force. The two most commonly used models are:

(1) *A Hookean spring* The spring force is $H\mathbf{Q}$ and the potential is given by $\frac{1}{2}H|\mathbf{Q}|^2$. Here H is the spring constant.

(2) *A finitely extensible nonlinear elastic (FENE) spring* The spring force is $H\mathbf{Q}/(1-(|\mathbf{Q}|/Q_0)^2)$ and the potential is given by $-\frac{1}{2}H|\mathbf{Q}|_0^2 \ln\left(1-\left(|\mathbf{Q}|/Q_0\right)^2\right)$. Here Q_0 is the maximal extension of the FENE spring.

The probability density function (pdf) $\psi(\mathbf{x},\mathbf{Q},t)$ of the dumbbells satisfies the Fokker–Planck equation, also known as the *Smoluchowski equation* [18, 28, 29],

$$\partial_t \psi + \nabla \cdot (\mathbf{u}\psi) + \nabla_{\mathbf{Q}} \cdot \left(\nabla\mathbf{u}\cdot\mathbf{Q} - \frac{2}{\zeta}\mathbf{F})\psi\right) = \frac{2k_B T}{\zeta}\Delta_{\mathbf{Q}}\psi + \frac{k_B T}{2\zeta}\Delta\psi.$$

Thus far we have discussed how to model the dynamics of the dumbbells in the solvent. Next, we will discuss how the dumbbells influence the dynamics of the solvent. To this end we need an expression for the polymer component of the stress, $\boldsymbol{\tau}_{\mathrm{p}}$. Imagine a plane immersed in the polymer fluid. There are two kinds of contributions that we need to consider [4]:

(1) the spring force due to the dumbbells that straddle the plane, which will be denoted as $\boldsymbol{\tau}_{\mathrm{p}}^{(c)}$;

(2) the momentum transfer caused by the crossing of the dumbbells through the plane, which will be denoted as $\boldsymbol{\tau}_{\mathrm{p}}^{(b)}$.

We first discuss $\boldsymbol{\tau}_{\mathrm{p}}^{(c)}$. Denote by n the number density of the polymer. Let us consider a cube of volume $1/n$; see Figure 5.2. Given a dumbbell with orientation $\mathbf{Q}$ in this region, the probability that it will cut the shaded plane is $(\mathbf{Q}\cdot\mathbf{n})/(1/n)^{1/3}$, where $\mathbf{n}$ is the unit normal vector of the plane. The force in the spring is $\mathbf{F}(\mathbf{Q})$. Averaging over $\mathbf{Q}$, we obtain

$$\int_{\mathbb{R}^3} \frac{\mathbf{Q}\cdot\mathbf{n}}{(1/n)^{1/3}}\mathbf{F}(\mathbf{Q})\psi(\mathbf{x},\mathbf{Q},t)\,d\mathbf{Q}.$$

The area of the plane within the cube is $n^{-2/3}$. Therefore the force per area is $n\langle(\mathbf{Q}\cdot\mathbf{n})\mathbf{F}(\mathbf{Q})\rangle = \mathbf{n}\cdot\boldsymbol{\tau}_{\mathrm{p}}^{(c)}$. Hence this part of the contribution to

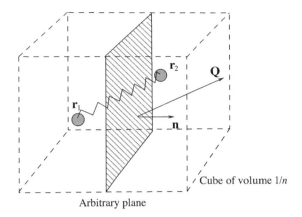

Figure 5.2. Illustration of the derivation of the formula for the stress $\boldsymbol{\tau}_{\mathrm{p}}^{(\mathrm{c})}$; the contribution from dumbbells that intersect with the control surface with orientation $\mathbf{Q}$ (courtesy of Tiejun Li).

the stress is

$$\boldsymbol{\tau}_{\mathrm{p}}^{(\mathrm{c})} = n\langle \mathbf{Q} \otimes \mathbf{F}(\mathbf{Q})\rangle, \qquad (5.1.10)$$

where $\langle \cdot \rangle$ denotes ensemble averaging in $\mathbf{Q}$ space.

Next, we turn to $\boldsymbol{\tau}_{\mathrm{p}}^{(\mathrm{b})}$. Consider a moving plane with velocity $\mathbf{u}$ (see Figure 5.3). Each dumbbell has two beads, which we label as 1 and 2 respectively. The number of 1-beads with velocity $\dot{\mathbf{r}}_1$ that cross the surface with area S per unit time will be $-n(\dot{\mathbf{r}}_1 - \mathbf{u}) \cdot \mathbf{n}S$. The amount of momentum transported as a result is

$$-n((\dot{\mathbf{r}}_1 - \mathbf{u}) \cdot \mathbf{n}S)m(\dot{\mathbf{r}}_1 - \mathbf{u}).$$

There is a similar expression for the 2-beads. The total momentum transported will be

$$-\iint \sum_{i=1}^{2} n((\dot{\mathbf{r}}_i - \mathbf{u}) \cdot \mathbf{n}S)m(\dot{\mathbf{r}}_i - \mathbf{u})\Psi(\dot{\mathbf{r}}_1, \dot{\mathbf{r}}_2)\, d\dot{\mathbf{r}}_1\, d\dot{\mathbf{r}}_2.$$

Assuming a Maxwellian distribution $\Psi(\dot{\mathbf{r}}_1, \dot{\mathbf{r}}_2)$ for the velocities of the beads, i.e.

$$\Psi(\dot{\mathbf{r}}_1, \dot{\mathbf{r}}_2) = \frac{1}{Z}\exp\left(-\frac{1}{2k_{\mathrm{B}}T}\left(m(\dot{\mathbf{r}}_1 - \mathbf{u})^2 + m(\dot{\mathbf{r}}_2 - \mathbf{u})^2\right)\right), \qquad (5.1.11)$$

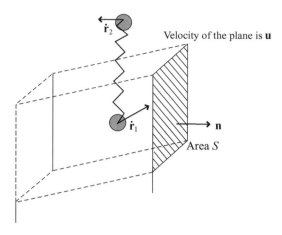

Figure 5.3. Illustration of the derivation of the formula for the stress $\tau_{\mathrm{p}}^{(b)}$ – the contribution from the transportation of momentum by the dumbbells (courtesy of Tiejun Li).

where $Z = \iint \exp[-(m(\dot{\mathbf{r}}_1 - \mathbf{u})^2 + m(\dot{\mathbf{r}}_2 - \mathbf{u})^2)/(2k_BT) \, d\dot{\mathbf{r}}_1]d\dot{\mathbf{r}}_2$, we can easily show that the mean momentum transported is equal to $-2nk_BT\mathbf{I} \cdot \mathbf{n}S = S\mathbf{n}\cdot\tau_{\mathrm{p}}^{(b)}$ and hence to $-2nk_BT\mathbf{I}$. Of this, $-nk_BT\mathbf{I}$ is the contribution to the pressure. The remaining part gives

$$\tau_{\mathrm{p}}^{(b)} = -nk_BT\mathbf{I}. \tag{5.1.12}$$

Combining (5.1.10) and (5.1.12) we obtain *Kramers' expression for the polymeric stress* in the dumbbell model:

$$\tau_{\mathrm{p}} = -nk_BT\mathbf{I} + n\langle\mathbf{F}(\mathbf{Q}) \otimes \mathbf{Q}\rangle. \tag{5.1.13}$$

It is standard practice to drop the term $-nk_BT\mathbf{I}$, since it can be lumped into the pressure.

To nondimensionalize the model we use

$$\mathbf{x} \to L_0\mathbf{x}, \quad \mathbf{Q} \to l_0\mathbf{Q}, \quad \mathbf{u} \to U_0\mathbf{u}, \quad t \to t, \quad p \to P_0p,$$

where L_0 and U_0 are typical length and velocity scales for the fluid flow and

$$l_0 = \sqrt{\frac{k_BT}{H}}, \quad T_0 = \frac{L_0}{U_0}, \quad P_0 = \rho_0U_0^2.$$

In addition, we define

$$De = \frac{T_r}{T_0}, \quad Re = \frac{\rho_0U_0L_0}{\eta}, \quad \varepsilon = \frac{l_0}{L_0}, \quad \gamma = \frac{\eta_s}{\eta},$$

where $T_r = \zeta/(4H)$ is the relaxation time scale of the spring, ρ_0 is the density and η the total viscosity of the fluid. The quantity De is the *Deborah number* and Re and γ are the Reynolds number and the viscosity ratio respectively. A standard nondimensionalization procedure gives

$$\partial_t \mathbf{u} + (\mathbf{u} \cdot \nabla)\mathbf{u} + \nabla p = \frac{\gamma}{Re}\Delta \mathbf{u} + \frac{1-\gamma}{ReDe}\nabla \cdot \boldsymbol{\tau}_{\mathrm{p}}, \tag{5.1.14}$$

$$\nabla \cdot \mathbf{u} = 0, \tag{5.1.15}$$

$$\boldsymbol{\tau}_{\mathrm{p}} = \langle \mathbf{F}(\mathbf{Q}) \otimes \mathbf{Q} \rangle, \tag{5.1.16}$$

$$\frac{d\mathbf{x}}{dt} = \mathbf{u}(\mathbf{x}) + \frac{\varepsilon}{2\sqrt{De}}\dot{\mathbf{W}}_1(t), \tag{5.1.17}$$

$$\frac{d\mathbf{Q}}{dt} = \nabla \mathbf{u} \cdot \mathbf{Q} - \frac{1}{2De}\mathbf{F}(\mathbf{Q}) + \frac{1}{\sqrt{De}}\dot{\mathbf{W}}_2(t). \tag{5.1.18}$$

Here the ensemble average in (5.1.16) is defined with respect to the realizations of the white noises in (5.1.17) and (5.1.18). To simulate this model numerically one needs to start with an ensemble of Brownian particles, whose dynamics are governed by (5.1.17) and (5.1.18). One obtains a set of quantities $\{(\mathbf{x}_j, \mathbf{Q}_j)\}$ for the ensemble. The polymer stress $\boldsymbol{\tau}_{\mathrm{p}}$ can be computed by averaging the quantity in (5.1.16). Since ε in general is very small, it is customary to neglect the Brownian force in (5.1.17). A more precise derivation of these equations can be found in [2].

The system (5.1.14)–(5.1.18) is a model for the dynamics of a dilute polymer solution. One can see that the polymer is modeled by noninteracting Brownian particles. Their influence on each other is felt only through the flow. In addition, particles initiated at the same location as one another will always stay together. Thus, instead of tracking the position and conformation $(\mathbf{x}, \mathbf{Q})$ of each particle, one can model the stochastic dynamics of the conformation by the dynamics of a conformation field $\mathbf{Q}(\mathbf{x}, t)$, where $\mathbf{Q}(\mathbf{x}, t)$ is the conformation of a Brownian particle at location $\mathbf{x}$ and time t [41]. Combining (5.1.17) and (5.1.18) gives [41]

$$\partial_t \mathbf{Q} + (\mathbf{u} \cdot \nabla)\mathbf{Q} = \nabla \mathbf{u} \cdot \mathbf{Q} - \frac{1}{2De}\mathbf{F}(\mathbf{Q}) + \frac{1}{\sqrt{De}}\dot{\mathbf{W}}(t). \tag{5.1.19}$$

Note that in this model the noise term is independent of the space variable $\mathbf{x}$. The corresponding Smoluchowski equation is

$$\partial_t \psi + \nabla \cdot (\mathbf{u}\psi) + \nabla_{\mathbf{Q}} \cdot \left[\left(\nabla \mathbf{u} \cdot \mathbf{Q} - \frac{1}{2De}\mathbf{F}(\mathbf{Q}) \right) \psi \right] = \frac{1}{2De} \Delta_{\mathbf{Q}} \psi. \quad (5.1.20)$$

Equation (5.1.20) can be regarded as the Fokker–Planck equation for Brownian configuration fields. It can also be regarded as the kinetic equation for Brownian particles in the $(\mathbf{x}, \mathbf{Q})$ space. A dependence on velocity or momentum variables is absent, since we have assumed that the velocity is Maxwellian (see (5.1.11)).

The model at which we have just arrived is an example of the coupled macro–micro models: the macroscale Navier–Stokes-like system (5.1.14), (5.1.15) for the velocity field is coupled with the microscale Brownian dynamics model (5.1.17), (5.1.18) for the conformation dynamics of the polymers.

5.2 Extensions of the Cauchy–Born rule

In Section 4.2 we discussed how one can derive continuum nonlinear elasticity models from atomistic models using the Cauchy–Born rule. The discussion was limited to bulk crystals whose elastic energy takes the form

$$E(\{\mathbf{u}\}) = \int_\Omega W(\nabla \mathbf{u}) \, d\mathbf{x}.$$

In this section, we will discuss the extension to situations when higher-order effects, such as the effects of strain gradients, are important:

$$E(\{\mathbf{u}\}) = \int_\Omega W(\nabla \mathbf{u}, \nabla^2 \mathbf{u}, \dots) \, d\mathbf{x}. \quad (5.2.1)$$

This is the case when the crystal is small, or at least one dimension of the crystal is small, as is the case for rods, sheets or thin films.

The essence of the Cauchy–Born rule is that, locally, one can approximate the displacement field by a homogeneous deformation, i.e. a linear approximation:

$$\mathbf{u}(\mathbf{x}) \approx \tilde{\mathbf{u}}_i(\mathbf{x}) = \mathbf{u}(\mathbf{x}_i) + (\nabla \mathbf{u})(\mathbf{x}_i)(\mathbf{x} - \mathbf{x}_i), \quad \tilde{\mathbf{y}}_j = \mathbf{x}_j + \tilde{\mathbf{u}}_i(\mathbf{x}_j), \quad (5.2.2)$$

for each $\mathbf{x}_i$, where $\mathbf{x}_j$ is close to $\mathbf{x}_i$, say, within the interaction range of $\mathbf{x}_i$. This expression is then used in the atomistic energy to give a continuum energy density that depends only on $\nabla\mathbf{u}$.

5.2.1 High-order, exponential and local Cauchy–Born rules

Equation (5.2.2) uses only the leading-order approximation in the Taylor expansion. A straightforward extension is to use a higher-order approximation:

$$\mathbf{u}(\mathbf{x}) \approx \tilde{\mathbf{u}}_i(\mathbf{x}) = \mathbf{u}(\mathbf{x}_i) + (\nabla\mathbf{u})(\mathbf{x}_i)(\mathbf{x} - \mathbf{x}_i) \qquad (5.2.3)$$
$$+ \tfrac{1}{2}(\nabla^2\mathbf{u})(\mathbf{x}_i) : (\mathbf{x} - \mathbf{x}_i) \otimes (\mathbf{x} - \mathbf{x}_i). \qquad (5.2.4)$$

If we proceed in the same way as before with this new approximation we obtain a stored energy density that depends on both $\nabla\mathbf{u}$ and $\nabla^2\mathbf{u}$ [7, 19]:

$$W = W_{\mathrm{HCB}}(\nabla\mathbf{u}(\mathbf{x}), \nabla^2\mathbf{u}(\mathbf{x})),$$

This is an example of the *high-order Cauchy–Born rule.*

When curvature effects are important, as is the case for rods, thin films and sheets, the Cauchy–Born rules should be modified accordingly. One approach is to use the exponential map which maps curved objects, such as surfaces, to their tangent spaces. The other approach is to make use of the local Taylor expansion. The resulting Cauchy–Born approximations are called the *exponential Cauchy–Born rule* [1] and the *local Cauchy–Born rule* [45] respectively. As before, their main ingredient is a kinematic approximation to the positions of the atoms in the neighborhood of a given atom. Instead of describing the abstract procedure, we will illustrate these ideas using some simple examples.

5.2.2 An example of a one-dimensional chain

Consider a one-dimensional chain of atoms on the xy-plane. Before deformation the atoms are on a straight line, with positions $(x_j, 0)^{\mathrm{T}}$, $x_j = ja$. To be specific, we will assume that the energy of the system takes the form

$$V = \sum_{i,j,k} V(\mathbf{y}_i, \mathbf{y}_j, \mathbf{y}_k) = \sum_i V_i, \qquad (5.2.5)$$

where

$$V_i = \sum_{j,k} V(\mathbf{y}_i, \mathbf{y}_j, \mathbf{y}_k),$$

and atoms j, k are the nearest or more distant neighbors of atom i. We will first discuss the exponential Cauchy–Born rule. Assume that the chain is deformed to a smooth curve in the plane. Assume further that the local deformation gradient (here the stretching or compression ratio) at x_i is F and the curvature after deformation is κ. In order to find the approximate position of the neighboring atoms after the chain is deformed, we will assume without loss of generality that the atom at x_i has zero displacement. We will consider the effects of stretching or compression and of curvature separately. On the one hand, after stretching or compression, the atom at $(x_j, 0)^{\mathrm{T}}$ is mapped to $(= x_i + (1 + F)(x_j - x_i), 0)^{\mathrm{T}}$. On the other hand, curvature alone will map $(x_j, 0)^{\mathrm{T}}$ to

$$\begin{pmatrix} x_i + \sin(\kappa(x_j - x_i))/\kappa \\ [1 - \cos(\kappa(x_j - x_i))]/\kappa \end{pmatrix}.$$

The composition of the two maps gives us:

$$\mathbf{y}_j = \begin{pmatrix} x_i + \sin((1 + F)\kappa(x_j - x_i))/\kappa \\ [1 - \cos((1 + F)\kappa(x_j - x_i))]/\kappa \end{pmatrix}. \tag{5.2.6}$$

Substituting into (5.2.5) we obtain the energy density as a function of F and κ:

$$W(F, \kappa)(x_i) = \frac{1}{a} \sum_{j,k} V(\mathbf{y}_i, \mathbf{y}_j, \mathbf{y}_k),$$

where $\mathbf{y}_i = (x_i, 0)^{\mathrm{T}}$ and $\mathbf{y}_j$ and $\mathbf{y}_k$ are given by (5.2.6).

Next we consider the local Cauchy–Born rule. We now specify the displacement by the vector field $\mathbf{u} = (u(x), v(x))^{\mathrm{T}}$. At x_i we approximate the deformed position of the neighboring atoms by

$$\mathbf{y}_j = (x_i, 0)^{\mathrm{T}} + \mathbf{u}(x_i) + \frac{d\mathbf{u}}{dx}(x_i)(x_j - x_i) + \frac{1}{2}\frac{d^2\mathbf{u}}{dx^2}(x_i)(x_j - x_i)^2. \tag{5.2.7}$$

Substituting into (5.2.5), we obtain the energy density as a function of $(d\mathbf{u}/dx)(x_i)$ and $(d^2\mathbf{u}/dx^2)(x_i)$:

$$W\left(\frac{d\mathbf{u}}{dx}(x_i), \frac{d^2\mathbf{u}}{dx^2}(x_i)\right) = \frac{1}{a} \sum_{j,k} V(\tilde{\mathbf{y}}_i, \tilde{\mathbf{y}}_j, \tilde{\mathbf{y}}_k),$$

where again $\mathbf{y}_i = (x_i, 0)^{\mathrm{T}}$ and $\mathbf{y}_j$ and $\mathbf{y}_k$ are given by (5.2.7).

We can relate these two results using the simple formulas

$$F = \left| \mathbf{e}_1 + \frac{d\mathbf{u}}{dx} \right| - 1,$$

$$\kappa = \frac{\left| \left(\mathbf{e}_1 + \dfrac{d\mathbf{u}}{dx} \right) \times \dfrac{d^2\mathbf{u}}{dx^2} \right|}{\left| \mathbf{e}_1 + \dfrac{d\mathbf{u}}{dx} \right|^3}$$

where $\mathbf{e}_1 = (1,0)^{\mathrm{T}}$.

5.2.3 Sheets and nanotubes

Sheets are two-dimensional surfaces. Examples of sheet-like structures include the graphene sheet and the carbon nanotube. These structures can sustain very large elastic deformations. To model such large deformations, we need an accurate model of nonlinear elasticity. Naturally, one way of acquiring such a model is to derive it from an accurate atomistic model.

For simplicity, we will assume that the reference configuration is a flat sheet that occupies the x_1x_2-plane. After deformation, atoms on the sheet are mapped as follows:

$$\begin{pmatrix} x_1 \\ x_2 \\ 0 \end{pmatrix} \longrightarrow \begin{pmatrix} y_1 \\ y_2 \\ y_3 \end{pmatrix} = \begin{pmatrix} x_1 \\ x_2 \\ 0 \end{pmatrix} + \begin{pmatrix} u_1(x_1, x_2) \\ u_2(x_1, x_2) \\ u_3(x_1, x_2) \end{pmatrix}.$$

For nanotubes the reference configuration should be a cylinder, in fact. However, this does not change much the discussion that follows.

As before we start with an accurate atomistic model, which for simplicity will be expressed as a three-body potential. We will look for a set of kinematic approximations that convert the atomistic model to a continuum model.

The exponential Cauchy–Born rule As before, to apply the exponential Cauchy–Born rule [1], we proceed in two steps:

(1) *Step 1* deforming the tangent plane;
(2) *Step 2* using the exponential map to take into account the curvature effect.

In practice, the exponential map is approximated by a combination of two consecutive deformations along the two principal curvature directions. The two principal curvatures, κ_1 and κ_2, can be calculated by diagonalizing the curvature tensor $\boldsymbol{\kappa}$. Without loss of generality, we can assume that $\boldsymbol{\kappa}$ takes the simple diagonal form $\boldsymbol{\kappa} = \mathrm{diag}\{\kappa_1, \kappa_2\}$. In this way, the kinematic approximation can be expressed using a three-step procedure.

(1) Deform the tangent plane using the two-dimensional deformation gradient tensor $\mathbf{F} = \nabla\mathbf{u}$ as in the standard Cauchy–Born rule:

$$\mathbf{w} = \begin{pmatrix} w_1 \\ w_2 \end{pmatrix} = (\mathbf{I} + \mathbf{F})\mathbf{x} = (\mathbf{I} + \mathbf{F}) \begin{pmatrix} x_1 \\ x_2 \end{pmatrix}.$$

(2) Roll the plane into a cylinder with curvature κ_1 along the first principal curvature direction to obtain the first curvature correction:

$$\Delta\mathbf{w}^1 = \begin{pmatrix} (\sin(\kappa_1 w_1))/\kappa_1 - w_1 \\ 0 \\ (1 - \cos(\kappa_1 w_1))/\kappa_1 \end{pmatrix}.$$

(3) Roll the plane into a cylinder with curvature κ_2 along the second principal curvature direction to obtain the second curvature correction:

$$\Delta\mathbf{w}^2 = \begin{pmatrix} 0 \\ (\sin(\kappa_2 w_2))/\kappa_2 - w_2 \\ (1 - \cos(\kappa_2 w_2))/\kappa_2 \end{pmatrix}.$$

The deformed position of $\mathbf{y}$ is then given by

$$\begin{aligned}
\mathbf{y} &= \mathbf{w} + \Delta\mathbf{w}^1 + \Delta\mathbf{w}^2 \\
&= \begin{pmatrix} w_1 \eta(\kappa_1 w_1) \\ w_2 \eta(\kappa_2 w_2) \\ \frac{1}{2}\kappa_1 w_1^2 \eta^2(\frac{1}{2}\kappa_1 w_1) + \frac{1}{2}\kappa_2 w_2^2 \eta^2(\frac{1}{2}\kappa_2 w_2) \end{pmatrix}
\end{aligned} \tag{5.2.8}$$

where $\eta(x) = (\sin x)/x$ and the two-dimensional vector $\mathbf{w} = (w_1, w_2)^{\mathrm{T}}$ is identified with the three-dimensional vector $\mathbf{w} = (w_1, w_2, 0)^{\mathrm{T}}$.

The exponential Cauchy–Born rule was used extensively by Arroyo and Belytschko to study the mechanical deformation of both single- and multi-walled nanotubes (see [1]). An example of their results is shown in Figure 5.4.

The local Cauchy–Born rule At each point $\mathbf{x}_i = (x_{i,1}, x_{i,2})^{\mathrm{T}}$ on the sheet we approximate the displaced position of neighboring atoms via a Taylor expansion:

$$\begin{pmatrix} y_{j,1} \\ y_{j,2} \\ y_{j,3} \end{pmatrix} = \begin{pmatrix} x_{j,1} \\ x_{j,2} \\ 0 \end{pmatrix} + \begin{pmatrix} u_1(x_{j,1}, x_{j,2}) \\ u_2(x_{j,1}, x_{j,2}) \\ u_3(x_{j,1}, x_{j,2}) \end{pmatrix} \approx \begin{pmatrix} \tilde{y}_{j,1} \\ \tilde{y}_{j,2} \\ \tilde{y}_{j,3} \end{pmatrix}$$

$$= \mathbf{x}_i + \mathbf{u}(x_{i,1}, x_{i,2}) + \frac{\partial \mathbf{u}(x_{i,1}, x_{i,2})}{\partial x_1}(x_{j,1} - x_{i,1})$$

$$+ \frac{\partial \mathbf{u}(x_{i,1}, x_{i,2})}{\partial x_2}(x_{j,2} - x_{i,2}) + \frac{1}{2}\frac{\partial^2 \mathbf{u}(x_{i,1}, x_{i,2})}{\partial x_1^2}(x_{j,1} - x_{i,1})^2$$

$$+ \frac{\partial^2 \mathbf{u}(x_{i,1}, x_{i,2})}{\partial x_1 \partial x_2}(x_{j,1} - x_{i,1})(x_{j,2} - x_{i,2})$$

$$+ \frac{1}{2}\frac{\partial^2 \mathbf{u}(x_{i,1}, x_{i,2})}{\partial x_2^2}(x_{j,2} - x_{i,2})^2, \qquad (5.2.9)$$

where $\mathbf{u} = (u_1, u_2, u_3)^{\mathrm{T}}$ is the displacement field and $\mathbf{x}_i = (x_{i,1}, x_{i,2})^{\mathrm{T}}$ is the reference position of the ith atom. Substituting this into the atomistic model, we obtain the stored energy density at $\mathbf{x}_i$:

$$W_{\mathrm{LCB}}(\nabla \mathbf{u}, \nabla^2 \mathbf{u}) = \frac{1}{D_0}\sum_{j,k} V(\mathbf{y}_i, \mathbf{y}_j, \mathbf{y}_k),$$

where D_0 is the area of a unit cell and $\mathbf{y}_i = (x_{i,1}, x_{i,2}, 0)^{\mathrm{T}}$.

Note that this procedure can be used for both simple and complex lattices.

As an example, let us consider a graphene sheet, which is a two-dimensional complex lattice obtained from the union of two triangular lattices. In equilibrium, one set of basis vectors for the triangular lattice is given by

$$\boldsymbol{A}_1 = a(1, 0), \quad \boldsymbol{A}_2 = a\left(\frac{1}{2}, \frac{\sqrt{3}}{2}\right),$$

where a is the lattice constant, and the shift vector is

$$\mathbf{p} = a\left(\frac{1}{2}, \frac{\sqrt{3}}{6}\right).$$

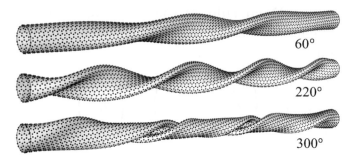

Figure 5.4. For comparison, the results of the MD simulation and those of the continuum model for a twisted nanotube using the exponential Cauchy–Born rule. The dots represent the results of the MD simulation. The gray background is computed from the exponential Cauchy–Born rule ([1], courtesy of Ted Belytschko).

This example is treated in [1] and [45] using the Tersoff potential as the atomistic model. Figure 5.4 shows results for a twisted nanotube, modeled using the atomistic model as well as the continuum model obtained using the exponential Cauchy–Born rule. Figure 5.5 shows results for the nanotube under compression, modeled using the atomistic model and the continuum model obtained using the local Cauchy–Born rule. Very good agreement is found in both cases. This suggests that mechanical deformation of nanotubes can be described well by continuum theories in this regime. The value of such a viewpoint has also been demonstrated using considerably simplified continuum models [44].

5.3 The moving contact line problem

When drops of water or of another liquid are placed on a solid surface, a contact line forms at the intersection between the liquid, the air and the solid. More generally, the same situation arises when one is considering the dynamics of immiscible fluids in a channel [12, 14], as in the case of secondary oil recovery when oil is drained out of the pores by some other liquid. Like the problem of crack propagation, contact line dynamics has often been used as an example to illustrate the need for formulating coupled continuum–atomistic models, since it has been realized for some time that classical continuum models are inadequate near the contact line and atomistic models should give a more accurate representation [21, 34]. While developing coupled atomistic–continuum models is certainly a very important avenue of research, in this section we will illustrate that

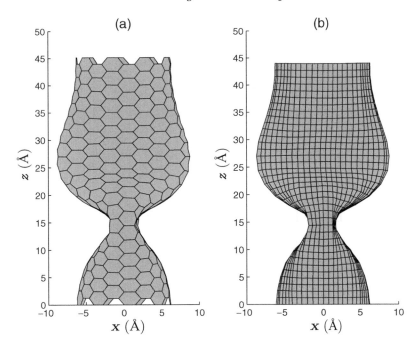

Figure 5.5. The results of the MD simulation (left) and the continuum model (right) for compressed nanotube using the local Cauchy–Born rule (from Yang and E [45]).

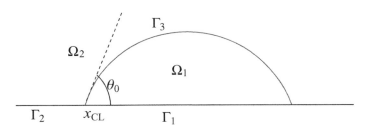

Figure 5.6. Two fluid phases, Ω_1 and Ω_2, on a substrate. The point x_{CL} represents the contact line and Γ_i $(i = 1, 2, 3)$ are the interfaces; θ_0 is the contact angle. (Courtesy of Weiqing Ren.)

multiscale ideas can also be used in a different way, to help informulating better continuum models.

5.3.1 Classical continuum theory

The setup is shown in Figure 5.6. The two fluid phases and the substrate are separated by three interfaces, Γ_1, Γ_2 and Γ_3. We denote the surface tension coefficient of Γ_i by γ_i, $i = 1, 2, 3$.

The equilibrium problem and the static configuration of the contact line were understood a long time ago through the work of Laplace, Young and Gauss [11]. At equilibrium the total energy of the system is the sum of the interfacial energies:

$$E = \sum_{i=1,2,3} \gamma_i \, |\Gamma_i| \, .$$

We will only consider the case of *partial wetting*, i.e.

$$|\gamma_1 - \gamma_2| < \gamma_3.$$

In this case, the Euler–Lagrange equation for the functional E is given by

$$\gamma_1 - \gamma_2 + \gamma_3 \cos \theta_0 = 0. \qquad (5.3.1)$$

This is *Young's relation* for the case of partial wetting, and it defines the *static contact angle* θ_0.

Now let us turn to the dynamic problem. To be specific, let us consider the situation of two immiscible fluids in a channel separated by an interface. To model the dynamics of such a system, the most obvious approach is to use the Navier–Stokes equation in the fluid phases:

$$\rho_i \left(\partial_t \mathbf{u} + \mathbf{u} \cdot \nabla \mathbf{u} \right) = -\nabla p + \mu_i \Delta \mathbf{u}, \quad \text{in } \Omega_i,$$
$$\nabla \cdot \mathbf{u} = 0, \qquad (5.3.2)$$

together with the no-slip boundary condition on the solid surface:

$$\mathbf{u} = \mathbf{u}_{\mathrm{w}}.$$

Here ρ_i and $\mu_i, i = 1, 2$, are the density and viscosity of the two fluid phases respectively and $\mathbf{u}_{\mathrm{w}}$ is the velocity of the wall. At the fluid–fluid interface we use the standard Laplace–Young interfacial conditions:

$$\gamma_3 \kappa = \mathbf{n} \cdot [\boldsymbol{\tau}] \cdot \mathbf{n}, \quad \mathbf{t} \cdot [\boldsymbol{\tau}] \cdot \mathbf{n} = \mathbf{0}, \qquad (5.3.3)$$

where $\mathbf{t}$ and $\mathbf{n}$ denote the unit tangent and normal vectors to the interface respectively, $\boldsymbol{\tau}$ denotes the total stress (the viscous stress and the pressure) in the fluids, κ is the mean curvature of the fluid–fluid interface and $[\cdot]$ denotes the jump of the quantity inside the bracket across the interface.

It was discovered long ago that in the case of partial wetting this model predicts a singularity at the contact line, known as the *Huh–Scriven singularity* [22]. This is most clearly seen in two dimensions. Let (r, ϕ) be the polar coordinates centered at the contact line. Then a simple Taylor expansion at the contact line gives, for $r \ll 1$,

$$\psi \sim r\left((C\phi + D)\cos\phi + (E\phi + F)\sin\phi\right)$$

for the stream function ψ, where C, D, E, F are constants. The velocity is given in terms of the stream function via $\mathbf{u} = (-\partial_y\psi, \partial_x\psi)$. Therefore we obtain

$$\nabla\mathbf{u} \sim \frac{1}{r}, \qquad \int |\nabla\mathbf{u}|^2\, d\mathbf{x} = +\infty.$$

For this reason, questions have been raised on the validity of the conventional hydrodynamics formalism, i.e. the Navier–Stokes equation with a no-slip boundary condition, for this problem. Many proposals have been made with the aim of removing this singularity (for a review, see [34]). While they have all succeeded in doing so, it is not clear whether any of them has a solid physical foundation.

Detailed molecular dynamics studies have revealed the fact that near the contact line there is a *slip region*, in which the no-slip boundary condition is violated [24, 38]. The existence of such a slip region is not surprising. In fact in a Couette flow setting (i.e. channel flow driven by walls moving in opposite directions), if the system reaches a steady state then the contact line has to experience complete slip, i.e. the relative slip velocity between the fluids and the solid surface is equal to the velocity of the solid surface. What is more important is that these molecular dynamics studies have opened up the door to a detailed analysis of what is going on near the contact line, at the atomistic level.

Even though molecular dynamics is able to offer much-needed details about the contact line problem, it is not very attractive as a theoretical tool compared with the continuum models. It is much less productive to think about the behavior of the individual molecules than that of the continuum fields. Therefore, from a theoretical viewpoint, it is still of interest to develop better continuum models. We will demonstrate that multiscale ideas are indeed quite useful in this respect.

5.3.2 Improved continuum models

The strategy that we will discuss consists of two steps. The first step is to find the simplest form of the constitutive relations, including the boundary conditions, from thermodynamic considerations. The second step is to measure, using MD, the detailed functional dependence of the constitutive relations using molecular dynamics. Such a strategy was pursued in [36].

Three constitutive relations needed:

(1) *The constitutive relation in the bulk*, which describes the response of the fluid in the bulk. It has been argued that, even for simple fluids, the linear constitutive relation, which is usually quite accurate for bulk fluids, may not be accurate enough near the contact line owing to the presence of large velocity gradients there [38, 40]. The careful molecular dynamics studies reported in [31] do not support this argument, however. Instead, it was found that, at least for simple fluids, a linear constitutive relation holds quite well right around the contact line. In other words, even though the velocity gradient is rather large, it is not large enough to induce significant nonlinear effects in the bulk response. Therefore we will assume for simplicity that linear constitutive relations are sufficient in the bulk.

(2) *The constitutive relation at the fluid–solid interface*. It has been observed that the standard no-slip boundary condition no longer holds in a region near the contact line [24, 25, 31, 34, 38, 40]. This region is the slip region. The linear friction law at the surface gives rise to the Navier boundary condition. However, nonlinear effects might be important there.

(3) *The constitutive relation at the contact line*, which determines the relationship between the contact angle and the contact line velocity. An obvious possibility is to set the dynamic contact angle to be the same as the equilibrium contact angle. However, this effectively suppresses energy dissipation at the contact line; dissipative processes at the contact line can be quite important [5]. Therefore, in general one has to allow the deviation of the dynamic contact angle from its equilibrium value.

To derive the needed constitutive relation, we will follow the principles discussed in Chapter 4 and ask the following question: what are the simplest forms of boundary conditions that are consistent with the second law of thermodynamics?

For simplicity, we will focus on the case of two-dimensional flows. The total (free) energy of the system is given by

$$E = \sum_{i=1,2} \frac{1}{2} \int_{\Omega_i} \rho_i |\mathbf{u}|^2 \, d\mathbf{x} + \sum_{i=1,2,3} \gamma_i |\Gamma_i|. \tag{5.3.4}$$

A straightforward calculation gives

$$\frac{dE}{dt} = -\sum_{i=1,2} \int_{\Omega_i} \eta_i |\nabla \mathbf{u}|^2 \, d\mathbf{x} + \sum_{i=1,2} \int_{\Gamma_i} \mathbf{u} \cdot \boldsymbol{\tau} \mathbf{n} \, dS + \gamma_3 \left(\cos \theta_{\mathrm{d}} - \cos \theta_0 \right) u_{\mathrm{CL}},$$

$$\tag{5.3.5}$$

where θ_{d} is the dynamic contact angle, θ_0 is the static contact angle defined in (5.3.1), u_{CL} is the velocity of the contact line and $\mathbf{n}$ is the unit normal vector of the solid surface.

Let u_{s} and $\boldsymbol{\tau}_{\mathrm{s}}$ be the slip velocity and the shear stress respectively:

$$u_{\mathrm{s}} = (\mathbf{u} - \mathbf{u}_{\mathrm{w}}) \cdot \mathbf{t}, \qquad \tau_{\mathrm{s}} = \mathbf{t} \cdot \boldsymbol{\tau} \mathbf{n}. \tag{5.3.6}$$

Let

$$\tau_{\mathrm{Y}} = \gamma_3 \left(\cos \theta_{\mathrm{d}} - \cos \theta_0 \right). \tag{5.3.7}$$

Here, τ_{Y} is the so-called *unbalanced* or *uncompensated Young's stress* [5, 12, 32, 34]. It represents the force on the contact line that arises from the deviation of the fluid interface from its equilibrium configuration. With this notation, we can rewrite (5.3.5) as

$$\frac{dE}{dt} = -\sum_{i=1,2} \int_{\Omega_i} \eta_i |\nabla \mathbf{u}|^2 \, d\mathbf{x} + \sum_{i=1,2} \int_{\Gamma_i} u_{\mathrm{s}} \tau_{\mathrm{s}} \, dS + u_{\mathrm{CL}} \tau_{\mathrm{Y}}.$$

According to the second law of thermodynamics, the rate of energy dissipation dE/dt is necessarily nonpositive for arbitrary realizations of $\mathbf{u}$. It follows easily that each of the three contributions has to be nonpositive. We now examine the implication of this basic requirement and the constraints it places on the form of the constitutive relations.

First, we look at the fluid–solid interfaces Γ_1 and Γ_2. There we must have

$$\mathbf{u} \cdot \mathbf{n} = 0. \tag{5.3.8}$$

This implies that the shear stress should depend only on u_{s}. Assuming that this dependence is local, then τ_{s} is a function of the local values of

u_s and its derivatives. To the lowest order, we simply have:

$$\tau_s = f(u_s). \tag{5.3.9}$$

Since $\int_{\Gamma_i} u_s \tau_s \, dS \leqslant 0$ $(i = 1, 2)$ for an arbitrary realization of $\mathbf{u}$, the function f must satisfy $wf(w) \leq 0$.

Next, we turn to the contact line. From thermodynamic considerations, τ_Y is necessarily a functional of the contact line velocity u_{CL}. The simplest assumption is that it is a local function of u_{CL}:

$$\tau_Y = f_{CL}(u_{CL}). \tag{5.3.10}$$

The function f_{CL} has to satisfy $uf_{CL}(u) \leq 0$.

The Navier–Stokes equation together with the boundary conditions (5.3.3), (5.3.8), (5.3.9) and (5.3.10) constitute our continuum model for the moving contact line problem. The constitutive relations in (5.3.9) and (5.3.10) are not yet fully specified; the exact functional dependence of f and f_{CL} has to be obtained through other means. For this continuum model, the energy dissipation relation has a very clean form:

$$\frac{dE}{dt} = -\sum_{i=1,2} \int_{\Omega_i} \eta_i \left| \nabla \mathbf{u} \right|^2 dx + \sum_{i=1,2} \int_{\Gamma_i} u_s f(u_s) \, dS + u_{CL} f_{CL}(u_{CL}).$$
$$\tag{5.3.11}$$

The three terms on the right-hand side represent the energies dissipated in the bulk fluid, at the solid surface and at the contact line respectively.

So far, we have followed the standard procedure in generalized thermodynamics or nonequilibrium thermodynamics [13, 30]. At this point we have two options. One is to continue following the standard procedure and write down linear relations between the generalized fluxes (here u_s and u_{CL}) and the generalized forces (here τ_s and τ_Y). The other is to use microscopic models to measure f and f_{CL} directly.

If we continue to follow the standard procedure in non-equilibrium thermodynamics, then the next step is to propose linear constitutive relations of the form (see Section 4.1)

$$\tau_s = -\beta_i u_s, \quad i = 1, 2, \tag{5.3.12}$$

and

$$\gamma_3 \left(\cos \theta_d - \cos \theta_0 \right) = -\beta_{CL} u_{CL}; \tag{5.3.13}$$

here $\beta_i, i = 1, 2$, and β_{CL} are the friction coefficients at the solid surface for the two fluid phases and for the contact line respectively. Equation (5.3.12) is the well-known *Navier boundary condition*. Putting things together, we obtain the following continuum model for the moving contact line problem.

(1) In the bulk we have

$$
\begin{cases}
\rho_i(\partial_t \mathbf{u} + (\mathbf{u} \cdot \nabla)\mathbf{u}) = -\nabla p + \mu_i \nabla^2 \mathbf{u} + \mathbf{f}, \\
\nabla \cdot \mathbf{u} \qquad\qquad = 0.
\end{cases}
\tag{5.3.14}
$$

(2) The fluid interface Γ is advected by the velocity field, and

$$
\begin{aligned}
\mathbf{n} \cdot [\boldsymbol{\tau}]\,\mathbf{n} &= \gamma_3 \kappa, \\
\mathbf{t} \cdot [\boldsymbol{\tau}]\,\mathbf{n} &= 0,
\end{aligned}
\tag{5.3.15}
$$

where $\boldsymbol{\tau} = -p\mathbf{I} + \mu(\nabla\mathbf{u} + \nabla\mathbf{u}^{\mathrm{T}})$ is the stress tensor and κ is the mean curvature of the interface.

(3) On the solid surface, we have

$$
\beta_i u_{\mathrm{s}} = \mathbf{t} \cdot \boldsymbol{\tau}\mathbf{n}, \quad \mathbf{u} \cdot \mathbf{n} = 0.
\tag{5.3.16}
$$

(4) At the contact line, we have

$$
\beta_{\mathrm{CL}} u_{\mathrm{CL}} = -\gamma_3 \left(\cos\theta_{\mathrm{d}} - \cos\theta_0\right).
\tag{5.3.17}
$$

This is the model put forward by Ren and E [34].

Some remarks about the physical significance of the parameter β_{CL} are in order.

(1) It is a three-phase friction coefficient.
(2) It has the dimension of viscosity. This should be contrasted with β_i, which has the dimension of viscosity/length; and this length has to be a microscopic length. Consequently, the Navier slip boundary condition is only significant in a slip region of microscopic size. In comparison, β_{CL} should be of macroscopic significance.

One can write (5.3.17) as

$$
\cos\theta_{\mathrm{d}} - \cos\theta_0 = -(\beta_{\mathrm{CL}}/\gamma_3) u_{\mathrm{CL}}.
\tag{5.3.18}
$$

This can be viewed as a kinetic correction to the condition

$$\theta_{\mathrm{d}} = \theta_0. \tag{5.3.19}$$

Equation (5.3.19) states that the dynamic contact angle is equal to the equilibrium contact angle. This is the analog of the *Gibbs–Thomson condition* in the theory of crystal growth [20]. In this regard, (5.3.18) is the analog of the corrected Gibbs–Thomson relation, i.e. with kinetic effects taken into account [20].

It is interesting to note that if the condition (5.3.19) is used together with the Navier boundary condition, then the Huh–Scriven singularity is removed but there is still a weak singularity [37]. Naturally we would expect that if we used (5.3.18) instead of (5.3.19) then this weak singularity would also be removed. This is still an open question.

Since any function is approximately linear in a small enough interval, these linear constitutive relations will be adequate in appropriate regimes. What is important is how big these regimes are and what happens outside them. To answer these questions, we resort to microscopic models.

5.3.3 Measuring the boundary conditions using molecular dynamics

Thermodynamic considerations tell us what the boundary conditions should depend on but not the detailed functional dependence. This latter information should be obtained from laboratory experiments, or, as we explain now, numerical experiments using microscopic simulations such as molecular dynamics.

The results of the previous section tell us that f and f_{CL} depend on the slip velocity. Therefore, we need to set up an MD system such that, at the statistical steady state, it has a constant slip velocity. This can be done using a MD system in the Couette flow geometry, shown in Figure 5.7, where the fluid particles are confined between two solid walls, and the walls move in opposite directions at a constant speed U relative to the x-axis [24, 31, 34, 38, 39]. A standard way of modeling immiscible fluids is to use a modified form of the Lennard-Jones (LJ) potential:

$$V^{\mathrm{LJ}}(r) = 4\varepsilon \left(\left(\frac{\sigma}{r} \right)^{12} - \zeta \left(\frac{\sigma}{r} \right)^6 \right)$$

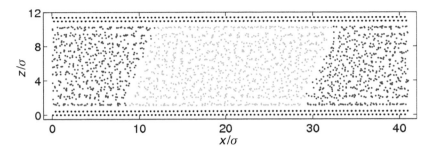

Figure 5.7. An instantaneous configuration of a two-phase fluid system obtained from a molecular dynamics simulation. The two regular lines of dots represent the particles of the wall. The system is driven by the motion of the two confining walls in opposite directions along the x-axis.

and take $\zeta = 1$ for particles of the same species and $\zeta = -1$ for particles of different species. Different values of ε can be used to model the fluid–fluid and fluid–solid interaction strengths. The simplest way to model the solid walls is to use a few layers of atoms in a close-packed structure, i.e. the face-centered cubic lattice in the [111] direction, with some number density ρ_w. For the purposes of illustration, we will consider only the particular case when the two fluid phases each have the same density and the same interaction strength with the solid wall. In this case, we have $\gamma_1 = \gamma_2$ and $\theta_0 = 90°$.

Standard techniques of molecular dynamics can be used to bring the system to a statistical steady state. The friction force along the wall between the fluid and solid particles can be computed at different values of U. A typical profile of the friction force along the wall is shown in Figure 5.8. Notice that the friction force drops sharply in the fluid–fluid interfacial region. This region is defined as the contact line region. The average friction force in this region is recorded as the friction force at the contact line and is plotted as a function of U in Figure 5.8. Here ε_{wf} denotes the value of ε in the LJ potential for the interaction between the fluid and the solid particles of the wall. The corresponding values for the fluid–fluid and solid–solid interaction are equal and are denoted by ε.

From these results one can see that the friction law is linear at small values of U and becomes nonlinear at large values. The friction force reaches a maximum at a certain critical contact-line speed (relative to the wall), then decreases as the speed increases further. The friction force

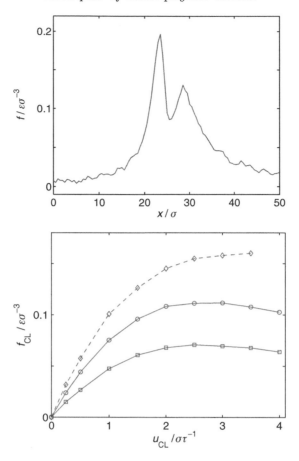

Figure 5.8. Upper panel: a typical profile of the friction force along the solid wall. Lower panel: the friction force in the contact line region versus the slip velocity at the contact line for different fluid and solid densities as well as different interaction parameters: $\rho_{\mathrm{w}} = 8\rho_{\mathrm{f}}, 12\rho_{\mathrm{f}}$ for the broken curve and solid curves respectively; $\varepsilon_{\mathrm{wf}} = 0.2\varepsilon$ (diamonds), $\varepsilon_{\mathrm{wf}} = 0.3\varepsilon$ (squares) and $\varepsilon_{\mathrm{wf}} = 0.5\varepsilon$ (circles). Here ρ_{w} and ρ_{f} are the particle densities in the wall and in the liquid respectively.

is smaller for systems with larger solid number densities. We also see that the critical contact-line speed, at which the friction force starts to decrease, also becomes smaller for systems with larger solid densities.

The combination of macroscopic thermodynamic considerations and molecular dynamics studies allows us to establish a fairly complete and fairly solid continuum model for this problem. However, it should also be noted that the accuracy of such a procedure relies on the accuracy of the microscopic model. From a quantitative viewpoint, we have considered a rather simplistic situation here. For systems of practical interest it is usually nontrivial to come up with accurate microscopic models.

5.4 Notes

In this chapter we have discussed how multiscale ideas can be used to develop better models, not just numerical algorithms. The material discussed in the section on polymer fluids is quite standard and can be found in [4], for example. Our presentation follows that of the review article by Li and Zhang [26]. We have only touched upon the situation of dilute polymer solutions. For a more general introduction to polymer structure and dynamics we refer to [3, 4, 10, 29].

Clearly the philosophy of coupling the macroscopic model with some description of the microscopic constituents can be applied to a large variety of problems. We have already mentioned the example of Car–Parrinello molecular dynamics in which the dynamics of the nuclei is coupled with a description of the electronic structure. Another interesting example is the analysis of material failure by the coupling of a macroscopic model with a description of the dynamics of the microdamages [43].

Developing continuum elasticity models from atomistic models has also been a subject of intensive research. We refer to [1, 7, 17] for more discussions. Using continuum concepts to study the mechanics of nanotubes has also attracted a great deal of attention; see for example [44].

As remarked earlier, the moving contact line problem has often been used to illustrate the need to develop coupled atomistic–continuum simulation strategies [21, 27, 33]. The viewpoint adopted here is different: instead of coupling continuum and atomistic models on the fly, we demonstrated a sequential coupling strategy in which the needed boundary conditions are obtained beforehand using molecular dynamics. This was possible since we limited the form of the boundary conditions to some particular class of functions that depend on very few variables (the friction force depends only on the local slip velocity). The advantage of this approach is that it produces a complete continuum model with which we can do analysis, not just simulation. It also reveals some physically interesting quantities such as the three-phase friction coefficient. The disadvantage is that it is limited to the special forms of constitutive relations we considered. Therefore in some situations it may not be adequate. This is an example of the top-down approach in the spirit of the heterogeneous multiscale method (HMM), to be discussed in more detail in the next chapter.

An alternative is the bottom-up approach: one performs extensive MD simulations and tries to extract the effective continuum model by mining the MD data. In the context of the moving contact line problem, very careful work on obtaining the boundary condition at the contact line region from MD data was reported in [32]. Note that the generality of this approach is also limited by the setup of the MD. Since the MD reported in [32] also used a Couette flow geometry, their boundary conditions were limited to the form considered here.

Finally, a word about the Huh–Scriven singularity. Although the slip boundary condition helps to remove this singularity, it should be noted that the no-slip boundary condition is recovered in the effective macroscopic models at length scales comparable with the system size [35]. Therefore, at the macroscopic scale the Huh–Scriven singularity seems to be unavoidable.

References for Chapter 5

[1] M. Arroyo and T. Belytschko, "Finite element methods for the nonlinear mechanics of crystalline sheets and nanotubes," *Int. J. Num. Meth. Engrg.*, vol. 59, pp. 419–256, 2004.

[2] J.W. Barrat and E. Süli, "Existence of global weak solutions to some regularized kinetic models for dilute polymers," *Multiscale Model. Simul.*, vol. 6, no. 2, 506–546, 2007.

[3] R.B. Bird, R.C. Armstrong and O. Hassager, *Dynamics of Polymeric Liquids, Vol. 1: Fluid Mechanics*, John Wiley, 1987.

[4] R.B. Bird, C.F. Curtiss, R.C. Armstrong and O. Hassager, *Dynamics of Polymeric Liquids, Vol. 2: Kinetic Theory*, John Wiley, 1987.

[5] T.D. Blake, "Dynamic contact angles and wetting kinetics," in *Wettability*, Marcel Dekker, 1993.

[6] T.D. Blake and J.M. Haynes, "Kinetics of liquid/liquid displacement," *J. Colloid Interface Sci.*, vol. 30, pp. 421–423, 1969.

[7] X. Blanc, C. Le Bris and P.-L. Lions, "From molecular models to continuum mechanics," *Arch. Rat. Mech. Anal.*, vol. 164, pp. 341–381, 2002.

[8] D. Bonn, J. Eggers, J. Indekeu, J. Meunier and E. Rolley, "Wetting and spreading," *Rev. Mod. Phys.*, vol. 81, pp. 739–805, 2009.

[9] R. Car and M. Parrinello, "Unified approach for molecular dynamics and density-functional theory," *Phys. Rev. Lett.*, vol. 55, pp. 2471–2474, 1985.

[10] M. Doi and S.F. Edwards, *The Theory of Polymer Dynamics*, Oxford University Press, 1986.

[11] P.-G. de Gennes, "Wetting: statics and dynamics," *Rev. Mod. Phys.*, vol. 57, pp. 827–863, 1985.

[12] P.-G. de Gennes, F. Brochard-Wyart, and D. Quere, *Capillarity and Wetting Phenomena: Drops, Bubbles, Pearls, Waves*, Springer-Verlag, 2003.

[13] S.R. de Groot and P. Mazur, *Non-Equilibrium Thermodynamics*, Dover Publications, 1984.

[14] V.E. Dussan, "On the spreading of liquids on solid surfaces: static and dynamic contact lines," *Ann. Rev. Fluid Mech.*, vol. 11, pp. 371–400, 1979.

[15] W. E and P.B. Ming, "Cauchy–Born rule and the stability of the crystalline solids: static problems," *Arch. Rat. Mech. Anal.*, vol. 183, pp. 241–297, 2007.

[16] W. E and P.B. Ming, "Cauchy–Born rule and the stability of the crystalline solids: dynamic problems," *Acta Math. Appl. Sin.*, Engl. Ser., vol. 23, pp. 529–550, 2007.

[17] G. Friesecke and R.D. James, "A scheme for the passage from atomic to continuum theory for thin films, nanotubes and nanorods," *J. Mech. Phys. Solids*, vol. 48, pp. 1519–1540, 2000.

[18] C.W. Gardiner, *Handbook of Stochastic Methods*, 2nd edition, Springer-Verlag, 1997.

[19] X. Guo, J.B. Wang and H.W. Zhang, "Mechanical properties of single-walled carbon nanotubes based on higher order Cauchy–Born rule," *Int. J. Solids Struct.*, vol. 43, pp. 1276–1290, 2006.

[20] M.E. Gurtin, *Thermodynamics of Evolving Phase Boundaries in the Plane*, Oxford University Press, 1993.

[21] N. Hadjiconstantinou, "Hybrid atomistic-continuum formulations and the moving contact-line problems," *J. Comp. Phys*, vol. 154, pp. 245–265, 1999.

[22] C. Huh and L.E. Scriven, "Hydrodynamic model of steady movement of a solid/liquid/fluid contact line," *J. Colloid Interface Sci.*, vol. 35, pp. 85–101, 1971.

[23] D. Jacqmin, "Contact-line dynamics of a diffusive fluid interface", *J. Fluid. Mech.*, vol. 402, pp. 57–88, 2000.

[24] J. Koplik, J.R. Banavar and J.F. Willemsen, "Molecular dynamics of Poiseuille flow and moving contact lines," *Phys. Rev. Lett.*, vol. 60, pp. 1282–1285, 1988.

[25] J. Koplik, J.R. Banavar and J.F. Willemsen, "Molecular dynamics of fluid flow at solid surfaces," *Phys. Fluids A*, vol. 1, pp. 781–794, 1989.

[26] T. Li and P.W. Zhang, "Mathematical analysis of multi-scale models of complex fluids," *Comm. Math. Sci.*, vol. 5, no.1, pp. 1–51, 2007.

[27] S.T. O'Connell and P.A. Thompson, "Molecular dynamics–continuum hybrid computations: a tool for studying complex fluid flows," *Phys. Rev. E*, vol. 52, no. 6, pp. R5792–R5795, 1995.

[28] B. Oksendal, *Stochastic Differential Equations*, 4th edition, Springer-Verlag, 1995.

[29] H.C. Öttinger, *Stochastic Processes in Polymeric Liquids*, Springer-Verlag, 1996.

[30] H.C. Öttinger, *Beyond Equilibrium Thermodynamics*, Wiley Interscience, 2005.

[31] T. Qian, X.P. Wang and P. Sheng, "Molecular scale contact line hydrodynamics of immiscible flows," *Phys. Rev. E*, vol. 68, pp. 016 306-1–016 306-15, 2003.

[32] T. Qian, X.P. Wang and P. Sheng, "Molecular hydrodynamics of the moving contact line in two-phase immiscible flows," *Commun. Comput. Phys.*, vol. 1, no. 1, pp. 1–52, 2006.

[33] W. Ren and W. E, "Heterogeneous multiscale method for the modeling of complex fluids and micro-fluidics," *J. Comp. Phys.*, vol. 204, pp. 1–26, 2005.

[34] W. Ren and W. E, "Boundary conditions for moving contact line problem," *Phys. Fluids*, vol. 19, pp. 022 101–022 101-15, 2007.

[35] W. Ren and W. E, "Hysteresis and effective models for the contact line problem on heterogeneous surfaces," preprint, 2010.

[36] W. Ren, D. Hu and W. E, "Continuum models for the contact line problem," *Phys. Fluids*, vol. 22, pp. 102 103–102 121, 2010.

[37] B. Schweizer, "A well-posed model for dynamic contact angles," *Nonlinear Anal. Theory Methods Appl.*, vol. 43, pp. 109–125, 2001.

[38] P.A. Thompson and M.O. Robbins, "Simulations of contact-line motion: slip and the dynamic contact angle," *Phys. Rev. Lett.*, vol. 63, pp. 766–769, 1989.

[39] P.A. Thompson and S.M. Troian, "A general boundary condition for liquid flow at solid surfaces," *Nature*, vol. 389, pp. 360–362, 1997.

[40] P.A. Thompson, W.B. Brinckerhoff and M.O. Robbins, "Microscopic studies of static and dynamics contact angles," *J. Adhes. Sci. Technol.*, vol. 7, pp. 535–554, 1993.

[41] B.H. van den Brule, A.P. van Heel and M.A. Hulsen, "Brownian configuration fields: a new method for simulating viscoelastic flow," *Macromol. Symp.*, vol. 121, pp. 205–217, 1997.

[42] G. Voth, ed., *Coarse-Graining of Condensed Phase and Biomolecular Systems*, CRC Press, 2009.

[43] H. Wang, Y. Bai, M. Xia and F. Ke, "Microdamage evolution, energy dissipation and their trans-scale effects on macroscopic failure," *Mech. Mater.*, vol. 38, pp. 57–67, 2006.

[44] B.I. Yakobson, C.J. Brabec, and J. Bernholc, "Nanomechanics of carbon tubes: instabilities beyond linear response," *Phys. Rev. Lett.*, vol. 76, pp. 2511–2514, 1996.

[45] J.Z. Yang and W. E, "Generalized Cauchy–Born rules for elastic deformation of sheets, plates, and rods: derivation of continuum models from atomistic models," *Phys. Rev. B*, vol. 74, pp. 184 110-1–184 110-11, 2006.

[46] P. Zhang, Y. Huang, P.H. Geubelle, P.A. Klein and K.C. Hwang, "The elastic modulus of single-wall carbon nanotubes: a continuum analysis incorporating interatomic potentials," *Int. J. Solids Struct.*, vol. 39, pp. 3893–3906, 2002.

6

Capturing the macroscale behavior

In many areas of science and engineering we often face the following dilemma. On the one hand we are interested in the macroscale behavior of a system but the empirical macroscale models we have at our disposal are inadequate for various reasons. On the other hand, the microscopic models are far from being practical, either computationally or analytically. The most well-known examples of the above dilemma include the following.

(1) In molecular dynamics, the accuracy of the empirical force fields is often limited. To improve accuracy we need to make use of electronic structure models such as those in density functional theory.

(2) When we are modeling the dynamics of real gases, the empirical equations of state or other constitutive relations are often too crude. Microscopic models such as kinetic models can be made more accurate but they are also harder to deal with.

(3) Continuum models of non-Newtonian fluids using empirical constitutive relations have had limited success [3]. However, it is practically impossible to use microscopic models such as molecular dynamics to analyze the large-scale flows of engineering interest.

In the language of Chapter 1 these are type B problems, or problems that require extended coupling between the macro- and microscale models.

Traditionally, such problems are treated using sequential multiscale modeling. However, as we remarked in Chapter 1, this is not feasible if the unknown components of the macroscale model depend on many variables. A good example is molecular dynamics. The interatomic forces should in principle depend on the positions of all the atoms in the system, but it is impractical to precompute the interatomic forces as functions of the atomic positions involved if the system has more than ten atoms. Therefore, in this chapter we will focus on concurrent-coupling techniques.

Many different numerical methods have been developed to deal with these problems. Examples include Car–Parrinello molecular dynamics (see (1) above) [8], kinetic schemes for studying gas dynamics (which partly address (2)) [16] and the quasi-continuum method for studying the deformation of solids [53, 73]. All these methods share the following features.

(1) They allow us to model the macroscale quantities of interest by making use of the appropriate microscale models instead of ad hoc macroscale models.

(2) Computational complexity is reduced by exploring the disparity between the macro- and microscales in the problem. In the Car–Parrinello method this is done by modifying the value of the fictitious mass parameter for the Kohn–Sham orbitals in the formulation. In the Knap–Ortiz version of the quasicontinuum method [53], it is done by calculating the energies using only small clusters of atoms instead of all the atoms. In kinetic schemes, it is done by solving the kinetic equation locally near the cell boundaries.

These successes and those of more traditional multiscale algorithms such as the multigrid method have given impetus to establishing general frameworks for such multiscale methods [5, 18, 26, 52]. The hope is that, as in the case of finite difference and finite element methods for solving differential equations, a general framework might lead to general designing principles and guidelines for carrying out error analysis. In [5], Achi Brandt reviewed a general strategy for extending the multigrid method and renormalization group analysis to general multi-physics problems. The new strategy in principle allows the use of atomistic models, such as Monte Carlo models or molecular dynamics, at the finest level of a multigrid hierarchy. It does not initially require explicit macroscale models. In fact Brandt remarked that one might be able to construct an effective macroscale model from the data accumulated during the computation. In addition one can exploit scale separation by restricting the microscopic model to small windows, and consequently a few sweeps might be enough to equilibrate the microscopic model in the interior of the windows. As in traditional multigrid methods, the extended multigrid method follows an interpolation–equilibration–restriction strategy, except that the simulations done at the macro and micro levels can be of very different natures; for example, we could have continuum simulations at the macro level and atomistic simulations at the micro level.

In the heterogeneous multiscale method (HMM) [18, 26], one assumes that the form of the macroscale model is roughly known but some details of the model are missing. On the basis of this information one selects a suitable macroscale solver. Owing to the fact that the macro model is not known explicitly, the microscale model is used during the computation to supply whatever data are needed in the macro solver but are missing from the macro model. Scale separation is exploited by observing that, when estimating the missing data using the microscale model, the computational domain is totally decoupled from the physical domain for the macroscale solver, and it only has to be large enough to guarantee the required accuracy for the data. For the same reason there is no direct communication between the different microscopic simulations that are carried out to estimate data at different locations in the macroscopic computational domain. All communications are done through the macro solver.

The general philosophy of the equation-free approach is also very similar to that of the extended multigrid method and HMM. The basic strategy for implementing this philosophy is to link together simulations of the microscopic models on small spatiotemporal domains in order to mimic the behavior of a system at large scale. This is done through interpolation in space, and extrapolation in time, of ensemble-averaged macroscale variables obtained from the microscopic simulations.

There are obvious similarities between these approaches. A most important similarity is that they all use a multigrid style of coupling, namely at each macro time step or macro iteration step the microscale solver needs to be re-initialized. This can be a rather difficult task, particularly when the macroscale model is a continuum model and the microscale model is discrete. The seamless strategy introduced by E, Ren and Vanden-Eijnden [31] bypasses such a requirement. The basic idea is to modify artificially the time scale of the microscale problem and solve the macroscale model and the modified microscale model simultaneously.

6.1 Some classical examples

6.1.1 Car–Parrinello molecular dynamics

In *Car–Parrinello molecular dynamics* (CPMD) [8], the macroscale quantities of interest are the positions and velocities of the atoms (or nuclei), which obey Newton's second law:

$$M_I \ddot{\mathbf{R}}_I = -\nabla_{\mathbf{R}_I} V.$$

Here M_I and $\mathbf{R}_I$ are respectively the mass and position of the Ith atom. The unknown component of the model is the interatomic potential V or force field $-\nabla_{\mathbf{R}_I} V$. We assume that we have at our disposal a sufficiently accurate electronic structure model, such as a density functional theory model. Since the atoms interact via Coulomb forces, we should be able to compute the force on the atoms once we know the electronic structure. Car and Parrinello devised a very elegant way of doing this, by coupling molecular dynamics with electronic structure models "on-the-fly" [8].

We begin with some remarks about the electronic structure model. In theory we could start from first principles using the quantum many-body problem. In practice one often uses various approximate models, such as Kohn–Sham density functional theory (see Section 4.4). We will rewrite the Kohn–Sham functional for a system of N orthornormal wave functions $\{\psi_n\}_{n=1}^N$ as follows (using atomic units):

$$I_{\mathrm{KS}}\{\psi_n; \mathbf{R}_I\} = \sum_{n=1}^N \int \frac{1}{2} |\nabla \psi_n(\mathbf{x})|^2 \, d\mathbf{x} + J[\rho] + E_{\mathrm{xc}}[\rho], \qquad (6.1.1)$$

where, as before, $\rho(\mathbf{x})$ is the electron density,

$$J[\rho] = \frac{1}{2} \int \int \frac{(\rho(\mathbf{x}) - m(\mathbf{x}))(\rho(\mathbf{y}) - m(\mathbf{y}))}{|\mathbf{x} - \mathbf{y}|} \, d\mathbf{x} \, d\mathbf{y}$$

and E_{xc} is the exchange-correlation energy. The variables $\{\mathbf{R}_I\}$ enter through a pseudo-potential m:

$$m(\mathbf{x}) = \sum_I m_I^{\mathrm{a}}(\mathbf{x} - \mathbf{R}_I),$$

where m_I^{a} is the pseudo-potential for the Ith atom.

Since nuclei are much heavier than electrons, one often makes the adiabatic approximation that the electrons are in the ground state determined by the positions of the nuclei. Under this approximation, the electronic structure is slaved by the state of the nuclei. Denote by $\{\psi_n^*\}$ the set of wave functions that minimize I_{KS}. Obviously $\{\psi_n^*\}$ depends on $\{\mathbf{R}_I\}$. Let

$$E_{\text{KS}}\{\mathbf{R}_I\} = I_{\text{KS}}\{\psi_n^*; \mathbf{R}_I\}$$

The *Hellman–Feynman theorem* states that [61]

$$\nabla_{\mathbf{R}_I} E_{\text{KS}}\{\mathbf{R}_I\} = \nabla_{\mathbf{R}_I} I_{\text{KS}}\{\psi_n^*; \mathbf{R}_I\},$$

i.e. when calculating $\nabla_{\mathbf{R}_I} E_{\text{KS}}$ one can ignore the dependence of $\{\psi_n^*\}$ on $\{\mathbf{R}_I\}$. This is a tremendous technical simplification, and it follows from a simple observation in calculus. Let $f(x, y)$ be a smooth function. For each fixed value of x (which plays the role of $\{\mathbf{R}_I\}$), denote by $y^*(x)$ the minimizer of $g(y) = f(x, y)$. Then $(\partial/\partial y) f(x, y^*(x)) = 0$. Therefore, we have

$$\frac{d}{dx} f(x, y^*(x)) = \frac{\partial}{\partial x} f(x, y^*(x)) + \frac{\partial}{\partial y} f(x, y^*(x)) \frac{d}{dx} y^*(x) = \frac{\partial}{\partial x} f(x, y^*(x)).$$

Under the adiabatic approximation one can define the *Born–Oppenheimer dynamics*:

$$M_I \ddot{\mathbf{R}}_I = -\nabla_{\mathbf{R}_I} I_{\text{KS}}\{\psi_n^*; \mathbf{R}_I\}, \tag{6.1.2}$$

$$\{\psi_n^*, n = 1, \dots, N\} \text{ is the set of minimizers of } I_{KS}. \tag{6.1.3}$$

To implement such a Born–Oppenheimer dynamics, one may proceed as follows.

(1) Select an integrator for the molecular dynamics equation (6.1.2), for example, the Verlet scheme [37].
(2) Calculate the forces on the nuclei by solving the electronic structure problem (6.1.3) using an iterative method.

This procedure requires the electronic structure problem to be solved accurately at each time step.

Instead of following strictly the Born–Oppenheimer dynamics, Car and Parrinello developed a much more seamless approach. They worked with an extended phase space for both the nuclear positions and the Kohn–Sham wave functions, and introduced the extended Lagrangian

$$\mathcal{L}\left\{\mathbf{R}_I, \psi_n, \dot{\mathbf{R}}_I, \dot{\psi}_n\right\}$$
$$= \frac{1}{2} \sum_I M_I |\dot{\mathbf{R}}_I|^2 + \frac{1}{2} \sum_n \mu \int |\dot{\psi}_n(\mathbf{x})|^2 \, d\mathbf{x} - I_{\text{KS}}\{\mathbf{R}_I, \psi_n\}, \tag{6.1.4}$$

where μ is the "mass" for the Kohn–Sham wave functions. The Car–Parrinello molecular dynamics (CPMD) is obtained by following standard procedures in classical mechanics for this extended Lagrangian, subject to the constraint that $\{\psi_n\}$ is a orthonormal set of wave functions:

$$M_I \ddot{\mathbf{R}}_I = -\nabla_{\mathbf{R}_I} I_{\text{KS}}, \qquad (6.1.5)$$

$$\mu \ddot{\psi}_n = -\frac{\delta E}{\delta \psi_n} + \sum_m \Lambda_{nm} \psi_m.$$

Here the Λ_{mn} are Lagrange multipliers for the orthonormality constraint on the wave functions. This formulation has the advantage that the electrons and nuclei are treated on an equal footing. Standard CPMD may also contain dissipative terms; here we have neglected them.

So far we have only considered the multi-physics aspect of CPMD. There is also a multiscale aspect associated with the disparity between the time scales for the nuclei and the electrons caused by the disparity between their masses. The natural choice for the value of the parameter μ should be the value of the electron mass, which is at least three orders of magnitude smaller than the nuclear mass. However, since we are only interested in the dynamics of the nuclei, we may use other values of μ as long as we still obtain an accurate enough approximation for the dynamics of the nuclei. The Born–Oppenheimer approximation is obtained in this context when μ is set to 0. Car and Parrinello adopted an opposite strategy, which is often more convenient in practice, namely to let μ equal some fictitious value much larger than the electron mass. The actual value used is determined by the accuracy requirement.

6.1.2 The quasicontinuum method

Our next example is the (local) *quasicontinuum (QC) method* [53, 73]. In this example our interest is in the macroscale deformation of solids. As discussed in Chapter 4, in a continuum approach this is done by solving a variational problem for the displacement field $\mathbf{u}$, $\Omega \to R^3$, where Ω is the domain that defines the undeformed configuration of the solid,

$$\min \int_\Omega \Big(W(\nabla \mathbf{u}(\mathbf{x})) - \mathbf{f}(\mathbf{x}) \cdot \mathbf{u}(\mathbf{x}) \Big) \, d\mathbf{x},$$

and $\mathbf{f}$ is the applied force. This requires knowing W, the stored energy density. Traditionally W is obtained from empirical considerations. The main purpose of QC is to bypass such empirical strategies and instead use a sufficiently accurate atomistic model.

At the outset, we know that local QC is a piecewise-linear finite element method on a coarse mesh. The mesh is generated by selecting a set of *representative atoms* (rep-atoms), denoted by $\{\mathbf{x}_\alpha\}$, and then forming a triangulation using these rep-atoms. Once the displacement of the rep-atoms is known, the displacement of the rest of the atoms is determined by linear interpolation. Obviously there is a one-to-one correspondence between the displacement of the rep-atoms and the trial functions in the finite element space over the triangulation. The key question is how to compute the energy of a trial function. Two different approaches have . been proposed to deal with this problem.

The first is to use the Cauchy–Born rule (see Section 4.2). Given any trial function $\mathbf{u}$ in the finite element space, since the deformation gradient $\mathbf{A} = \nabla\mathbf{u}$ is uniform within each element one may approximate the energy of each element using the energy density of an infinite crystal that is uniformly deformed with deformation gradient $\mathbf{A}$ at that element. Denote by $\mathcal{E}(\mathbf{A})$ the strain energy density obtained this way. The approximate total energy is then given by

$$E_{\text{tot}}(\mathbf{u}) = \sum_K \mathcal{E}(\mathbf{A}|_K)|K|, \qquad\qquad (6.1.6)$$

where the sum is over all elements; $|K|$ denotes the volume of the element K and $\mathbf{A}|_K$ is the deformation gradient of $\mathbf{u}$ on K.

The second approach is to compute the energy associated with each representative atom by performing a direct summation of the interatomic potential over a small cluster of atoms around the rep-atom [32, 53]; see Figure 6.1. The position $\mathbf{y}$ of an atom in the cluster is given in terms of the trial function $\mathbf{u}$ by $\mathbf{y} = \mathbf{x}+\mathbf{u}(\mathbf{x})$, where $\mathbf{x}$ is the position of the atom in the equilibrium configuration. Knowing all the positions of the atoms, the total energy in the cluster can be computed using the atomistic model. Denote by E_α the average energy of an atom in the cluster around the rep-atom indexed by α. The total energy is computed approximately using

$$E_{\text{tot}}(\mathbf{u}) = \sum_{\alpha=1}^{N_{\text{rep}}} n_\alpha E_\alpha, \qquad\qquad (6.1.7)$$

where $\{n_\alpha\}$ is a set of suitably chosen weights. Roughly speaking, if the mesh itself is smoothly varying then n_α should be the number of atoms

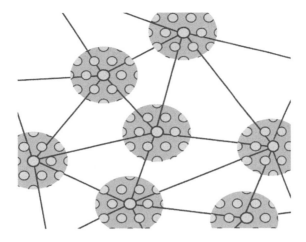

Figure 6.1. Schematic illustration of the cluster summation rule (6.1.8) in QC. Only the atoms in the small clusters need to be visited during the computation. (Courtesy of M. Ortiz.)

that the rep-atom at $\mathbf{x}_\alpha$ is taken to represent. In general one might have to use different weights for different sectors of the clusters (a sector is the intersection of a cluster with an element):

$$E_{\text{tot}}(\mathbf{u}) = \sum_{\alpha,j} n_{\alpha,j} E_{\alpha,j}, \qquad (6.1.8)$$

where $E_{\alpha,j}$ is the energy density for the jth sector of the αth rep-atom; see also [59].

Once we have an expression for the total energy of any trial function, the displacement field can be found by minimizing this energy functional subject to the appropriate boundary conditions.

We have only discussed one component of QC. Another important component is the use of adaptive mesh refinement to resolve the atomistic features near defects (type A problems). This will be discussed in the next chapter.

6.1.3 Kinetic schemes

Our next example is a class of numerical schemes for gas-dynamics simulation that use only the kinetic model. Such schemes are called kinetic schemes (see for example [16, 62, 63, 70, 78]; see also the related work on lattice Boltzmann methods [9, 71]). Here the macroscopic quantities of interest are the density, pressure and velocity fields of the gas. The starting point is a kinetic model, such as the Boltzmann equation (see Section 4.3):

$$\partial_t f + (\mathbf{v} \cdot \nabla) f = \frac{1}{\varepsilon} C(f), \tag{6.1.9}$$

where $f = f(\mathbf{x}, \mathbf{v}, t)$ is the one-particle phase-space distribution function, which is also our microscale state variable, $C(f)$ is the collision term and ε is the Knudsen number. The macroscale state variables U are the usual hydrodynamic variables of mass, momentum and energy densities, which are related to the microscale state variable f by

$$\rho = \int f(\mathbf{v}) d\mathbf{v}, \quad \rho \mathbf{u} = \int f(\mathbf{v}) \mathbf{v} d\mathbf{v}, \quad E = \int f(\mathbf{v}) \frac{|\mathbf{v}|^2}{2} d\mathbf{v}. \tag{6.1.10}$$

From the Boltzmann equation, we have

$$\partial_t \begin{pmatrix} \rho \\ \rho \mathbf{u} \\ E \end{pmatrix} + \nabla \cdot \mathbf{F} = 0, \tag{6.1.11}$$

where

$$\mathbf{F} = \int_{\mathbb{R}^3} f \begin{pmatrix} \mathbf{v} \\ \mathbf{v} \otimes \mathbf{v} \\ \frac{1}{2} |\mathbf{v}|^2 \mathbf{v} \end{pmatrix} d\mathbf{v}. \tag{6.1.12}$$

When $\varepsilon \ll 1$, the distribution function f is close to the local equilibrium states (or local Maxwellians), defined as

$$M(\mathbf{x}, \mathbf{v}, t) = \frac{\rho(\mathbf{x}, t)}{(2\pi k_B T(\mathbf{x}, t))^{3/2}} \exp \left(-\frac{|\mathbf{v} - \mathbf{u}(\mathbf{x}, t)|^2}{2k_B T(\mathbf{x}, t)} \right), \tag{6.1.13}$$

where T is the absolute temperature.

For simplicity, we will focus on the one-dimensional case. We first divide the computational domain in the physical space into cells of size Δx. We will denote by x_j the center of the jth cell and by $x_{j+1/2}$ the cell boundary between the jth and $(j+1)$th cells. In a first-order method we represent the solution as piecewise constants, i.e.

$$\mathbf{U}_j = (\rho_j, \rho_j u_j, E_j), \quad x \in (x_{j-1/2}, x_{j+1/2}].$$

A finite volume scheme takes the form

$$
\begin{cases}
\rho_j^{n+1} - \rho_j^n + \dfrac{\Delta t}{\Delta x}\left(F_{j+1/2}^{(1)} - F_{j-1/2}^{(1)}\right) & = 0, \\[2mm]
(\rho u)_j^{n+1} - (\rho u)_j^n + \dfrac{\Delta t}{\Delta x}\left(F_{j+1/2}^{(2)} - F_{j-1/2}^{(2)}\right) & = 0, \\[2mm]
E_j^{n+1} - E_j^n + \dfrac{\Delta t}{\Delta x}\left(F_{j+1/2}^{(3)} - F_{j-1/2}^{(3)}\right) & = 0,
\end{cases}
\tag{6.1.14}
$$

where $\mathbf{F}_{j+1/2} = (F_{j+1/2}^{(1)}, F_{j+1/2}^{(2)}, F_{j+1/2}^{(3)})^{\mathrm{T}}$ is the numerical flux at cell boundary $x_{j+1/2}$.

The fluxes $\mathbf{F}_{j+1/2}$ are computed by solving the kinetic equation, using the following three-step procedure.

(1) Initialize the kinetic equation using the local Maxwellian with parameters (ρ, u, T) given by $\mathbf{U}_j$.
(2) Solve the kinetic equation locally around cell boundaries where the values of the fluxes are needed. The kinetic equation is solved by splitting. Onc first solves the transport equation, which gives $f(x, v, t) = f(x - v(t - t_n), v, t_n)$. Then one takes into account the collision term, by projecting the solution onto local Maxwellians.
(3) Use (6.1.12) to compute the numerical fluxes.

Omitting the details, we arrive at the following expression for the numerical fluxes [16]:

$$
\mathbf{F}_{j+1/2} = \mathbf{F}^+(\mathbf{U}_j) + \mathbf{F}^-(\mathbf{U}_{j+1}),
\tag{6.1.15}
$$

with

$$
\mathbf{F}^\pm(\mathbf{U}) = \begin{pmatrix}
\rho u A^\pm(S) \pm \dfrac{\rho}{\sqrt{2\pi\beta}} B(S) \\[3mm]
(p + \rho u^2) A^\pm(S) \pm \dfrac{\rho u}{\sqrt{2\pi\beta}} B(S) \\[3mm]
(pu + \rho u e) A^\pm(S) \pm \dfrac{\rho}{\sqrt{2\pi\beta}}\left(\dfrac{p}{2\rho} + e\right) B(S)
\end{pmatrix}
\tag{6.1.16}
$$

where $\beta = 1/(k_B T)$,

$$
S = \frac{u}{\sqrt{2k_B T}}, \quad A^\pm(S) = \frac{1 \pm \operatorname{erf}(S)}{2},
$$

$$
B(S) = e^{-S^2}, \quad p = \rho k_B T, \quad e = \frac{1}{2}u^2 + k_B T;
$$

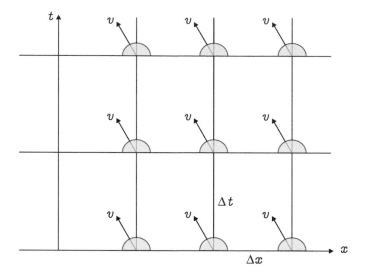

Figure 6.2. Schematic for the derivation in kinetic schemes. A finite volume method is imposed in the xt domain and the kinetic equation is solved (e.g. analytically) over the shaded regions to give the fluxes needed in the finite volume method. The vectors v indicate the extra velocity variable in the kinetic model; this represents the microstructure for the problem (6.1.9).

here erf is the error function. This is the simplest kinetic scheme. A schematic is shown in Figure 6.2.

In this particular example the end product of the whole process may seem to be very much the same as that of standard finite volume methods. Nevertheless, the philosophy embodied in this process is quite significant and is useful for situations when the kinetic model or the local equilibrium distributions are more complicated.

6.1.4 Cloud-resolving convection parametrization

Cloud-resolving convection parametrization (CRCP), also known as super-parametrization, is a technique introduced by Grabowski and Smolarkiewicz [43] in which the large-scale tropical dynamics of the atmosphere is captured by embedding cloud scale models at each grid point of a large-scale model. The macroscale model is a large eddy simulation model. However, in contrast with standard large-eddy simulation, the effects of the subgrid scales are modeled not by empirical models but rather through the introduction of explicit cloud-scale models. We will omit the details here and refer the interested reader to the original article [43]. For more recent developments on this subject, see [77].

6.2 The multigrid and equation-free approaches

Given the success of these different multiscale methods, it is natural to ask whether we can formulate a general framework. When solving differential equations, such general strategies have proven to be very useful. A good example is the finite element method, which has provided not only a general designing principle for numerical algorithms but also general guidelines for error analysis. In recent years, several attempts have been made to construct general strategies for multiscale methods. In the following, we will discuss three examples of such general strategies: the extended multigrid method, the heterogeneous multiscale method and the equation-free approach.

6.2.1 Extended multigrid method

One of the first attempts to construct a general framework for multiscale modeling was the extension of the *multigrid method* by Achi Brandt. In its original form [4, 45], the multigrid method was an efficient way of solving the algebraic equations obtained from discretizing PDEs (see Section 3.1). The objective was to find accurate solutions of a PDE. Since the 1990s, Brandt and others have discussed the possibility of extending the traditional multi-grid method in a number of directions [5].

(1) The models used at the different scales can be quite general and quite different in nature. For example, one may use Monte Carlo methods or molecular dynamics at small scales, and traditional continuum models at large scales.

(2) The traditional multigrid method is an efficient solver for the fine-scale details of the problem. In the extended version, one might be interested only in capturing the macroscale behavior. This is particularly useful in cases when closed-form macroscale models are not available, and it is desirable to have numerical algorithms that are capable of capturing the macroscale behavior directly, using microscale models. In this case, one might even be able to reconstruct the effective macroscale model at the end of the computation. To quote from [5]: "At [a] sufficient[ly] coarse level, this

entire algorithm effectively produces *macroscopic 'equations'* for the simulated system ... This can yield a macroscopic numerical description for the fluid even for those cases where the traditional derivation of closed-form differential equations is inapplicable."

(3) One may limit the microscopic simulation at the fine levels to small spatial-temporal domains. As was remarked by Brandt: "a few sweeps are enough, due to the fast CMC (conditional Monte Carlo) equilibration. This fast equilibration also implies that the interpolation can be done just over a restricted *subdomain*, serving as [a] *window*: in the window interior fine-level equilibration is reached."

(4) Renormalization group ideas are integrated into the new multigrid method.

The general procedure is quite similar to standard multigrid methods. At each cycle one repeats the following steps at the different levels of grids [5].

(1) *Interpolation.* Use the current values of the state variables at the coarse level to initialize or constraint the (micro)model at the fine level.

(2) *Equilibration.* Run the fine-level model (on small windows) for a number of steps.

(3) *Restriction (projection).* Project the variables at the fine level back to the coarse level.

This is a two-level procedure. Generalization to multiple levels is straightforward. Unlike in traditional multigrid methods, in which the successive coarse and fine levels are simply nested finite difference or finite element grids, here the coarse and fine levels can be very different and very far apart.

Brandt also described applications to many different areas, including electronic structure analysis, the solution of integrodifferential equations, the modeling of high-frequency wave propagation, Monte Carlo methods in statistical mechanics, complex fluids and image processing [5].

Brandt's picture is very broad and very appealing. However, at an algorithmic level, it is quite unclear to the present author how one can carry out such a program in practice, even though much has been written on the subject (see, for example, [5] and the references therein). We will return to this discussion later in the chapter.

6.2.2 The equation-free approach

The equation-free approach is another framework for addressing the same kinds of question and is based on a similar philosophy. It consists of a set of techniques including coarse bifurcation analysis, projective integrators, the gap-tooth scheme and patch dynamics [52]. At an abstract level, the equation-free approach is built on a three-stage lift–evolve–restrict coarse time-stepper procedure [52]. Given a macroscopic state $U(t)$ at some time t, the coarse time-stepper consists of the following basic elements [52].

(1) *Lift.* Transform the macroscale initial data to one or more consistent microscopic realizations.

(2) *Evolve.* Use the microscopic simulator (the detailed time-stepper) to evolve these realizations for the desired short macroscopic time τ, generating the value(s) $u(\tau, t)$.

(3) *Restrict.* Obtain the restriction of u and define the coarse time-stepper solution as $\mathcal{T}_c^\tau U(t) = \mathcal{M} u(\tau, t)$.

Here $\mathcal{M}$ is the restriction operator that maps microstate variables to macro-state variables.

So far this is very similar to a two-level interpolate–equilibrate–restrict multigrid procedure. The additional component of the equation-free approach is to exploit scale separation in the problem using interpolation (in space) and extrapolation (in time).

Coarse projective integration In coarse projective integrators, one generates an ensemble of microscopic solutions for short times by running the microscopic solver with an ensemble of initial data that are consistent with the current macrostate. One evaluates the average values of the coarse variables over this ensemble. The time derivatives for the coarse variables are computed using these averaged values, and these coarse time derivatives are used to extrapolate the coarse variables over a much larger time step.

Consider the example

$$\begin{cases} \dot{\mathbf{x}}^\varepsilon = f(\mathbf{x}^\varepsilon, \mathbf{y}^\varepsilon, t), & \mathbf{x}^\varepsilon(0) = \mathbf{x}, \\ \dot{\mathbf{y}}^\varepsilon = \dfrac{1}{\varepsilon} g(\mathbf{x}^\varepsilon, \mathbf{y}^\varepsilon, t), & \mathbf{y}^\varepsilon(0) = \mathbf{y}, \end{cases} \qquad (6.2.1)$$

where $(\mathbf{x}, \mathbf{y}) \in \mathbb{R}^n \times \mathbb{R}^m$. Here the macroscale variable of interest is $\mathbf{x}$. For this problem, the projective integrator proceeds in two steps.

(1) *Relaxation step.* In the simplest case of a forward Euler scheme with micro time step δt, we have

$$\begin{cases} \hat{\mathbf{x}}^{n,m+1} = \hat{\mathbf{x}}^{n,m} + \delta t f(\hat{\mathbf{x}}^{n,m}, \hat{\mathbf{y}}^{n,m}, t^n + m\delta t), \\ \hat{\mathbf{y}}^{n,m+1} = \hat{\mathbf{y}}^{n,m} + \dfrac{\delta t}{\varepsilon} g(\hat{\mathbf{x}}^{n,m}, \hat{\mathbf{y}}^{n,m}, t^n + m\delta t), \end{cases} \qquad (6.2.2)$$

with some initial data $\hat{\mathbf{x}}_{n,0}$ and $\hat{\mathbf{y}}_{n,0}$, $m = 0, 1, \ldots, M-1$.

(2) *Extrapolation step.* Compute the approximate value of $\dot{\mathbf{x}}_n$ and then extrapolate:

$$\dot{\mathbf{x}}^n = \frac{\mathbf{x}^{n,M} - \mathbf{x}^{n,M-1}}{\delta t},$$
$$\mathbf{x}^{n+1} = \mathbf{x}^n + \Delta t \, \dot{\mathbf{x}}^n.$$

If one wants to construct higher-order schemes, one uses higher-order extrapolation:

$$\ddot{\mathbf{x}}^n = \frac{\mathbf{x}^{n,M} - 2\mathbf{x}^{n,M-1} + \mathbf{x}^{n,M-2}}{(\delta t)^2},$$
$$\mathbf{x}^{n+1} = \mathbf{x}^n + \Delta t \dot{\mathbf{x}}^n + \tfrac{1}{2}(\Delta t)^2 \ddot{\mathbf{x}}^n$$

and so on.

A main advantage of projective integrators is that they are relatively simple in the cases when they apply.

Dealing with the space variable Brandt noticed that owing to the fast equilibration of the microscopic model, it might be enough to conduct the microscopic simulations on "small windows." The "gap-tooth scheme" builds on a similar intuition.

The basic idea of the gap-tooth scheme is to "use the microscopic rules themselves, in smaller parts of the domain and, through computational averaging within the subdomains, followed by interpolation, to evaluate the coarse field $U(t, x)$, the time-stepper, and the time derivative field over the entire domain [52].

"Given a finite-dimensional representation $\{U_j^N\}$ of the coarse solution (e.g. nodal values, cell averages, spectral coefficients, coefficients for finite elements or empirical basis functions) the steps of the gap-tooth scheme are the following.

(1) *Boundary conditions.* Construct boundary conditions, for each small box, based on the coarse representation $\{U_j^N\}$.

(2) *Lift.* Use lifting to map the coarse representation $\{U_j^N\}$ to [the] initial data for each small box.

(3) *Evolve.* Solve the detailed equation (the microscale model) for time $t \in [0, \tau]$ in each small box $y \in [0, h] \equiv [x_j - h/2, x_j + h/2]$ with the boundary conditions and initial data given by steps (1) and (2).

(4) *Restrict.* Define the representation of the coarse solution at the next time level by restricting the solutions of the detailed equation in the boxes at $t = \tau$."

One interesting application of the gap-tooth scheme was presented in [39], in which a particle model was used to capture the macroscale dynamics of the viscous Burgers equation. The microscale model is a one-dimensional Brownian dynamics model:

$$\dot{x}_j(t) = u_j^\delta(t) + \dot{w}_j(t), \tag{6.2.3}$$

where $x_j(t)$ is the position of the jth particle at time t, the $\{\dot{w}_j\}$ are independent white noises and $u_j^\delta(t)$ is the drift velocity, which is defined to be the density of particles over an interval of length δ centered at $x_j(t)$. To apply the gap-tooth scheme, microscale simulation boxes of size δ are placed uniformly on the real axis. Particle simulations of (6.2.3) are carried out inside the boxes. At each macroscopic time step n, particles are initialized according to the known local (macroscopic) density at that time step. The boundary condition for the particle simulation is imposed as follows. For the kth box, there is an outgoing flux (the number of particles that exit the box per unit time) and there also is an incoming flux, at both the left- and right-hand sides of the box. These are denoted by $J_k^{\ell,+}$, $J_k^{\ell,-}$, $J_k^{r,+}$, and $J_k^{r,-}$ respectively. The outgoing fluxes $J_k^{r,+}$ and $J_k^{\ell,-}$ can be measured from the simulation. The incoming fluxes $J_k^{\ell,+}$, $J_k^{r,-}$ have to be prescribed. One possibility is to set

$$J_k^{\ell,+} = \alpha J_{k-1}^{r,+} + (1 - \alpha)J_k^{r,+},$$
$$J_k^{r,-} = \alpha J_{k+1}^{\ell,-} + (1 - \alpha)J_k^{\ell,-},$$

where $\alpha = \delta/\Delta x$, δ is the size of the box and Δx is the distance between the centers of adjacent boxes. To implement this boundary condition, particles that come out of a box are either "teletransported" to the next box or returned to the same box from the other side, according to the probability distribution specified by the boundary condition shown above. This procedure corresponds to a linear interpolation. Quadratic interpolations are also discussed in [39].

To update the particle density at the next time step, $t_{n+1} = (n + 1)\Delta t$, the particle densities inside the boxes at t_n are computed, and interpolated in space to give the macroscopic density profile at the new time step.

Patch dynamics Patch dynamics is a combination of projective integrators and the gap-tooth scheme. It consists of the following [52].

(1) *Short time steps.* Repeat the gap-tooth time-stepper a few times, i.e. compute patch boundary conditions, lift to the patches, evolve microscopically and then restrict.

(2) *Extrapolate.* Advance the coarse fields for a long time step into the future through projective integration. This involves estimation of the time derivatives for the coarse field variables, using the successively reported coarse fields, followed by a large projective step.

An example of patch dynamics will be presented in Section 6.7.

6.3 The heterogeneous multiscale method

6.3.1 The main components of the method

We now turn to the framework of the *heterogeneous multiscale method (HMM)*. The general setup is as follows. At the macroscopic level, we have an incomplete macroscale model:

$$\partial_t U = L(U; D), \tag{6.3.1}$$

where D denotes the data needed for the macroscale model to be complete. For example, for complex fluids U might be the macroscopic velocity field and D might be the stress tensor. In addition, we also have a microscopic model:

$$\partial_t u = \mathcal{L}(u; U). \tag{6.3.2}$$

Here the macroscale variable U may enter the system via constraints. We may also write the macro and micro models abstractly as

$$F(U, D) = 0, \tag{6.3.3}$$
$$f(u, d) = 0, \quad d = d(U),$$

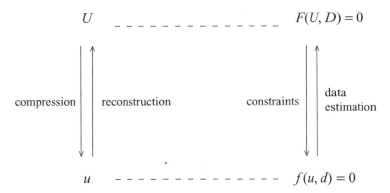

Figure 6.3. Schematics of the HMM framework. The macroscale model is solved with data D obtained from the (local) microscale model, which is then constrained by the (local) values of the macroscale variables U.

where d represents the data needed to set up the microscale model; see Figure 6.3. For example, if the microscale model is the NVT ensemble of molecular dynamics, d might be the temperature value.

Equations (6.3.1) or (6.3.3) represent the knowledge we have about the possible form of the effective macroscale model.

The general philosophy of HMM is to solve the incomplete macroscale model by extracting the needed data from the microscale model. The coupling between the macro and micro models is done in such a way that the macro state provides the constraints for setting up the micro model and the micro model provides the needed constitutive data D for the macro model. The two main components of HMM are as follows.

(1) *A macroscopic solver.* On the basis of whatever knowledge is available about the macroscale behavior of the system, we make an assumption about the form of the macroscale model from which we select a suitable macroscale solver. For example, if we are dealing with a variational problem, we may use a finite element method as the macroscale solver.

(2) *A procedure for estimating the missing macroscale data D using the microscale model.* This is typically done in two steps.

   (a) Constrained microscale simulation. At each point where some macroscale data is needed, perform a series of

microscopic simulations which are constrained to be consistent with the local value of the macro variable.

(b) Data processing. Use the results from the microscopic simulations to extract the macroscale data needed in the macroscale solver.

Consider for example the case when the microscale model is an elliptic
equation with multiscale coefficients:

$$-\nabla \cdot (a^\varepsilon(\mathbf{x})\nabla)u^\varepsilon(\mathbf{x}) = f(\mathbf{x}). \tag{6.3.4}$$

Assume that the macroscale model is of the form

$$-\nabla \cdot (A(\mathbf{x})\nabla)U(\mathbf{x}) = f(\mathbf{x}).$$

Naturally, for the macroscale solver we choose standard finite element
methods, e.g. the piecewise linear finite element method, over a coarse
mesh. The datum that needs to be estimated is the stiffness matrix for
the finite element method. If $A = A(\mathbf{x})$ were known then we would simply follow standard practice and use numerical quadrature to compute
the elements in the stiffness matrix. Since A is not known, we set up a
microscale simulation around each quadrature point in order to estimate
the needed function value at that quadrature point. The details of this
procedure will be discussed in Chapter 8.

As a second example consider an incompressible polymeric fluid flow,
for which the macroscale model is a continuum model for the macroscale
velocity field $\mathbf{U}$ in the form

$$\rho_0(\partial_t\mathbf{U} + (\mathbf{U} \cdot \nabla)\mathbf{U}) = \nabla \cdot \boldsymbol{\sigma},$$
$$\nabla \cdot \mathbf{U} = 0.$$

These are simply statements of the conservation of momentum and mass
for a fluid of constant density ρ_0. The unknown data D correspond to the
stress $\boldsymbol{\sigma}$.

Let us assume that the micro model is a molecular dynamics model for
the particles that make up the fluid:

$$m_j \frac{d^2\mathbf{y}_j}{dt^2} = \mathbf{f}_j, \quad j = 1, \ldots, N.$$

Here m_j and $\mathbf{y}_j$ are respectively the mass and position of the jth particle
and $\mathbf{f}_j$ is the force acting on the jth particle.

Given that the macroscale model is in the form of an incompressible
flow equation, it is natural to select the projection method as the macro

solver [11]. In the implementation of the projection method, we will need the values of $\boldsymbol{\sigma}$ at the appropriate grid points. These are the data that need to be estimated. At this point, we have to make an assumption about what $\boldsymbol{\sigma}$ depends on since this enters in the constraints that we put on the microscale model. Let us assume that

$$\boldsymbol{\sigma} = \boldsymbol{\sigma}(\nabla \mathbf{U}).$$

We will constrain the molecular dynamics in such a way that the average strain rate is given by the value of $\nabla \mathbf{U}$ at the relevant grid point. In general, implementing such constraints is the most difficult step in HMM. For the present example, one possible strategy is discussed in [65].

From the molecular dynamics results, we need to extract the values of the needed components of the stress. For this purpose, we need a formula that expresses the stress in terms of the molecular dynamics output. This can be obtained by modifying the Irving–Kirkwood formula [50]. These details will be explained in Section 6.6.

In summary, three main ingredients are required in HMM:

(1) a macro solver, here the projection method;
(2) a micro solver, here constrained molecular dynamics;
(3) a data estimator, here the modified Irving–Kirkwood formula.

We can write down the HMM procedure formally as follows. At each macro time step:

(1) given the current state of the macro variables U^n, re-initialize the micro-variables, setting

$$u^{n,0} = RU^n; \tag{6.3.5}$$

(2) evolve the micro variables for a number of micro time-steps, via

$$u^{n,m+1} = \mathcal{S}_{\delta t}(u^{n,m}; U^n), \quad m = 0, 1, \ldots, M-1; \tag{6.3.6}$$

(3) estimate D using

$$D^n = \mathcal{D}_M(u^{n,0}, u^{n,1}, \ldots, u^{n,M}); \tag{6.3.7}$$

(4) evolve the macro variables for one macro time-step using the macro-solver, via

$$U^{n+1} = S_{\Delta t}(U^n; D^n). \tag{6.3.8}$$

Here R is a reconstruction operator which plays the same role as the interpolation or prolongation operators in the multigrid method, $S_{\delta t}$ is the micro solver, which also depends on U^n through the constraints, as indicated, and $\mathcal{D}_M$ is some data processing operator which in general involves spatial–temporal ensemble averaging; it is sometimes referred to as the data estimator. Finally $S_{\Delta t}$ is the macro solver.

For dynamic problems there are two important time scales that we need to consider. The first, denoted by t_M, is the time scale for the dynamics of the macro variables. The second, denoted by τ_ε, is the relaxation time for the microscopic model. We will need to distinguish two different cases. In the first case the two time scales are comparable, i.e. $\tau_\varepsilon \sim t_M$. Then, from the viewpoint of numerical efficiency, there is not much room for maneuvre as far as time scales are concerned: we just have to evolve the microscale model along with the macroscale model. In the second case $\tau_\varepsilon \ll t_M$. This is the case on which we will focus. The general guidelines in this case are as follows.

(1) Choose Δt to resolve accurately the t_M time scale.

(2) Choose M such that $M\delta t$ covers the τ_ε time scale sufficiently to allow equilibration to take place in the micro model.

Next we consider an example in some detail.

6.3.2 Simulating gas dynamics using molecular dynamics

Instead of using kinetic theory as in kinetic schemes, we will discuss how to perform gas dynamics simulations using molecular dynamics models. It is not clear at this point in time how relevant this is in real applications. We will simply use it as an example to illustrate HMM (see Figure 6.4).

The microscale model is molecular dynamics, which we will write as

$$\frac{d\mathbf{x}_j}{dt} = \mathbf{v}_j, \quad m_j \frac{d\mathbf{v}_j}{dt} = \sum_k \mathbf{f}_{jk}.$$

Here $\mathbf{f}_{jk} = -\nabla_{\mathbf{x}_j} V_0(|\mathbf{x}_j - \mathbf{x}_k|)$ is the force on the jth atom from the kth atom. For simplicity we have assumed that there are only two-body interactions. Now recall that if we define

$$\tilde{\rho}(\mathbf{x}, t) = \sum_j m_j \delta(\mathbf{x} - \mathbf{x}_j(t)),$$

$$\tilde{\mathbf{m}}(\mathbf{x}, t) = \sum_j m_j \mathbf{v}_j(t) \delta(\mathbf{x} - \mathbf{x}_j(t)),$$

$$\tilde{E}(\mathbf{x}, t) = \sum_j \tfrac{1}{2} m_j |\mathbf{v}_j(t)|^2 \delta(\mathbf{x} - \mathbf{x}_j(t))$$

$$+ \tfrac{1}{2} \sum_j \left(\sum_{i \neq j} V_0(|\mathbf{x}_j(t) - \mathbf{x}_i(t)|) \right) \delta(\mathbf{x} - \mathbf{x}_j(t))$$

then we have

$$\begin{aligned}
\partial_t \tilde{\rho} + \nabla_{\mathbf{x}} \cdot \tilde{\mathbf{m}} &= 0, \\
\partial_t \tilde{\mathbf{m}} + \nabla_{\mathbf{x}} \cdot \tilde{\boldsymbol{\sigma}} &= 0, \\
\partial_t \tilde{E} + \nabla_{\mathbf{x}} \cdot \tilde{\mathbf{J}} &= 0,
\end{aligned} \qquad (6.3.9)$$

where $\tilde{\boldsymbol{\sigma}}$ and $\tilde{\mathbf{J}}$ are defined in terms of $\{\mathbf{x}_j, \mathbf{v}_j\}$ (see Section 4.2). These equations are the analogs of the conservation laws used in deriving the continuum models of gas dynamics and provide our starting point for coupling the macroscale and microscale models.

Since the macroscale model is a system of conservation laws, it is natural to use the finite volume method as the macro solver. There are many different versions of the finite volume method (see for example [56]). We will choose the version that uses the least amount of information about the details of the model since we do not know much about these details. For example, a central type of scheme such as the Lax–Friedrichs scheme is preferred over schemes that rely on characteristic decomposition, such as the Roe scheme. Thus we will pick the Lax–Friedrichs scheme as the macro solver:

$$U_{j+1/2}^{n+1} = \frac{U_j^n + U_{j+1}^n}{2} - \frac{\Delta t}{\Delta x} \left(F_{j+1}^n - F_j^n \right).$$

The data to be estimated are the fluxes. This is done by performing MD simulations at places where the numerical fluxes are needed.

The Lax–Friedrichs scheme has another feature that helps to simplify the data estimation step: it is defined on a staggered grid. Consequently, to estimate F_j^n we need to set up an MD with the constraints that the

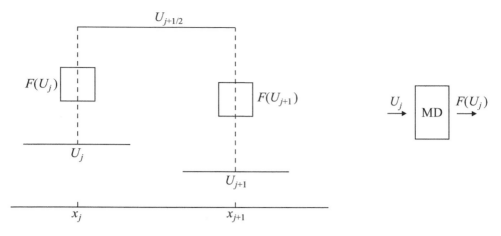

Figure 6.4. The HMM procedure for capturing the macroscopic behavior of gas dynamics using molecular dynamics (MD): on the left, the macroscale solver, here the Lax–Friedrichs scheme; on the right, the data estimation procedure. This setup takes advantage of the staggered grid used in the macroscale solver. (Courtesy of Xiantao Li.)

average mass, momentum and energy densities are equal to the U_j^n. This can be achieved simply by initializing the MD with such constraints and then using the periodic boundary condition. Note that these densities are conserved quantities of the MD.

As the MD proceeds we extract the needed data, namely $\tilde{\sigma}$ and $\tilde{J}$, by performing statistical averaging using the Irving–Kirkwood type of formula discussed in Section 4.2. Some typical results are shown in Figure 6.5. One can see that after a limited number of steps, the estimated values of $\tilde{\sigma}$ and $\tilde{J}$ have already reached reasonable accuracy; this is independent of how large the macroscale system is.

Shown in Figure 6.6 is an example of numerical results produced using the HMM just described for a slightly different application, namely one-dimensional thermoelasticity. This is a Riemann problem mimicking the setup of a shock tube problem, i.e. the macroscale initial condition is piecewise constant, $U_0 = U_1$ in the left half of the domain and $U_0 = U_{\mathrm{r}}$ in the right half of the domain. If we neglect the viscous and thermal dissipation then the macroscale solution should be self-similar. This fact is used in the figure: the MD results are computed for a small system and for a short time and then rescaled in space and time for comparison with the HMM results, which were computed for much larger systems and much longer times.

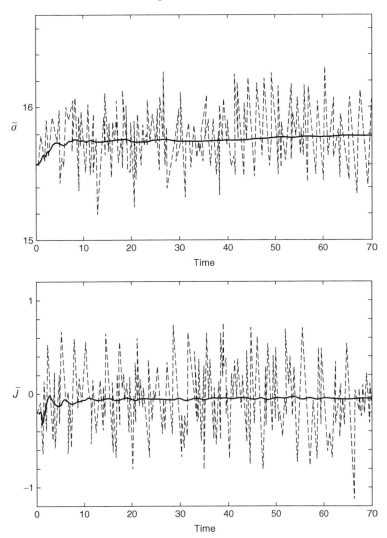

Figure 6.5. Estimated values of the stress and heat flux (solid lines) as a function of the molecular dynamics time scale. The broken lines show the instantaneous values of the stress and heat flux computed using an Irving–Kirkwood type of formula; see Section 4.2. The solid lines give the time-averaged values using the instantaneous data up to the corresponding times.

6.3.3 The classical examples from the HMM viewpoint

Let us now reexamine the examples in Section 6.1 from an HMM perspective.

The Born–Oppenheimer dynamics can be viewed as a special case of HMM in which the macro solver is the molecular dynamics algorithm

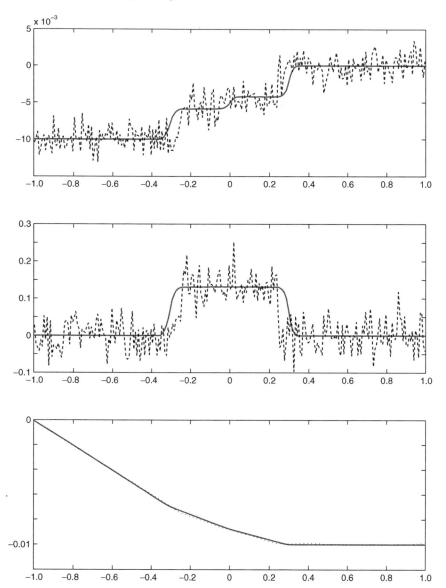

Figure 6.6. Numerical solutions obtained using HMM computations and solutions obtained using direct MD. Solid lines, computed solutions; broken lines, full atom MD simulations (one realization). Top, the strain; middle, the velocity; bottom, the displacement.

chosen for (6.1.2), the data to be estimated are the force components acting on the nuclei, and data estimation is done through an iterative algorithm for the electronic structure model. For local QC the macro solver is the linear finite element method and the data to be estimated are the energy associated with the trial function or the forces on the representative atoms. Data estimation can be done either using the Cauchy–Born rule or the cluster-summation rule. For the kinetic scheme, the macro solver is the finite volume method. The data to be estimated are the numerical fluxes. Data estimation is done by solving the kinetic equation locally around the point of interest.

These algorithms were proposed for vastly different applications, so it is quite interesting that they can be considered from a unified viewpoint, one that also adds some perspectives of its own.

(1) As will be discussed below (see Section 6.7), the HMM framework provides a general strategy for analyzing the stability and accuracy of multiscale methods. This can be used to analyze the algorithms discussed in Section 6.1. One example, the analysis of the local QC, is carried out in [23] using this viewpoint.

(2) The HMM provides an alternative perspective on how to improve these existing methods. For example, if we are interested in developing high-order kinetic schemes, we might start with a high-order macro solver and solve the kinetic model to a high level of accuracy. This may not be so different from what experts in kinetic schemes would do anyway. But it does make the ideas a bit more systematic.

There is one very notable exception, though, and that is the Car–Parrinello molecular dynamics (CPMD). It has a rather different flavor from that of the HMM we have discussed so far. The coupling used in CPMD is much more seamless than the multigrid style of coupling used in HMM, which requires one to go back and forth explicitly between the macro- and microstates. We will return to this question later when we discuss seamless coupling strategies (see Section 6.5).

6.3.4 Modifying traditional algorithms to handle multiscale problems

The heterogeneous multiscale method can also be viewed as a way of modifying traditional numerical algorithms for the purpose of handling efficiently multiscale problems. To illustrate this point, we will discuss how to modify the fast multipole method to handle a multiscale charge distribution.

As was discussed in Section 3.2, the fast multipole method (FMM) is among the most efficient algorithms for evaluating the Coulomb potential due to a distribution of charges [44]. Consider the following problem:

$$\phi^\varepsilon(\mathbf{x}) = \int_\Omega \frac{q^\varepsilon(\mathbf{y})}{|\mathbf{x} - \mathbf{y}|} \, d\mathbf{y}, \qquad (6.3.10)$$

where $q^\varepsilon(\mathbf{y}) = q(\mathbf{y}, \mathbf{y}/\varepsilon)$ is a smooth function which is periodic in the second variable with period $I = [0,1]^3$, and $\varepsilon \ll 1$. Problems of this type arise in the electronic structure analysis of crystalline materials [21]. Applying FMM directly to the evaluation of ϕ^ε will require a cost of $\mathcal{O}(\varepsilon^{-3})$, since the smallest boxes used in FMM should resolve the smallest scale of q^ε, which is $\mathcal{O}(\varepsilon)$. A much more efficient approach is to combine FMM with HMM [47].

In the spirit of HMM we will select FMM as the macroscale solver; the size of the smallest boxes in FMM is decided by the need to resolve the large-scale features of the charge distribution but not the small-scale features. The data needed in FMM are the coefficients of the multipole expansion:

$$M_{k,j}^p = \int_{C_{k,j}} q\left(\mathbf{y}, \frac{\mathbf{y}}{\varepsilon}\right)(\mathbf{y} - \mathbf{x}_{k,j})^p \, d\mathbf{y},$$

where $(C_{k,j}, \mathbf{x}_{k,j})$ denotes the jth box and box center at the kth level. These integrals can be evaluated, approximately using, for example,

$$M_{k,j}^p \simeq \int_{C_{k,j}} \int_I q(\mathbf{y}, \mathbf{z})(\mathbf{y} - \mathbf{x}_{k,j})^p \, d\mathbf{y} \, d\mathbf{z}$$

The total cost of this approach is $\mathcal{O}(1)$.

6.4 Some general remarks

6.4.1 Similarities and differences

There are obvious similarities between the extended multigrid method, HMM and the equation-free approach.

(1) They all have the same objective, namely, developing algorithms that capture the macroscale behavior of complex systems with the help of microscopic models, thus bypassing empirical macroscopic models, yet without the need to resolve all microscopic details.

Table 6.1. *The terminologies used in the extended multigrid method, HMM and the equation-free approach*

	Macro to micro	Micro to macro
Extended multigrid	interpolation	restriction (projection)
HMM	reconstruction	compression
Equation-free	lifting	restriction

Table 6.2. *The basic structures of the extended multigrid method and the equation-free approach*

Extended multigrid	Equation-free
interpolation	lifting
equilibration	evolution (equilibration)
restriction (projection)	restriction
	extrapolation

(2) They all use a multigrid style of coupling, namely, the macro and micro variables are converted back and forth at each macro time step or macro iteration step, even though somewhat different terminologies are used (see Table 6.1).

(3) They all explore the possibility of restricting microscale model simulation to small domains (boxes, windows) for short times (or few sweeps).

Moreover, as was mentioned earlier, the basic structure for the extended multigrid method is remarkably close to that of the equation-free approach, as can be seen in Table 6.2. Note in particular that the purpose of the "evolution" step in the equation-free approach is to equilibrate.

There are also some differences. As was pointed out already, compared with the multi-grid approach, the equation-free approach uses extrapolation as a way of exploring time-scale separation. Between HMM and equation-free it is quite clear that, at a philosophical level, HMM is a top-down approach, it is a more "equation-based" technique, although the "equation," i.e., the preconceived form of the macroscale model, is incomplete. In contrast, the equation-free approach seems to be more of a bottom-up strategy. Its purpose is to set up microscopic simulations on small windows to mimic a simulation over the whole macroscopic physical domain. One clear example is the gap-tooth scheme in [39] discussed earlier. The main thrust there is to couple the different microscopic simulations on the small boxes in order to mimic a microscopic simulation performed over the entire physical domain, using basically the smoothness of the macro variables.

At a technical level, if we write the effective macroscale model in the form

$$\partial_t U = F(\{U\})$$

then the difference between HMM and the equation-free approach is that equation-free focuses on the left-hand side whereas HMM focuses on the right-hand side. Here we have written $F(\{U\})$ instead of $F(U)$ to emphasize that the right-hand side may depend on more variables than just the pointwise value of U: for example, it may depend on the gradients of U. Equation-free can be viewed as a strategy in which the microscale models are used to compute the approximate value of $\partial_t U$, and then this value is used to advance U over macro time steps. In HMM, however, the first question one asks is: what can we say about the structure of $F(\{U\})$? This information is then used to select a macro solver. For example, in the case when the effective macroscale model is a stochastic differential equation, $F(\{U\})$ is a sum of two terms, a drift term and a noise term; HMM treats the two terms separately [75].

If we do not know anything about the structure of $F(\{U\})$ then evaluating $F(\{U\})$ is the same as evaluating $\partial_t U$ directly. In this case, HMM still provides the flexibility of selecting a suitable macro solver.

6.4.2 Difficulties with the three approaches

Difficulties with the multi-grid method While its general philosophy is very attractive, the algorithmic details for the extended multi-grid method are sparse. In particular, it is not clear whether Brandt aims at constructing linear or sublinear scaling algorithms. The discussions about "small windows" and "few sweeps" suggest that the goal is to construct sublinear scaling algorithms. But his insistence on complete generality raises some doubts: clearly it is not possible to construct sublinear scaling algorithms that work for general problems.

Difficulties with HMM The most significant shortcoming of HMM is that it is based on a preconceived macroscale model. If the form of the macroscale model is chosen incorrectly, one cannot expect the resulting HMM procedure to produce accurate results. For example, if the effective macroscale model should be a stochastic ODE but one makes the assumption that it is a deterministic ODE then the stochastic component of the macroscale solution will not be captured correctly by an HMM based on such an assumption.

There is a important reason for starting with the macro solver: Even for problems for which we do have a sufficiently accurate macroscale model, finding an effective numerical algorithm for that macroscale model may still be a significant task. Indeed this has been the focus of the computational mathematics community for more than 50 years. One example is Euler's equation in gas dynamics, whose solutions typically contain shocks, i.e. discontinuities [56]. In this case the numerical algorithms have to satisfy certain constraints in order to be able to solve Euler's equation accurately. Obviously this should also be a concern for multiscale methods.

For practical problems of interest, we often have some accumulated knowledge about what the macroscale model should be like. Such information can be used when making assumptions about the macroscale model used in HMM. In cases when one makes a wrong assumption, one can still argue that HMM produces an "optimal approximation" for the macroscale behavior of the solution in the class of models considered. In this sense, HMM is a way of addressing the following question: what is the best one can do given the knowledge we have about the problem at all scales?

Difficulties with the equation-free approach There is still some confusion about the basic philosophy of the equation-free approach. For this reason, we have carefully followed the original papers in our above presentation of this method. To give an example about the kind of confusion we are concerned with, let us consider the example of analyzing the free energy profile of a complex system by computing the probability density of some coarse-grained variables (see for example [34]). The probability density $f = f(q)$ is assumed to satisfy a Fokker–Planck equation of the form

$$\partial_q(\partial_q[D(q)f(q)] - V(q)f(q)) = 0; \qquad (6.4.1)$$

here q is some coarse-grained variable. Microscopic models are used to precompute the coefficients D and V. Equation (6.4.1) is then used to find the free energy density or for other purposes. This is a very useful approach, as was demonstrated in [34, 48] and numerous other examples. It also extends the standard free energy analysis by allowing a more general form of the diffusion coefficient D. However, from the viewpoint of multiscale modeling, it is anything but "equation-free": It is a typical example of the precomputing (or sequential coupling, see Chapter 1), equation-based technique. It has the standard merits and difficulties of all precomputing techniques. For example, it becomes unfeasible when the dimension of q is large.

In addition, as we will see later, the original patch dynamics may suffer from issues of numerical instability. To overcome these problems, a new version of the patch dynamics has been proposed in which a preconceived macroscale model and a macroscale solver is used to construct the lifting operator (see Section 6.7). The original appeal of the equation-free approach, namely that it does not require knowing much about the macroscale model, seems to be lost in this modified version.

In summary, finding a robust bottom-up approach that does not require making a priori assumptions on the form of the macroscale model is still an open problem. It is not clear at this stage whether it is possible to have a general approach of this kind and how useful it will be.

6.5 Seamless coupling

With the exception of Car–Parrinello molecular dynamics, the approaches discussed above all require one to convert back and forth between the macro- and microstates of the system. This can become rather difficult in actual implementations, particularly when one is constructing discrete microstates (which are needed in for example molecular dynamics) from continuous macroscale variables. The seamless strategy proposed in [31] is intended to bypass this difficult step.

To motivate the seamless algorithm, let us consider the trivial example of a stiff ODE:

$$
\begin{cases}
\dfrac{dx}{dt} = f(x, y), \\[2mm]
\dfrac{dy}{dt} = -\dfrac{1}{\varepsilon}(y - \varphi(x)).
\end{cases}
\tag{6.5.1}
$$

Using the notation of earlier sections, we have $U = x$, $u = (x, y)$. If we want an efficient algorithm for capturing the behavior of x without resolving the detailed behavior of y, we can simply increase the small parameter ε to ε', the size of which is determined by the accuracy requirement (the question of how the value of ε' affects the accuracy will be discussed later):

$$
\begin{cases}
\dfrac{dx}{dt} = f(x, y), \\
\dfrac{dy}{dt} = -\dfrac{1}{\varepsilon'}(y - \varphi(x)).
\end{cases}
\tag{6.5.2}
$$

This is then solved using standard ODE solvers. We will refer to this as "boosting."

We can look at this differently. Instead of changing the value of ε we can change the clock for the microscale model, i.e. if we use $\tau = t\varepsilon/\varepsilon'$ in the second equation in (6.5.2) then (6.5.2) can be written as

$$
\begin{cases}
\dfrac{dx}{dt} = f(x, y), \\
\dfrac{dy}{d\tau} = -\dfrac{1}{\varepsilon}(y - \varphi(x)).
\end{cases}
\tag{6.5.3}
$$

If we discretize this equation using standard ODE solvers but with different time-step sizes for the first and second equations in (6.5.3), we obtain the following algorithm:

$$
y^{n+1} = y^n - \frac{\delta\tau}{\varepsilon}(y^n - \varphi(x^n)),
\tag{6.5.4}
$$

$$
D^{n+1} = y^{n+1},
\tag{6.5.5}
$$

$$
x^{n+1} = x^n + \tilde{\Delta}t f(x^n, D^{n+1}).
\tag{6.5.6}
$$

Here $y^n \sim y(n\delta\tau)$ and $x^n \sim x(n\tilde{\Delta}t)$. The value of $\delta\tau$ is the time-step size we would use if we were attempting to solve (6.5.1) accurately. If (6.5.1) were the molecular dynamics equations then $\delta\tau$ would be the standard femtosecond time-step size. The quantity $\tilde{\Delta}t$ is the time step one would use for (6.5.2). It satisfies

$$
\frac{\tilde{\Delta}t}{\varepsilon'} = \frac{\delta\tau}{\varepsilon}.
$$

In general $\tilde{\Delta}t$ should be chosen such that one not only resolves the macro time scale but also allows the micro state to relax sufficiently, i.e. to adjust to the changing macroscale environment. For example, if Δt is the time-step size required for accurately resolving the macroscale dynamics and if τ_ε is the relaxation time of the microscopic model then we should choose $\tilde{\Delta} t = \Delta t / M$, where $M \gg \tau_\varepsilon / \delta \tau$.

The advantage of this second viewpoint is that it is quite general and does not require tuning parameters in the microscopic model. In a nutshell, the basic idea is as follows.

(1) Run the (constrained) micro solver using its own time-step $\delta \tau$.
(2) Run the macro solver at a pace that is slower than for a standard macro model: $\tilde{\Delta} t = \Delta t / M$.
(3) Exchange data between the micro and macro solvers at every step.

Intuitively, what one is doing is forcing the microscale model to accommodate the changes in the macroscale environment (here the change in x) at a much faster pace. For example, assume that the characteristic macro time scale is 1 second and the micro time scale is 1 femtosecond (10^{-15} second). In a brute-force calculation the micro model will run 10^{15} steps before the macroscale environment changes appreciably. The HMM procedure makes use of the separation of the time scales by running the micro model only until it is sufficiently relaxed, which requires many fewer (say M) steps than 10^{15}, and then extracting the data needed in order to evolve the macro system over a macro time-step of 1 second. Thus in effect HMM skips $10^{15} - M$ micro steps of calculation for the microscale model. It exchanges data between the macro and micro solvers after every 10^{15} micro time-step interval. The price one has to pay is that one has to re-initialize the microscale solver at each macro time-step, owing to the temporal gap created by skipping $10^{15} - M$ micro steps.

One can take a different viewpoint, however. If we define

$$\tilde{y}(k\tilde{\Delta} t + t^n) = y^{n,k}, \quad k = 1, \dots, M,$$

where $\tilde{\Delta} t = \Delta t / M$ and $y^{n,k}$ is the kth-step solution to the microscale model at the nth macro step in HMM, the variable $\tilde{y}$ is defined uniformly in the new rescaled time axis. By doing so we change the clock for the micro model (by a factor $\tilde{\Delta} t / \delta t$), but we no longer need to re-initialize the micro model every macro time step since the gap mentioned above no longer exists. Because the cost of the macro solver is typically very small compared with the cost of the micro solver, we may as well run the macro solver using a smaller time step (i.e. $1/M$ s) and exchange data

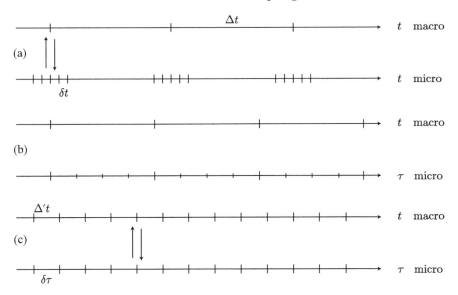

Figure 6.7. Illustration of (a) HMM (top panel) (b) rescaling the micro time scale (c) the seamless algorithm (bottom panel). In HMM, one macro time-step needs to be accompanied by many micro time-steps (for the micro solver) in order for the needed macro data to be estimated. In the seamless algorithm, one changes the clock for the micro solver but evolves the micro solver continuously.

every time-step. In this way the data exchanged tend to be smoother. It turns out that this has the added advantage that it also reduces the statistical error (see the numerical results presented in the next section). This is illustrated in Figure 6.7.

Now we turn to the general form of the seamless algorithm proposed in [31] (see also earlier related work in [36, 64]). Using the setup in Section 6.3, we can write the seamless algorithm as follows.

(1) Given the current state of the micro variables $u(\tau)$ and the macro variables $U(t)$, evolve the micro variables for one time step:

$$u(\tau + \delta\tau) = \mathcal{S}_{\delta\tau}(u(\tau); U(t)). \qquad (6.5.7)$$

(2) Estimate D:

$$D = \mathcal{D}(u(\tau + \delta\tau)). \qquad (6.5.8)$$

(3) Evolve the macro variables:

$$U(t + \tilde{\Delta}t) = S_{\tilde{\Delta}t}(U(t); D). \qquad (6.5.9)$$

In this algorithm, we alternate between the macro and micro solvers, each running with its own time-step (therefore the micro and macro solvers use different clocks). At every step the needed macroscale data is estimated from the results of the micro model (at that step) and is supplied to the macro solver. The new values of the macrostate variables are then used to constrain the micro solver.

Remarks 1. The seamless algorithm incurs an additional cost since we are now evolving the macro variables using much smaller time-steps. However, in most cases this additional cost is insignificant since the cost of the micro solver is still much bigger than the cost of the macro solver.

2. In HMM we normally use time-averaging to process the needed data. We may still do that in the new algorithm. However, it is not clear that this does in fact improve the accuracy. We will return to this issue later.

From the consideration of time scales alone, the computational savings in the seamless algorithm come from the fact that effectively the system evolves on the time step $\tilde{\Delta} t$. In the case when the time scales are disparate, $\tilde{\Delta} t$ can be much larger than $\delta\tau$. Therefore one can define the savings factor:

$$C_{\mathrm{S}} = \frac{\tilde{\Delta} t}{\delta\tau} = \frac{\Delta t}{M\delta\tau}. \tag{6.5.10}$$

As an example, let us consider the case when the microscopic model is molecular dynamics, and the time-step size is femtoseconds ($\delta\tau = 10^{-15}$ seconds). On the one hand, if one wants to simulate one second of physical time then one needs to compute for 10^{15} steps. On the other hand, assume that the relaxation time is of the order of picoseconds (10^{-12} seconds), corresponding to about 10^3 micro time-steps, then $M = 10^5$ is a reasonable choice. Simulating one second of physical time using the seamless algorithm requires 10^5 steps. This is a factor 10^{10} saving. The price to pay is that we no longer obtain accurate information at the level of the microscopic details – we can only hope to get accurate information for the macrostate variables.

Example 1: Stochastic ODEs with multiple time scales Consider the stochastic ODE

$$\begin{cases} \dfrac{dx}{dt} = f(x, y), \\ \dfrac{dy}{dt} = -\dfrac{1}{\varepsilon}(y - \varphi(x)) + \sqrt{\dfrac{2}{\varepsilon}}\dot{w}, \end{cases} \tag{6.5.11}$$

where $\dot{w}(t)$ is a standard white noise. The averaging theorems suggest that the effective macroscale equation should be of the form of an ODE:

$$\frac{dx}{dt} = F(x).$$

The HMM procedure with forward Euler as the macro solver proceeds as follows.

(1) Initialize the micro solver, e.g. set $y^{n,0} = y^{n-1,M}$.
(2) Apply the micro solver for M micro steps:

$$y^{n,m+1} = y^{n,m} - \frac{\delta t}{\varepsilon}(y^{n,m} - \varphi(x^n)) + \sqrt{\frac{\delta t}{\varepsilon}}\xi^{n,m}, \qquad (6.5.12)$$

with $m = 0, 1, \ldots, M-1$. Here the $\{\xi^{n,m}\}$ are independent normal random variables with mean 0 and variance 1.
(3) Estimate $F(x)$:

$$F^n = \frac{1}{M}\sum_{m=1}^{M} f(x^n, y^{n,m}). \qquad (6.5.13)$$

(4) Apply the macro solver:

$$x^{n+1} = x^n + \Delta t\, F^n. \qquad (6.5.14)$$

In contrast, the seamless algorithm based on the forward-Euler scheme is simply

$$y^{n+1} = y^n - \frac{\delta\tau}{\varepsilon}(y^n - \phi(x^n)) + \sqrt{\frac{\delta\tau}{\varepsilon}}\xi^n, \qquad (6.5.15)$$

$$x^{n+1} = x^n + \tilde{\Delta}t\, f(x^n, y^{n+1}), \qquad (6.5.16)$$

where the $\{\xi^n\}$ are independent normal random variables with mean 0 and variance 1. Note that for HMM we have $x^n \sim x(n\Delta t)$ but for the seamless algorithm we have $x^n \sim x(n\tilde{\Delta}t) = x(n\Delta t/M)$.

Example 2: The parabolic homogenization problem Consider

$$\partial_t u^\varepsilon = \partial_x\left(a\left(x, \frac{x}{\varepsilon}, t\right)\partial_x u^\varepsilon\right), \qquad (6.5.17)$$

where $a(x, y, t)$ is a smooth function and is periodic in y, say with period 1. The macroscale model is of the form

$$\partial_t U = \partial_x D, \tag{6.5.18}$$

$$D = \left\langle a\left(x, \frac{x}{\varepsilon}, t\right) \partial_x u^\varepsilon \right\rangle, \tag{6.5.19}$$

where $\langle \cdot \rangle$ means a spatial average and D is a flux.

As in HMM, if we choose a finite volume method as the macro solver then the flux D needs to be evaluated at the cell boundaries [1]. We will make the assumption that D depends on the local values of U and $\partial_x U$ only. Consequently, for the micro model we will impose the boundary condition that $u^\varepsilon(x, t) - Ax$ is periodic, where $A = \partial_x U$ is evaluated at the location of interest.

Denote the micro solver as

$$u^{n+1} = \mathcal{S}_{\delta\tau, \delta x}(u^n; A). \tag{6.5.20}$$

In HMM, assuming that we have the numerical approximation $\{U_j^n\}$ (where $t^n = n\Delta t, U_j^n \sim U(n\Delta t, j\Delta x)$) at the nth macro time step, we obtain the numerical approximation at the next macro time step as follows.

(1) For each j, let $A_j^n = (U_j^n - U_{j-1}^n)/\Delta x$.
(2) Re-initialize the micro solver in such a way that $u_j^0(x) - A_j^n x$ is periodic for each j.
(3) Apply the micro solver for M steps:

$$u_j^{n,m+1} = \mathcal{S}_{\delta\tau, \delta x}(u_j^{n,m}; A_j^n),$$

   with $m = 0, 1, \ldots, M - 1$.
(4) Compute

$$D_{j-1/2}^{n+1} = \left\langle a\left(x, \frac{x}{\varepsilon}, t^n\right) \partial_x u_j^{n,M} \right\rangle.$$

(5) Evolve the macrostate variables using

$$U_j^{n+1} = U_j^n + \Delta t \frac{D_{j+1/2}^{n+1} - D_{j-1/2}^{n+1}}{\Delta x}.$$

In comparison, given $\{U_j^n\}$ where U_j^n denotes the numerical approximation at time $t^n = n\tilde{\Delta}t$ inside the jth cell, the seamless strategy works as follows.

(1) For each j, let $A_j^n = (U_j^n - U_{j-1}^n)/\Delta x$.

(2) Evolve the microstate variables for one micro time-step:

$$u_j^{n+1} = \mathcal{S}_{\delta\tau,\delta x}(u_j^n; A_j^n).$$

(3) Compute

$$D_{j-1/2}^{n+1} = \left\langle a\left(x, \frac{x}{\varepsilon}, t^n\right) \partial_x u_j^{n+1}\right\rangle,$$

averaged over one period for the fast variable.

(4) Advance the macrostate for one reduced macro time-step:

$$U_j^{n+1} = U_j^n + \tilde{\Delta}t \frac{D_{j+1/2}^{n+1} - D_{j-1/2}^{n+1}}{\Delta x}.$$

It should be noted that, as in HMM, the seamless algorithm is also a top-down strategy based on a preselected macroscale solver. In this sense it can be regarded as a seamless version of HMM.

6.6 Application to fluids

The examples discussed above are all rather simplistic. As a more realistic example, we discuss how HMM and the seamless algorithm can be used to model the macroscopic fluid dynamics of complex molecules, under the assumption that the stress depends only on the rate of strain.

The basic formulation The microscopic model that we will use is a molecular dynamics model for chain molecules. Assume that we have N molecules, each molecule consisting of L beads connected by springs. Each bead moves according to Newton's equation:

$$m\frac{d^2\mathbf{x}_j}{d\tau^2} = -\frac{\partial V}{\partial \mathbf{x}_j}, \tag{6.6.1}$$

where $j = 1, \ldots, LN$ accounts for all the beads. The interaction potential consists of two parts.

(1) All beads interact via the Lennard-Jones (LJ) potential:

$$V^{\mathrm{LJ}}(r) = 4\varepsilon_0\left(\left(\frac{\sigma}{r}\right)^{12} - \left(\frac{\sigma}{r}\right)^6\right), \tag{6.6.2}$$

where r is the distance between the beads and ε_0 and σ are energy and length parameters respectively.

(2) There is an additional interaction between neighboring beads in each molecule via a spring force, modeled by the FENE potential (see Section 5.1):

$$V^{\text{FENE}}(r) = \begin{cases} \frac{1}{2}kr_0^2 \ln\left(1 - \left(\frac{r}{r_0}\right)^2\right) & \text{if } r < r_0, \\ \infty & \text{if } r \geq r_0. \end{cases} \tag{6.6.3}$$

In principle we should work with a set of compressible flow equations at the macro level and an NVE ensemble at the micro level, as discussed in Section 6.3, using equations (6.3.9) as the starting point for linking the macro and micro models. In practice we will make two approximations, for the sake of convenience. First, we will assume that the flow is incompressible at the macroscale. Second, we will work with the NVT ensemble in the molecular dynamics, i.e. we will impose constant temperature in the molecular dynamics [37] (see Section 4.2). This constraint is imposed by using various kinds of thermostat [37]. After making these two assumptions the conservation laws (6.3.9) with $\boldsymbol{\sigma}$ defined by the Irving–Kirkwood formula are no longer exact. However, we will continue to use the Irving–Kirkwood formula as the basis for extracting macroscopic stress from the MD data. This should be an acceptable approximation but its accuracy remains to be carefully validated. For simplicity only, we will also limit ourselves to the situation when the macroscale flow is two dimensional. The molecular dynamics, however, is done in three dimensions, and the simple periodic boundary condition is used in the third direction.

The macroscale model and the macro solver We will assume that the macroscopic density is a constant, ρ_0. Under this assumption the macroscopic model is of the form

$$\begin{cases} \rho_0\left(\partial_t \mathbf{U} + \nabla \cdot (\mathbf{U} \otimes \mathbf{U})\right) - \nabla \cdot \boldsymbol{\sigma}_{\text{s}} = 0, & \mathbf{x} \in \Omega, \\ \nabla \cdot \mathbf{U} = 0, \end{cases} \tag{6.6.4}$$

where $\mathbf{U}$ is the macroscopic velocity field and $\boldsymbol{\sigma}_{\text{s}}$ is the stress tensor. The data D that need to be supplied from the micro model are the components of the stress tensor $\boldsymbol{\sigma}_{\text{s}}$. We will make the assumption that the stress depends only on $\nabla \mathbf{U}$. As we remarked earlier (Section 6.3), since the macroscale model is in the form of the equations for incompressible flow it is natural to use the projection method as the macro solver [11]. Let us denote the time-step by Δt (or $\tilde{\Delta} t$ in the seamless method), and the numerical solution at time $t_n = n\Delta t$ by $\mathbf{U}^n$. In the projection method, we discretize the time derivative in the momentum equation using the forward Euler scheme:

$$\rho_0 \frac{\tilde{\mathbf{U}}^{n+1} - \mathbf{U}^n}{\Delta t} + \nabla \cdot (\rho_0 \mathbf{U}^n \otimes \mathbf{U}^n - \boldsymbol{\sigma}_{\mathrm{s}}^n) = 0. \qquad (6.6.5)$$

For the moment, the pressure as well as the incompressibility condition are neglected. Next, the velocity field $\mathbf{U}^{n+1}$ at the new time step, $t_{n+1} = (n+1)\Delta t$, is obtained by projecting $\tilde{\mathbf{U}}^{n+1}$ onto the divergence-free subspace:

$$\rho_0 \frac{\mathbf{U}^{n+1} - \tilde{\mathbf{U}}^{n+1}}{\Delta t} + \nabla P^{n+1} = 0, \qquad (6.6.6)$$

where the pressure field P^{n+1} is determined by the incompressibility condition

$$\nabla \cdot \mathbf{U}^{n+1} = 0.$$

In terms of the pressure field this becomes:

$$\Delta P^{n+1} = \frac{\rho_0}{\Delta t} \nabla \cdot \tilde{\mathbf{U}}^{n+1}, \qquad (6.6.7)$$

with a Neumann type of boundary condition.

The spatial derivatives in the equations above are discretized using a staggered grid, as shown in Figure 6.8. We denote the two components of the velocity field by U and V. In the staggered grid, U is defined at $\{(x_i, y_{j+1/2})\}$, V is defined at $\{(x_{i+1/2}, y_j)\}$ and P is defined at the center of each cell, $\{(x_{i+1/2}, y_{j+1/2})\}$. The diagonal components of σ_{s} are defined at $\{(x_{i+1/2}, y_{j+1/2})\}$, and the off-diagonal components are defined at $\{(x_i, y_j)\}$. The operators ∇ and Δ in equations (6.6.5)–(6.6.7) are discretized by standard central difference and the five-point formulas respectively.

The HMM algorithm Given an initial value $\mathbf{U}^0$ for the macro model, we set $n = 0$ and proceed as follows.

(1) Compute the velocity gradient $\mathbf{A}^n = \nabla \mathbf{U}^n$ at each grid point where the stress is needed.
(2) Initialize an MD system at each grid point.

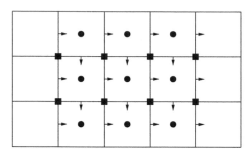

Figure 6.8. A staggered grid for the discretization of the spatial derivatives in the macro model of incompressible fluid dynamics. The components of the velocity field $\mathbf{U}$ are U, V. The component U is defined at the mid-point of the vertical edges, the component V is defined at the mid-point of the horizontal edges and P is defined at the centers of the cell. The diagonal components of the momentum flux are computed at the cell centers, indicated by the circles, and the off-diagonal components are computed at the grid points indicated by the squares.

(3) Evolve each MD system for M steps with a micro time-step $\delta\tau$. Each MD system is constrained by the local velocity gradient $\mathbf{A}^n$ through the boundary condition. More details about this constrained MD will be discussed later.

(4) Compute the stress $\boldsymbol{\sigma}_{\mathrm{s}}^n$ from the MD results. This is done using the Irving–Kirkwood formula after a sufficient number of relaxation steps. The stress is averaged over each MD box as in (6.6.12) (see below) and also over time to reduce the statistical error. More details will be given later.

(5) Evolve the macro model for one macro step, thus obtaining $\mathbf{U}^{n+1} \sim \mathbf{U}((n+1)\Delta t)$. This is done using the projection method (6.6.5)–(6.6.7) with a macro time-step Δt and the stress computed from MD.

(6) Set $n := n+1$, and go to step 1.

The seamless algorithm At each step, the procedure is as follows.

(1) Compute the velocity gradient $\mathbf{A}^n = \nabla \mathbf{U}^n$ at each grid point.

(2) Evolve each MD system and the equation for the MD box (6.6.9) (see below) for one micro time-step $\delta\tau$. For the MD one uses a periodic boundary condition with respect to an evolving box. The details of that will be discussed next.

(3) Compute the stress $\boldsymbol{\sigma}_s^n$ from the MD results. As in HMM, the stress is averaged over the MD box.

(4) Update the macro solution for one macro step $\tilde{\Delta}t$ to obtain $\mathbf{U}^{n+1} \sim \mathbf{U}((n+1)\tilde{\Delta}t)$. This is done using the scheme in (6.6.5)–(6.6.7) but with Δt replaced by the (smaller) time step $\tilde{\Delta}t$.

(5) Set $n := n + 1$, and go to step 1.

Compared with HMM, the main difference is that the micro solver (here the MD) runs continuously without re-initialization at each time-step. MD communicates with the macro solver at every time-step: the constraint (i.e. the velocity gradient) imposed on the MD changes at each MD step while the instantaneous stress computed from the MD step is used to evolve the macro velocity field. The simplest version of this algorithm for shear flows was introduced in [64].

Constrained MD and stress estimation Next we discuss how to set up the molecular dynamics in such a way to ensure that its average velocity field coincides with the local macroscale velocity field. We have assumed that the stress is a function of the rate of strain only. Therefore, associated with each macro grid point where the stress is needed, we make the approximation that the macroscale velocity field is a linear function:

$$\mathbf{U} = \mathbf{U}_0 + \mathbf{A}\mathbf{x}, \tag{6.6.8}$$

where $\mathbf{A}$ is the rate of strain tensor at the given macroscale grid point. According to our assumption, $\boldsymbol{\sigma}_s$ depends only on $\mathbf{A}$. Therefore, without loss of generality we can set $\mathbf{U}_0 = 0$. We will set up a molecular dynamics system which is constrained in such a way that its average strain rate coincides with $\mathbf{A}$. This constraint is imposed by using a periodic boundary condition for particle positions with respect to an evolving computational box. The vertices of the MD box move according to the dynamics:

$$\frac{d\mathbf{X}}{d\tau} = \mathbf{A}\mathbf{X}(\tau). \tag{6.6.9}$$

When a particle goes outside the simulation box, it is inserted back from the opposite side of the box and, at the same time, its velocity is modified according to the imposed velocity gradient: $\bar{\mathbf{v}}_i = \mathbf{v}_i + \mathbf{A}(\bar{\mathbf{x}}_i - \mathbf{x}_i)$. This is a generalization of the Lees–Edwards boundary condition for shear flows [2]. See [65] for more details.

From the results of the molecular dynamics we then compute the local stress using a modification of the Irving–Kirkwood formula. We split the particle velocity into two parts:

$$\mathbf{v}_i = \mathbf{v}_i' + \mathbf{U}(\mathbf{x}_i, \tau),$$

where $\mathbf{v}_i'$ is the fluctuating part of the velocity and $\mathbf{U}(\mathbf{x}) = \langle \sum_i \mathbf{v}_i \delta(\mathbf{x} - \mathbf{x}_i) \rangle$ is the mean velocity. Since the MD is constrained in such a way that the average velocity field is $\mathbf{U} = \mathbf{A}\mathbf{x}$, the fluctuating part of the velocity of the particles in (6.6.11) below is $\mathbf{v}_i' = \mathbf{v}_i - \mathbf{A}\mathbf{x}_i$. The average momentum flux $\tilde{\boldsymbol{\sigma}}$ can then be written as

$$\langle \tilde{\boldsymbol{\sigma}} \rangle = \langle \tilde{\rho} \rangle \mathbf{U} \otimes \mathbf{U} - \langle \tilde{\boldsymbol{\sigma}}_s \rangle, \tag{6.6.10}$$

where $\tilde{\boldsymbol{\sigma}}_s$ is given by

$$
\begin{aligned}
\tilde{\boldsymbol{\sigma}}_s = & -\sum_i m \mathbf{v}_i'(\tau) \otimes \mathbf{v}_i'(\tau) \delta(\mathbf{x} - \mathbf{x}_i(\tau)) \\
& - \tfrac{1}{2} \sum_i \sum_{j \neq i} (\mathbf{x}_i(\tau) - \mathbf{x}_j(\tau)) \otimes \mathbf{f}_{ij}(\tau) \\
& \times \int_0^1 \delta\left(\mathbf{x} - (1 - \lambda)\mathbf{x}_j(\tau) - \lambda \mathbf{x}_i(\tau)\right) d\lambda.
\end{aligned}
\tag{6.6.11}
$$

The ensemble average in (6.6.10) can be replaced or supplemented by a spatial average (or a spatial-temporal average, as in HMM):

$$\boldsymbol{\sigma}_s(n\tilde{\Delta}t) = \frac{1}{|\omega|} \int_\omega \tilde{\boldsymbol{\sigma}}_s(\mathbf{x}, n\delta\tau) \, d\mathbf{x}. \tag{6.6.12}$$

Here ω is the (time-dependent) MD simulation box and $|\omega|$ is the volume of ω.

There is a basic difference between HMM and the seamless algorithm regarding how the stress is obtained from the MD. In HMM, at each macro time step a constrained MD is performed and the stress is then computed following the dynamics. During the MD simulation, the matrix $\mathbf{A}$ in (6.6.9) is fixed at the value given by the macro state at that macro time-step. Usually the above formula is also averaged over time to reduce the statistical fluctuations. In the seamless algorithm, however, (6.6.9) as well as the MD equations are solved simultaneously with the macro model, and the data are exchanged at every time-step. Thus the matrix $\mathbf{A}$ changes at every step. This change is very slow, however, owing to the fact that $\tilde{\Delta}t$ is smaller by a factor M than the time-step Δt used in HMM.

Example: Driven cavity flow Next we present some results of the seamless algorithm for driven cavity flow. These results are taken from [31]. The results for HMM are quite similar. We will pay special attention to the effects of the fluctuations that are intrinsic to this type of algorithm. We consider two types of fluids: the first is composed of simple particles interacting via the Lennard-Jones (LJ) potential. The second is a polymer fluid, which, at the microscale, is modeled by the bead–spring model described earlier. Under normal conditions the LJ fluid is approximately Newtonian (i.e. the stress and the rate of strain have a linear relationship) and the macro flow behavior can be accurately described by the Navier–Stokes (NS) equation. Therefore, we will regard the solution of the Navier–Stokes equation as the exact solution and use it as a benchmark for the multiscale method. The details of the numerical parameters can be found in [31].

The numerical results for the LJ fluid are shown in Figure 6.9. The left-hand column of the figure shows the instantaneous streamlines obtained from the seamless multiscale method at several times. As a comparison, in the right-hand column we show numerical results for the Navier–Stokes equation at the same times. The parameters used in the Navier–Stokes equation are measured from separate MD simulations of the LJ fluid at the same density and temperature ($\rho = 0.81$, $\mu = 2.0$ in atomic units). From the figure we see that the results of the multiscale method agree very well with the solution of the Navier–Stokes equation.

To assess the performance of the seamless method further, we show in Figure 6.10 the x-component of the velocity as a function of time at two locations. The broken curves are the solution for the seamless method. As a comparison, we also plot the solution of the Navier–Stokes equation (solid curves). From the figures we see that the major difference in the two results is the fluctuation in the solution for the multiscale method. This is to be expected and is due to the statistical fluctuations in the stress tensor computed from MD. This is indeed a major difficulty for such multiscale methods. It can be improved in various ways, e.g. by employing ensemble averaging (i.e. using many MD replicas at each grid point), using a larger MD system (consequently the instantaneous stress will be averaged over larger space, see (6.6.12)) or by reducing the macro time-step (see the

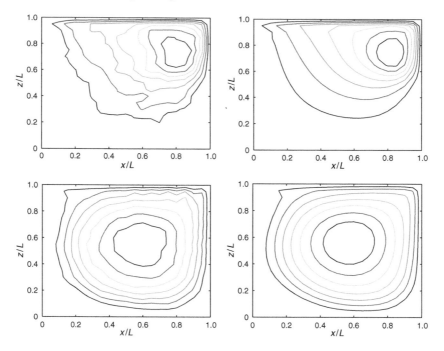

Figure 6.9. Streamlines of a Lennard-Jones fluid in a driven cavity flow at an early time ($t = 7.5 \times 10^3$ atomic units, upper figures) and at a late time ($t = 2.25 \times 10^4$ atomic units, lower figures). The left-hand column gives the results of the seamless multiscale method, and the right-hand column gives the numerical solution of the Navier–Stokes equation.

next example). Apart from the fluctuations, the multiscale result follows closely the solution of the Navier–Stokes equation.

Next we consider polymer fluids. Our MD system of interest contains 1000 polymers at each grid point; each polymer has 12 beads. The density of the beads is 0.81; the MD time-step is 0.002 (in atomic units). All the other parameters are the same as in the example for the LJ fluid. The numerical results are shown in Figure 6.11. In this example two different macro time-steps were used, $\tilde{\Delta}t = 0.5$ and $\tilde{\Delta}t = 0.25$; the results are shown in the two columns respectively. Comparing the two solutions we see that their overall behavior agrees very well. We also see that the solution obtained using the smaller macro time-step (the right-hand column) has fewer fluctuations. We can understand this in the following way. In the seamless method the stress tensor is implicitly averaged over time. Given a macro time T, the MD simulation in the seamless algorithm is carried out for a period $T\delta\tau/\tilde{\Delta}t$ in the clock of the macro solver;

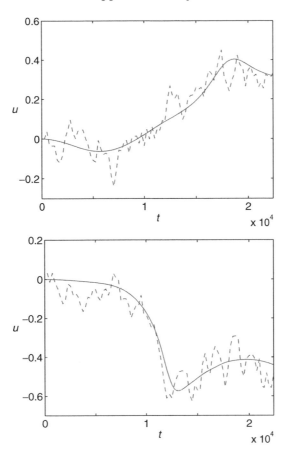

Figure 6.10. The x-component of the velocity at the locations $(1000, 1600)$ (upper panel) and $(1500, 500)$ (lower panel) as a function of time. The solid curves are solutions of the Navier–Stokes equation; the broken curves are solutions of the seamless multiscale method.

therefore reducing the macro time-step $\tilde{\Delta}t$ while keeping the micro time-step $\delta\tau$ fixed yields a longer MD simulation. Hence the numerical result has less statistical error. This is similar to HMM, in which the statistical error can be reduced by increasing the value of the parameter M, the number of MD steps in one macro time-step. This example shows that, in the seamless algorithm, the same effect can also be achieved by reducing the macro time-step. The difference is that in HMM the time averaging is done *explicitly* whereas in the seamless method it is done *implicitly*.

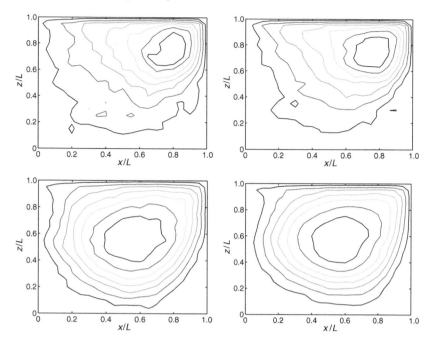

Figure 6.11. Streamlines of a polymer fluid in a driven cavity flow at an early time ($t = 7.5 \times 10^3$ atomic units, upper figures) and at a late time ($t = 2.25 \times 10^4$ atomic units, lower figures). The left-hand column gives the results of the seamless multiscale method with macro time-step $\Delta' t = 0.5$; the right-hand column gives the results for macro time-step $\Delta' t = 0.25$.

6.7 Stability, accuracy and efficiency

Error control for multiscale methods presents a new challenge to numerical analysis, particularly for problems involving multi-physics. Since multiscale methods are relatively new, and their errors typically involve several different contributions, error analysis is very much needed for understanding basic issues such as stability, accuracy and efficiency.

To begin with, what should we take as the exact or reference solution, to be compared with the numerical solution of a multiscale method? In general we should not take the detailed solution of the microscopic problem as the reference solution, since approximating that would require solving the full microscopic problem and therefore defeat the whole purpose. For type B problems, since we are only interested in the macroscale behavior of the solutions we can take Qu^ε as the reference solution, where u^ε is the solution to the microscopic model and Q is the compression (or projection) operator that maps a microscopic state to a macroscopic state. In practice, in order to obtain explicit analytical results for the behavior of the error, we have to limit ourselves to situations where we have explicit control of Qu^ε. These are situations for which limit theorems are available. These limit theorems allow us to approximate Qu^ε by $\bar{U}$, the solution to an effective macroscale model.

6.7.1 The heterogeneous multiscale method

The basic idea, as was explained in [18], is to compare the HMM solution with the solution of the selected macroscale solver for the effective macroscale model. Their difference is caused by an additional error in the HMM solution due to the error in the data estimation process. This new error term is called the HMM error and denoted by $e(\text{HMM})$. We will assume that both the HMM and the macro solver for the effective macroscale model can be expressed in the form

$$
\begin{aligned}
U_{\text{HMM}}^{n+1} &= U_{\text{HMM}}^n + \Delta t\, F^\varepsilon(U_{\text{HMM}}^n, U_{\text{HMM}}^{n-1}, \ldots), \\
\bar{U}_{\text{H}}^{n+1} &= \bar{U}_{\text{H}}^n + \Delta t \bar{F}(\bar{U}_{\text{H}}^n, \bar{U}_{\text{H}}^{n-1}, \ldots);
\end{aligned}
\tag{6.7.1}
$$

note that

$$
||Qu^\varepsilon - U_{\text{HMM}}|| \le ||Qu^\varepsilon - \bar{U}|| + ||\bar{U}_{\text{H}} - \bar{U}|| + ||U_{\text{HMM}} - \bar{U}_{\text{H}}||. \tag{6.7.2}
$$

Here $\bar{U}$ is the solution of the macroscale model, $\bar{U}_{\text{H}}$ is the numerical solution to the effective macroscale model computed using (6.7.1) and U_{HMM} is the HMM solution. The first term on the right-hand side of (6.7.2) is due to the error of the effective model and the second term to the error in the macroscale solver; the third term is the HMM error and is due to the error in the estimated data. Normally we expect that estimates of the following type will hold:

$$
||Qu^\varepsilon - \bar{U}|| \le C\varepsilon^\alpha, \tag{6.7.3}
$$

$$
||\bar{U} - \bar{U}_{\text{H}}|| \le C(\Delta t)^k, \tag{6.7.4}
$$

where k is the order of accuracy of the macro solver. In addition, define

$$
e(\text{HMM}) = \max_U ||\bar{F}(U^n, U^{n-1}, \ldots) - F^\varepsilon(U^n, U^{n-1}, \ldots)||.
$$

Then under general stability conditions one can show that [18]:

$$
||U_{\text{HMM}} - \bar{U}_{\text{H}}|| \le Ce(\text{HMM}) \tag{6.7.5}
$$

for some constant C. Therefore, we have

$$\|Qu^\varepsilon - U_{\mathrm{HMM}}\| \le C(\varepsilon^\alpha + (\Delta t)^k + e(\mathrm{HMM})).$$

The key to getting concrete error estimates and thereby giving guidelines to designing multiscale methods lies in the estimation of $e(\mathrm{HMM})$. This is specific to each problem.

Let us illustrate this general strategy using the example (6.5.1). Note that in the limit as $\varepsilon \to 0$ one expects y to tend to $\varphi(x)$, and hence that the effective dynamics for the slow variable x will be given by

$$\frac{d\bar{x}}{dt} = f(\bar{x}, \varphi(\bar{x})) = F(\bar{x}). \tag{6.7.6}$$

In fact it can be easily shown that if we denote by $(x_\varepsilon, y_\varepsilon)$ the solution of (6.5.1) and if $x_\varepsilon(0) = \bar{x}(0)$ then for any fixed $T > 0$ there is a constant C_1 independent of ε, such that

$$|x_\varepsilon(t) - \bar{x}(t)| \le C_1 \varepsilon$$

if $t \le T$. This is (6.7.3) for the present example.

Next, let us pick the forward Euler as the macro solver. If we use it to solve the effective model with time-step size Δt and if we denote the numerical solution by $\{\bar{x}_n\}$, then we have

$$|\bar{x}(t_n) - \bar{x}_n| \le C_2 \Delta t$$

if $t_n \le T$. This is the form taken by (6.7.4) for the present example.

The last component in our discussion of (6.5.1) is to estimate the HMM error. We can express HMM as:

$$x^{n+1} = x^n + \Delta t \tilde{f}(x^n)$$

for some function $\tilde{f}$ depending on the details of the HMM procedure. Let us assume that the microscale problem has been solved using forward Euler with time-step δt for M steps and that the last value is used as the estimated force for the macro solver, i.e. $\tilde{f}(x^n) = f(x^n, y^{n,M})$; then it is a simple matter to prove that [17]

$$e(\mathrm{HMM}) = |\tilde{f}(x^n) - F(x^n)| \le C|y^{n,M} - \varphi(x^n)| \le C_3 \left(\left|1 - \frac{\delta t}{\varepsilon}\right|^M + \frac{\delta t}{\varepsilon} \right). \tag{6.7.7}$$

Here the first term on the right-hand side accounts for the relaxation error for the fast variable and the second term accounts for the error in the micro-solver.

Putting all three components together, we have, for $t_n \leq T$,

$$|x_\varepsilon(t_n) - x^n| \leq C \left(\varepsilon + \Delta t + \left| 1 - \frac{\delta t}{\varepsilon} \right|^M + \frac{\delta t}{\varepsilon} \right). \qquad (6.7.8)$$

In the general case when kth-order stable macro and micro solvers are used, we have [17]:

$$|x_\varepsilon(t_n) - x^n| \leq C \left(\varepsilon + \Delta t^k + \left| A \left(\frac{\delta t}{\varepsilon} \right) \right|^M + \left(\frac{\delta t}{\varepsilon} \right)^\ell \right), \qquad (6.7.9)$$

where k and ℓ are the orders of accuracy for the macro and micro solvers respectively and $A(\lambda \delta t)$ is the amplification factor of the microscale solver for the ODE $\dot{y} = -\lambda y$.

This is what we obtain if we follow the standard procedure for analyzing an example using HMM, and it holds as long as the $\{y^{n,0}\}$ are bounded. However, if we take $y^{n,0} = y^{n-1,M}$ then (6.7.7) can be improved [26] to

$$e(\text{HMM}) \leq C \left(\frac{\varepsilon \Delta t}{M \delta t} + \Delta t \left| 1 - \frac{\delta t}{\varepsilon} \right|^M + \frac{\delta t}{\varepsilon} \right); \qquad (6.7.10)$$

similar results are obtained when higher-order macro and micro solvers are used.

In the same fashion, for stochastic ODEs (6.5.11) one can show that

$$\sup_{t \leq T} \mathbb{E} |x_\varepsilon(t) - \bar{x}(t)| \leq C \sqrt{\varepsilon},$$

where $\bar{x}$ is the solution to the limiting equation discussed in Section 6.5. If we use the re-initialization $y^{n,0} = y^{n-1,M}$ in HMM, we have, under suitable conditions [27, 26]

$$e(\text{HMM}) \leq C \left(\sqrt{\frac{\varepsilon \Delta t}{M \delta \tau}} + \left(\frac{\delta \tau}{\varepsilon} \right)^\ell \right), \qquad (6.7.11)$$

$$\sup_{t \leq T} \mathbb{E} |x_{\text{HMM}}(t) - x_\varepsilon(t)| \leq C \left(\sqrt{\varepsilon} + \sqrt{\frac{\varepsilon \Delta t}{M \delta \tau}} + \left(\frac{\delta \tau}{\varepsilon} \right)^\ell + \Delta t^k \right). \qquad (6.7.12)$$

Here k and ℓ are the (strong) orders of the macro and micro solvers respectively.

Error analysis of this type has also been carried out for finite element HMM applied to elliptic homogenization problems [30]. The structure of the error is very similar to what have we just discussed. The parabolic homogenization problem was been analyzed in [1, 60]. The error analysis and careful numerical studies for the hyperbolic homogenization problem were presented in [10].

The error analysis of HMM for the case when the effective macroscale model is stochastic was considered in [41, 40, 58].

Examples with discrete microscale models were tackled in [20]. The microscopic models can either be molecular dynamics or kinetic Monte Carlo methods. The macroscale model considered in [20] was based on either gas dynamics or general nonlinear conservation laws. It was proven in [20] that $e(\text{HMM})$ consists of three parts: the relaxation error, the error due to the finite size of the simulation box for the microscale model and the sampling error. Clearly, as the system size increases the finite size effect decreases but the relaxation error increases. Therefore this result gives some suggestions of how to choose the size of the domain and the duration for the microscale simulation. Unfortunately there are very few explicit results on the dependence of these errors on the box size. All known results were proved for lattice models. We refer to [20] for a discussion of this issue.

These examples are admittedly simple, compared with the real problems that we would like to handle. However, they do provide crucial insight into the structure of the error and they give an indication of the magnitude of each contribution.

6.7.2 The boosting algorithm

In principle, we should be able to develop a similar framework for the error analysis of the seamless algorithm. However, this program has not yet been implemented. Therefore, as in [31], we will focus on the simple examples of ODEs and SDEs discussed in Section 6.5. Denote by x_h the numerical solution for the macro variable x and by $\bar{x}$ the solution to the limiting equation (as $\varepsilon \to 0$). For these examples, the error for the seamless algorithm can be understood very simply using the observation mentioned earlier when we introduced the seamless algorithm: instead of viewing the micro and macro solvers as having different clocks, we can think of them as being run on the same clock but with a different value of ε that satisfies

$$\frac{\delta \tau}{\varepsilon} = \frac{\tilde{\Delta} t}{\varepsilon'}$$

or

$$\varepsilon' = \frac{\tilde{\Delta} t}{\delta \tau} \varepsilon. \tag{6.7.13}$$

Thus the seamless algorithm can be viewed as a standard algorithm applied to the modified problem with a modified parameter value. The error consists of two parts: that due to boosting the parameter value from ε to ε' and that due to solving the boosted model numerically. For the stiff ODE (6.2.1) we obtain the following: for any $T > 0$, there exists a constant C such that, for $t \leq T$,

$$|x_{\mathrm{h}}(t) - \bar{x}(t)| \leq C \left(\varepsilon' + \left(\frac{\tilde{\Delta} t}{\varepsilon'} \right)^{\ell} + \tilde{\Delta} t^{k} \right)$$

$$= C \left(\frac{\tilde{\Delta} t}{\delta \tau} \varepsilon + \left(\frac{\delta \tau}{\varepsilon} \right)^{\ell} + \tilde{\Delta} t^{k} \right). \tag{6.7.14}$$

In terms of the parameters entering the HMM algorithm, (6.7.14) can be written as

$$|x_{\mathrm{h}}(t) - x_{\varepsilon}(t)| \leq C \left(\varepsilon + \frac{\Delta t}{M} \frac{\varepsilon}{\delta \tau} + \left(\frac{\delta \tau}{\varepsilon} \right)^{\ell} + \left(\frac{\Delta t}{M} \right)^{k} \right). \tag{6.7.15}$$

This is to be compared with the error for the HMM algorithm given in (6.7.9) and (6.7.10). We can see that the accuracy is comparable.

Let us now consider the SDE (6.5.11). Using $\delta \tau / \varepsilon = \tilde{\Delta} t / \varepsilon'$, we see that (6.5.15) and (6.5.16) can be rewritten as

$$y^{n+1} = y^{n} - \frac{\tilde{\Delta} t}{\varepsilon'} (y^{n} - \varphi(x^{n})) + \sqrt{\frac{\tilde{\Delta} t}{\varepsilon'}} \xi^{n}, \tag{6.7.16}$$

$$x^{n+1} = x^{n} + \tilde{\Delta} t f(x^{n}, y^{n+1}), \tag{6.7.17}$$

which can be considered as a standard discretization of (6.5.11) with the parameter value of ε boosted to ε'. Therefore, the error for the seamless algorithm is controlled by [27]

$$\sup_{t \leq T} \mathbb{E}|x_\mathrm{h}(t) - \bar{x}(t)| \leq C \left(\sqrt{\varepsilon'} + \left(\frac{\tilde{\Delta} t}{\varepsilon'} \right)^\ell + (\Delta t)^k \right)$$

$$= C \left(\sqrt{\frac{\varepsilon \Delta t}{M \delta \tau}} + \left(\frac{\delta \tau}{\varepsilon} \right)^\ell + (\Delta t)^k \right), \tag{6.7.18}$$

$$\sup_{t \leq T} \mathbb{E}|x_\mathrm{h}(t) - x_\varepsilon(t)| \leq C \left(\sqrt{\varepsilon} + \sqrt{\frac{\varepsilon \Delta t}{M \delta \tau}} + \left(\frac{\delta \tau}{\varepsilon} \right)^\ell + (\Delta t)^k \right).$$

Comparing with (6.7.12), we see that the errors are comparable for this case also.

6.7.3 The equation-free approach

We begin with the stability of the projective integrators for stiff ODEs. This was analyzed in [38]. The essence of this analysis is very nicely illustrated by the following simple calculation presented in [35], in which ideas similar to that of the projective integrator were also proposed. For the sake of clarity we will follow the presentation in [35].

Consider the ODE

$$\begin{cases} \dfrac{dy}{dt} = -\dfrac{1}{\varepsilon} y, \\ y(0) = 1 \end{cases} \tag{6.7.19}$$

where $\varepsilon \ll 1$. We will use a composite scheme which consists of m steps of forward Euler with step size δt followed by one step of forward Euler with step size Δt, where $\delta t \ll \varepsilon$ and $\Delta t \gg \varepsilon$.

After one composite step the numerical solution is given by $y^1 = R_m y^0$, where the amplification factor R_m is given by

$$R_m = \left(1 - \frac{\delta t}{\varepsilon} \right)^m \left(1 - \frac{\Delta t}{\varepsilon} \right). \tag{6.7.20}$$

In order for the method to be stable, i.e. for $|R_m| \leq 1$, we simply need

$$m > -\frac{\log\left(\Delta t/\varepsilon - 1 \right)}{\log\left(1 - \delta t/\varepsilon \right)} \sim \frac{\varepsilon}{\delta t} \log\left(\Delta t/\varepsilon \right). \tag{6.7.21}$$

The values of Δt and $\delta t/\varepsilon$ are set by the error tolerance. Therefore, as $\varepsilon \to 0$ the total cost for advancing the system over an $\mathcal{O}(1)$ time interval is $\mathcal{O}(\log(1/\varepsilon))$, which may be compared with $\mathcal{O}(1/\varepsilon)$ using forward Euler with the same step sizes. This calculation illustrates well the effectiveness of solving stiff ODEs using both large and small time-steps.

Turning to the issue of spatial discretization in the equation-free approach, we will focus on patch dynamics since it contains features of both projective integrators and the gap-tooth scheme. We will restrict our attention to a one-dimensional problem and denote by $\{x_j = j\Delta x\}$ the macro grid points. Around each grid point there is a domain of size H over which the microscopic model is simulated and another domain of size h on which the microscale variables are averaged to yield the macroscale variables. Naturally, $h \leq H$. Given the values of the macro variables at the grid points, $\{U_j^n\}$, the lifted state is given by [52]

$$\tilde{u}_0(x_j) = \sum_{k=0}^{d} \frac{1}{k!} D_{k,j}(x - x_j)^k, \qquad (6.7.22)$$

where the $D_{k,j}$ are approximations to the derivatives of the macroscale profile at x_j, for example,

$$D_{2,j} = \frac{U_{j+1}^n - 2U_j^n + U_{j-1}^n}{\Delta x^2}, \quad D_{1,j} = \frac{U_{j+1}^n - U_{j-1}^n}{2\Delta x}, \quad D_{0,j} = U_j^n - \tfrac{1}{24}h^2 D_{2,j}.$$
$$(6.7.23)$$

Below we consider the case when $d = 2$ and the microscale model is the heat equation [52];

$$\partial_t u = \partial_x^2 u.$$

The effective macroscale equation is also a heat equation: $\partial_t U - \partial_x^2 U = 0$. Without loss of generality let $j = 0$ and let $\tilde{u}_0 = D_{0,0} + D_{1,0}x + \tfrac{1}{2}D_{2,0}x^2$. The solution to the microscale model after time δt is

$$\mathcal{S}_{\delta t}\tilde{u}_0(x) = D_{0,0} + D_{1,0}x + D_{2,0}(\tfrac{1}{2}x^2 + \delta t).$$

Denoting by $\mathcal{A}_h$ the averaging operator over the small domain (of size h), we have

$$\tilde{U}_{\delta t}^n = \mathcal{A}_h \mathcal{S}_{\delta t}\tilde{u}_0(x) = D_{0,0} + D_{2,0}\delta t + \tfrac{1}{24}D_{2,0}h^2 = U^n + D_{2,0}\delta t.$$

Simple first-order extrapolation gives the familiar scheme

$$U_0^{n+1} = U_0^n + \Delta t\, D_{2,0},$$

as was shown in [52]. This is both stable and consistent with the heat equation, which is the correct effective model at the macroscale.

Now let us turn to the case when the microscale model is the advection equation

$$\partial_t u + \partial_x u = 0.$$

In this case we have

$$\mathcal{S}_{\delta t}\tilde{u}_0(x) = D_{0,0} + D_{1,0}(x - \delta t) + \tfrac{1}{2}D_{2,0}(x - \delta t)^2.$$

Hence,

$$\tilde{U}_{\delta t}^n = \mathcal{A}_h \mathcal{S}_{\delta t}\tilde{u}_0(x) = D_{0,0} - D_{1,0}\delta t + \tfrac{1}{2}D_{0,0}\delta t^2 + \tfrac{1}{24}D_{0,0}h^2$$
$$= U^n - D_{1,0}\delta t + \tfrac{1}{2}D_{0,0}\delta t^2.$$

Simple first-order extrapolation yields

$$U_0^{n+1} = U_0^n + \Delta t \left(-D_{1,0} + \tfrac{1}{2}\delta t D_{2,0}\right).$$

Since $\delta t \ll \Delta t$, the last term is much smaller than the other terms and we are left essentially with the Richardson scheme,

$$U_0^{n+1} = U_0^n - \Delta t\, D_{1,0}.$$

This is unstable under the standard time-step size condition that $\Delta t \sim \Delta x$, owing to the central character of $D_{1,0}$.

There is also a consistency problem. Assume that the effective macroscale model is a fourth-order PDE. Without knowing this explicitly, the lifting operator may stop short of using any information on the fourth-order derivatives, which would lead to an inconsistent scheme. These issues are discussed in [25].

These examples are very simple but they illustrate a central difficulty with the bottom-up approach for multiscale modeling: since the microscale model is only solved on small domains for short times, it does not have enough influence on the overall scheme to bring out the character of the macroscopic process. Therefore the overall scheme does not encode enough information about the nature of the macroscale behavior. In particular, there is no guarantee of stability or consistency in such an approach.

These problems are addressed in a subsequent version of patch dynamics proposed in [68]. In this version, one assumes a macro model of the form

$$\partial_t U = F(U, \partial_x U, \ldots, \partial_x^d U, t),$$

to begin with, and then one selects a stable "method-of-lines discretization" of this macro model:

$$\partial_t U_i = F(U_i, D_i^1(U), \ldots, D_i^d(U), t).$$

Here $D^k(U)$ is some suitable finite difference discretization of $\partial_x^k U$. For example, for the advection equation one should use a one-sided discretization. The operator D^k is then used in the lifting operator:

$$\bar{u}_\varepsilon^i(x, t^n) = \sum_{k=0}^{d} D_i^k(\bar{U}^n) \frac{(x - x_i)^k}{k!}, \quad x \in \left[x_i - \frac{H}{2}, x_i + \frac{H}{2} \right].$$

The idea here is to use a stable macro solver to guide the construction of the lifting operator. This version of the patch dynamics does overcome the difficulties discussed above. However, it has lost its original appeal as an approach that does not require a preconceived form of the macroscale model. It now has the shortcomings of HMM but, unlike HMM, it uses the macro solver indirectly, through the lifting step. This adds to the complexity of the lifting operator without any obvious benefit in return. Indeed, in this new version the re-initialization process has to take into account not only consistency with the local values of the macro variables (which is the only requirement in HMM) but also some characteristics of the unknown macroscale model, such as the order of the effective macroscale PDE and the direction of the characteristic fields if the macroscale model happens to be a first-order PDE. The most unsettling aspect is that we do not know what else needs to be taken into account. This seems to be an issue for any kind of general bottom-up strategy.

6.8 Notes

Memory effects and time scale separation In much of the present chapter it has been assumed that there is a separation between the time scale for the coarse-grained variables and that for the remaining degrees of freedom. In many situations, particularly when we are analyzing the dynamics of macromolecules, it is unlikely that such a time scale separation really exists, yet it is still very desirable to consider a coarse-grained description. In this case the dynamics of the coarse-grained variables should in general have memory effects, as suggested by the Mori–Zwanzig formalism. How to handle such memory effects in practice is a question that should receive more attention.

Top-down versus bottom-up strategies An ideal multiscale method is one that relies solely on the microscale model, without the need to make a priori assumptions about the form of the macroscale model. Clearly such a strategy has to be a bottom-up strategy. Unfortunately, up to now there have been no such reliable, ideal, strategies of general interest. Several candidates have been proposed. The equation-free approach is one possibility. Another possible approach is provided by the renormalization group methods of Ceder and Curtarolo [14]. There the idea is to perform systematic coarse-graining to find the effective Hamiltonian at large scales. This is certainly an attractive strategy. But it is not clear how widely applicable it is. For example, it is not clear how to take into account dissipative processes such as heat conduction and viscous dissipation.

As we explained earlier, HMM is an example of a top-down strategy. As such, it represents a compromise between idealism and practicality. In some sense it is the most straightforward way of extending the sequential multiscale strategy to a concurrent setting.

The fiber bundle structure The philosophy used in HMM and the seamless algorithm can be better appreciated if we formulate multiscale problems using a fiber bundle structure, in which the local microstructure is represented by the fibers. The relevance of the concept of fiber bundles was noted in [18, 26]. A systematic treatment was presented in [22]. It is not clear that there is any real substance in such a viewpoint, but it is sometimes a convenient way to think about this class of multiscale problems. In addition, it gives rise to some interesting models (see [22]).

We start with a simple example. Consider the nonlinear homogenization problem

$$\partial_t u^\varepsilon = \nabla \cdot \left(\mathbf{a} \left(u^\varepsilon, \mathbf{x}, \frac{\mathbf{x}}{\varepsilon} \right) \nabla u^\varepsilon \right).$$

Here $\mathbf{a}(u, \mathbf{x}, \mathbf{z})$ is a smooth and uniformly positive definite tensor function, which is periodic in $\mathbf{z}$ with period Γ. For this problem the homogenized equation takes the form

$$\partial_t U = \nabla \cdot (\mathbf{A}(U, \mathbf{x}) \nabla U), \tag{6.8.1}$$

where

$$\mathbf{A}(U, \mathbf{x}) = \frac{1}{|\Gamma|} \int_{\Gamma} \mathbf{a}(U, \mathbf{x}, \mathbf{z})(\nabla_{\mathbf{z}}\, \chi(\mathbf{z}; U, \mathbf{x}) + \mathbf{I})\, d\mathbf{z}. \qquad (6.8.2)$$

Here $\mathbf{I}$ is the identity matrix and $\chi(\mathbf{z}; U, \mathbf{x})$ is the solution of

$$\nabla_{\mathbf{z}} \cdot (\mathbf{a}(U, \mathbf{x}, \mathbf{z})(\nabla_{\mathbf{z}}\, \chi + \mathbf{I})) = 0 \quad \text{in } \Gamma, \qquad (6.8.3)$$

with periodic boundary conditions.

Intuitively, we can describe the behavior of the solution as follows. At each point $\mathbf{x}$, $u^{\varepsilon}(\mathbf{x})$ is approximated closely by the value of $U(\mathbf{x})$ and the microstructure of u^{ε} is locally described by χ, which is the solution to the cell problem (6.8.3), is parametrized by $(U, \mathbf{x})$ and is a function over Γ. We can think of χ as being the fiber over $\mathbf{x}$ that describes the local microstructure.

Another example is the Cauchy–Born rule discussed in Chapter 5. Here the macroscale behavior is the large-scale deformation of the material and the microstructure is the local behavior of the crystal lattice at each point. It is more interesting to look at the case of complex lattices. Roughly speaking, the Cauchy–Born rule suggests the following picture about the deformation of the material. The underlying Bravais lattice undergoes a smooth deformation, described by the Cauchy–Born nonlinear elasticity model. The shift vectors $(\mathbf{p}_1, \dots, \mathbf{p}_s)$, which describe the local microstructure (within each unit cell), minimize the function $W(\mathbf{A}, \mathbf{p}_1, \dots, \mathbf{p}_s)$, where $\mathbf{A}$ is given by the deformation gradient for the Bravais lattice at the point of interest. Here the microstructure is parametrized by $\mathbf{A}$ and is a function of the indices for the shift vectors. The position of the shift vectors can undergo bifurcations, as is the case for structural phase transformation.

This last example becomes more interesting when the atomistic model is an electronic structure model and the electronic structure plays the role of the internal structure (the shift vectors). In this case the problem becomes very similar to the homogenization problem discussed in the first example, the local electronic structure being the fiber that describes the local microstructure.

To explore the fiber bundle structure further, let us note that there are two fiber bundles involved (see Figure 6.12). The first is a fiber bundle in the space of independent variables, for which the macroscale space–time

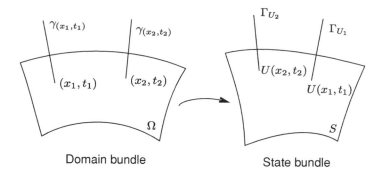

Figure 6.12. Fiber bundle structure; γ and Γ represent fibers.

domain of interest is the underlying base manifold, called the *domain base manifold*, and the domains for the additional fast variable $\mathbf{z}$ which describes the local microstructure are the fibers. The second is a fiber bundle that parametrizes the microstructure. The base manifold, the *state base manifold*, is the space of the parameters for the microstructure (e.g. the cell problem). The fibers are the space of functions over the fibers in the first fiber bundle.

A fiber bundle model generally consists of the following three components:

(1) a macroscale model

$$L(U; D(U)) = 0, \tag{6.8.4}$$

where $U = U(\mathbf{x})$ is a mapping from the domain base manifold Ω to the state base manifold S and D is the data that depend on the microstructure;

(2) the cell problem, which is a microstructure model over the fibers for any $\mathbf{x} \in \Omega$,

$$\mathcal{L}(u; U(\mathbf{x})) = 0, \tag{6.8.5}$$

where $U(\mathbf{x})$ enters as parameters (or constraints) for the microscale model;

(3) a relation that specifies the missing input to the macroscale model in terms of the solutions to the microstructure model,

$$D(U) = \mathcal{D}(U, u(\cdot; U)). \tag{6.8.6}$$

This complex language simply says that the cell problems are parametrized by U and the solutions to the cell problems give the missing data in the macro model.

For the homogenization example the first component is given by (6.8.1), the second by (6.8.3), and the third by (6.8.2).

This structure is useful in at least two ways. The first is that it gives us an underlying mathematical structure for the problems handled by HMM and the seamless algorithm. The second is that it yields very interesting models. For example, in the fiber bundle models the microstructure is usually slaved by the macrostate of the system. This is analogous to the Born–Oppenheimer model in *ab initio* molecular dynamics. One can also formulate the analog of the Car–Parrinello model by introducing relaxational dynamics for the fibers. This yields interesting models when applied to the examples discussed above. We refer to [22] for details and additional examples.

References for Chapter 6

[1] A. Abdulle and W. E, "Finite difference heterogeneous multi-scale method for homogenization problems," *J. Comput. Phys.*, vol. 191, no. 1, pp. 18–39, 2003.

[2] M.P. Allen and D.J. Tildesley, *Computer Simulation of Liquids*, Oxford University Press, 1987.

[3] R.B. Bird, R.C. Armstrong and O. Hassager, *Dynamics of Polymeric Liquids, Vol. 1, Fluid Dynamics*, John Wiley, 1987.

[4] A. Brandt, "Multi-level adaptive solutions to boundary value problems," *Math. Comp.*, vol. 31, no. 138, pp. 333–390, 1977.

[5] A. Brandt, "Multiscale scientific computation: review 2001," in *Multiscale and Multiresolution Methods: Theory and Applications, Lecture Notes in Computational Science and Engineering*, T.J. Barth *et al.*, eds., vol. 20, pp. 3–96, Springer-Verlag, 2002.

[6] Y. Cao, D. Gillespie and L. Petzold, "Multiscale stochastic simulation algorithm with stochastic partial equilibrium assumption for chemically reacting systems," *J. Comp. Phys.*, vol. 206, pp. 395–411, 2005.

[7] Y. Cao, D. Gillespie and L. Petzold, "The slow scale stochastic simulation algorithm," *J. Chem. Phys.*, vol. 122, pp. 14 116-1–14 116-18, 2005.

[8] R. Car and M. Parrinello, "Unified approach for molecular dynamics and density-functional theory," *Phys. Rev. Lett.*, vol. 55, no. 22, pp. 2471–2474, 1985.

[9] S. Chen and G.D. Doolen, "Lattice Boltzmann methods for fluid flows," *Ann. Rev. Fluid Mech.*, vol. 30, pp. 329–364, 1998.

[10] S. Chen, W. E and C.W. Shu, "The heterogeneous multiscale method based on the discontinuous Galerkin method for hyperbolic and parabolic problems," *SIAM MMS*, vol. 3, pp. 871–894, 2005.

[11] A.J. Chorin, "A numerical method for solving incompressible viscous flow problems," *J. Comput. Phys.*, vol. 2, pp. 12–26, 1967.

[12] A.J. Chorin, O.H. Hald and R. Kupferman, "Optimal prediction with memory," *Physica D*, vol. 166, pp. 239–257, 2002.

[13] A.J. Chorin, A.P. Kast and R. Kupferman, "Optimal prediction of underresolved dynamics," *Proc. Nat. Acad. Sci. USA*, vol. 95, no. 8, pp. 4094–4098, 1998.

[14] S. Curtarolo and G. Ceder, "Dynamics of an inhomogeneously coarse grained multiscale system," *Phys. Rev. Lett.*, vol. 88, pp. 255 504-1– 255 504-4, 2002.

[15] G. Dahlquist and A. Björck, *Numerical Methods*, Prentice-Hall, 1974.

[16] S.M. Deshpande, "Kinetic flux splitting schemes," in *Computational Fluid Dynamics Review*, John Wiley & Sons, 1995.

[17] W. E, "Analysis of the heterogeneous multiscale method for ordinary differential equations," *Comm. Math. Sci.*, vol. 1, no. 3, pp. 423–436, 2003.

[18] W. E and B. Engquist, "The heterogeneous multi-scale methods," *Comm. Math. Sci.*, vol. 1, pp. 87–133, 2003.

[19] W. E and B. Engquist, "Multiscale modeling and computation," *Not. AMS*, vol. 50, no. 9, pp. 1062–1070, 2003.

[20] W. E and X. Li, "Analysis of the heterogeneous multiscale method for gas dynamics," *Meth. Appl. Anal.*, vol. 11, pp. 557–572, 2004.

[21] W. E and J. Lu, "The continuum limit and the QM-continuum approximation of quantum mechanical models of solids," *Comm. Math. Sci.*, vol. 5, no. 3, pp. 679–696, 2007.

[22] W. E and J. Lu, "Seamless multiscale modeling via dynamics on fiber bundles," *Comm. Math. Sci.*, vol. 5, no. 3, pp. 649–663, 2007.

[23] W. E and P.B. Ming, "Analysis of the multiscale methods," *J. Comput. Math.*, vol. 22, pp. 210–219, 2004.

[24] W. E and E. Vanden-Eijnden, "Metastability, conformation dynamics, and transition pathways in complex systems," in *Multiscale Modeling and Simulation*, S. Attinger and P. Koumoutsakos, eds., Lecture Notes in Computational Science and Engineering, vol. 39, Springer-Verlag, 2004.

[25] W. E and E. Vanden-Eijnden, "Some critical issues for the 'equation-free' approach to multiscale modeling," unpublished, 2008.

[26] W. E, B. Engquist, X. Li, W. Ren and E. Vanden-Eijnden, "Heterogeneous multiscale methods: a review," *Comm. Comput. Phys.*, vol. 3, no. 3, pp. 367–450, 2007.

[27] W. E, D. Liu and E. Vanden-Eijnden, "Analysis of multiscale methods for stochastic differential equations," *Comm. Pure Appl. Math.*, vol. 58, no. 11, pp. 1544–1585, 2005.

[28] W. E, D. Liu and E. Vanden-Eijnden, "Nested stochastic simulation algorithm for chemical kinetic systems with disparate rates," *J. Chem. Phys.*, vol. 123, pp. 194 107-1–194 107-8, 2005.

[29] W. E, D. Liu and E. Vanden-Eijnden, "Nested stochastic simulation algorithms for chemical kinetic systems with multiple time scales," *J. Comput. Phys.*, vol. 221, no. 1, pp. 158–180, 2007.

[30] W. E, P.B. Ming and P.W. Zhang, "Analysis of the heterogeneous multiscale method for elliptic homogenization problems," *J. Amer. Math. Soc.*, vol. 18, no. 1, pp. 121–156, 2005.

[31] W. E, W. Ren and E. Vanden-Eijnden, "A general seamless strategy for multiscale modeling," *J. Comput. Phys.*, vol. 228, pp. 5437–5453, 2009.

[32] B. Eidel and A. Stukowski, "A variational formulation of the quasicontinuum method based on energy sampling in clusters," *J. Mech. Phys. Solids*, vol. 57, no. 1, pp. 87–108, 2009.

[33] B. Engquist and R. Tsai, "Heterogeneous multiscale method for a class of stiff ODEs," *Math. Comp.*, vol. 74, pp. 1707–1742, 2005.

[34] R. Erban, I.G. Kevrekidis, D. Adalsteinsson and T.C. Elston, "Gene regulatory networks: a coarse-grained, equation-free approach to multiscale computations," *J. Chem. Phys.*, vol. 124, pp. 084 106-1–084 106-17, 2006.

[35] K. Eriksson, C. Johnson and A. Logg, "Explicit time-stepping for stiff ODEs," *SIAM J. Sci. Comput.*, vol. 25, no. 4, pp. 1142–1157, 2003.

[36] I. Fatkullin and E. Vanden-Eijnden, "A computational strategy for multiscale systems with applications to Lorenz 96 model," *J. Comput. Phys.*, vol. 200, no. 2, pp. 605–638, 2004.

[37] D. Frenkel and B. Smit, *Understanding Molecular Simulation: From Algorithms to Applications*, 2nd edition, Academic Press, 2001.

[38] C.W. Gear and I.G. Kevrekidis, "Projective methods for stiff differential equations: problems with gaps in their eigenvalue spectrum," *SIAM J. Sci. Comput.*, vol. 24, pp. 1091–1106, 2003.

[39] C.W. Gear, J. Li and I.G. Kevrekidis, "The gap-tooth scheme for particle simulations," *Phys. Lett. A*, vol. 316, pp. 190–195, 2003.

[40] D. Givon and I.G. Kevrekidis, "Multiscale integration schemes for jump-diffusion systems," *SIAM MMS*, vol. 7, pp. 495–516, 2008.

[41] D. Givon, I. G. Kevrekidis, and R. Kupferman, "Strong convergence of projective integration schemes for singularly perturbed stochastic differential systems," *Comm. Math. Sci.* vol. 4, pp. 707–729, 2006.

[42] J. Goodman and A. Sokal, "Multigrid Monte Carlo method: conceptual foundations," *Phys. Rev. D*, vol. 40, no. 6, pp. 2035–2071, 1989.

[43] W. Grabowski, "Coupling cloud processes with the large-scale dynamics using the cloud-resolving convection parameterization (CRCP)," *J. Atmos. Sci.*, vol. 58, pp. 978–997, 2001.

[44] L. Greengard, "Fast algorithms for classical physics," *Science*, vol. 265, pp. 909–914, 1994.

[45] W. Hackbusch, "Convergence of multigrid iterations applied to difference equations," *Math. Comp.*, vol. 34, no. 150, pp. 425–440, 1980.

[46] E.L. Haseltine and J.B. Rawlings, "Approximate simulation of coupled fast and slow reactions for stochastic kinetics," *J. Chem. Phys.*, vol. 117, pp. 6959–6969, 2002.

[47] J. Huang, private communication.

[48] G. Hummer and Y. Kevrekidis, "Coarse molecular dynamics of a peptide fragment: free energy, kinetics, and long-time dynamics computations," *J. Chem. Phys.*, vol. 118, no. 23, pp. 10 762–10 773, 2003.

[49] M. Iannuzzi, A. Laio and M. Parrinello, "Efficient exploration of reaction potential energy surfaces using Car–Parrinello molecular dynamics," *Phys. Rev. Lett.*, vol. 90, pp. 238302-1–238302-4, 2003.

[50] J.H. Irving and J. G. Kirkwood, "The statistical mechanical theory of transport processes IV," *J. Chem. Phys.*, vol. 18, pp. 817–829, 1950.

[51] M. A. Katsoulakis, P. Plechac and L. Rey-Bellet, "Numerical and statistical methods for the coarse-graining of many-particle stochastic systems," *J. Sci. Comput.*, vol. 27, pp. 43–71, 2008.

[52] I.G. Kevrekidis, C.W. Gear, J.M. Hyman, P.G. Kevrekidis, O. Runborg and C. Theodoropoulos, "Equation-free, coarse-grained multiscale computation: enabling microscopic simulators to perform system-level analysis," *Comm. Math. Sci.*, vol. 1, no. 4, pp. 715–762, 2003.

[53] J. Knap and M. Ortiz, "An analysis of the quasicontinuum method," *J. Mech. Phys. Solids*, vol. 49, no. 9, pp. 1899–1923, 2001.

[54] K. Kremer, F. Müller-Plathe, "Multiscale problems in polymer science: simulation approaches," *MRS Bull.*, vol. 26, pp. 205–210, 2001.

[55] A. Laio and M. Parrinello, "Escaping free-energy minima," *Proc. Nat. Acad. Sci. USA*, vol. 99, pp. 12 562–12 566, 2002.

[56] R. LeVeque, *Numerical Methods for Conservation Laws*, Birkhauser, 1990.

[57] X. Li, private communication.

[58] D. Liu, "Analysis of multiscale methods for stochastic dynamical systems with multiple time scales," *Mult. Model Simul.*, vol. 8, pp. 944–964, 2010.

[59] M. Luskin and C. Ortner, "An analysis of node-based cluster summation rules in the quasi-continuum method," *SIAM J. Numer. Anal.*, vol. 47, pp. 3070–3086, 2009.

[60] P.B. Ming and P.W. Zhang, "Analysis of the heterogeneous multiscale method for parabolic homogenization problems," *Math. Comp.*, vol. 76, pp. 153–177, 2007.

[61] R.G. Parr and W. Yang, *Density Functional Theory of Atoms and Molecules*, Oxford University Press, 1989.

[62] B. Perthame and F. Coron, "Numerical passage from kinetic to fluid equations," *SIAM J. Numer. Anal.*, vol. 28, pp. 26–42, 1991.

[63] D.I. Pullin, "Direct simulation methods for compressible inviscid ideal gas flow," *J. Comput. Phys.*, vol. 34, pp. 231–244, 1980.

[64] W. Ren, "Seamless multiscale modeling of complex fluids using fiber bundle dynamics," *Comm. Math. Sci.*, vol. 5, pp. 1027–1037, 2007.

[65] W. Ren and W. E, "Heterogeneous multiscale method for the modeling of complex fluids and microfluidics," *J. Comput. Phys.*, vol. 204, no. 1, pp. 1–26, 2005.

[66] R.E. Rudd and J.Q. Broughton, "Coarse-grained molecular dynamics and the atomic limit of finite elements," *Phys. Rev. B*, vol. 58, no. 10, pp. R5893–R5896, 1998.

[67] R.E. Rudd and J.Q. Broughton, "Coarse-grained molecular dynamics: nonlinear finite element and finite temperature," *Phys. Rev. B*, vol. 72, pp. 144 104–144 136, 2005.

[68] G. Samaey, D. Roose, and I.G. Kevrekidis, "Finite difference patch dynamics for advection homogenization problems," in *Model Reduction and Coarse-Graining Approaches for Multiscale Phenomena, Part II*, A. N. Gorban, ed., pp. 225–246, Springer-Verlag, 2006.

[69] A. Samant and D.G. Vlachos, "Overcoming stiffness in stochastic simulation stemming from partial equilibrium: a multiscale Monte Carlo algorithm," *J. Chem. Phys.*, vol. 123, pp. 144 114-1–144 114-8, 2005.

[70] R.H. Sanders and K.H. Prendergast, "On the origin of the 3-kiloparsec arm," *Astrophys. J.*, vol. 188, pp. 489–500, 1974.

[71] S. Succi, O. Filippova, G. Mith and E. Kaxiras, "Applying the lattice Boltzmann equation to multiscale fluid problems," *Comp. Sci. Eng.*, vol. 3, pp. 26, 2001.

[72] W. Sweldens, "The lifting scheme: a construction of second generation wavelets," *SIAM J. Math. Anal.*, vol. 29, no. 2, pp. 511–546, 1997.

[73] E.B. Tadmor, M. Ortiz and R. Phillips, "Quasicontinuum analysis of defects in crystals," *Phil. Mag.*, vol. 73, pp. 1529–1563, 1996.

[74] M. Tuckerman, "Molecular dynamics algorithm for multiple time scales: systems with disparate masses," *J. Chem. Phys.*, vol. 94, pp. 1465–1469, 1991.

[75] E. Vanden-Eijnden, "Numerical techniques for multiscale dynamical systems with stochastic effects," *Comm. Math. Sci.*, vol. 1, no. 2, pp. 385–391, 2003.

[76] E. Vanden-Eijnden, "On HMM-like integrators and projective integration methods for systems with multiple time scales," *Comm. Math. Sci.*, vol. 5, pp. 495–505, 2007.

[77] Y. Xing, W. Grabowski and A.J. Majda, "New efficient sparse space–time algorithms for superparameterization on mesoscales," *Mon. Weather Rev.*, vol. 137, pp. 4307–4324, 2009.

[78] K. Xu and K.H. Prendergast, "Numerical Navier–Stokes solutions from gas kinetic theory," *J. Comput. Phys.*, vol. 114, pp. 9–17, 1994.

7

Resolving local events or singularities

We now turn to type A problems. These are problems for which a macroscopic model is sufficiently accurate except in isolated regions where some important events take place and a microscale model is needed to resolve the details of these local events. Examples of such local events include chemical reactions, interfaces, shocks, boundary layers, cracks, contact lines, triple junctions and dislocations. Conventional macroscale models fail to resolve the important details of these events. Yet, other than for these local events a full microscopic description is unnecessary. Therefore a coupled macro–micro modeling approach is preferred in order to capture both the macroscale behavior of the system and the important details of such local events or singularities.

One of the earliest and most well-known examples of this type of approach is the coupled quantum mechanics and molecular mechanics (QM–MM) method proposed by Warshel and Levitt during the 1970s for modeling the chemical reactions of biomolecules [52] (see also [51]). In this approach, quantum mechanics models are used in regions where the chemical reaction takes place, and classical molecular mechanics models are used elsewhere. Such a combined QM–MM approach has now become a standard tool in chemistry and biology. This kind of philosophy has been applied to a variety of problems in different disciplines.

In the general setting, a locally coupled macro–micro modeling strategy involves the following major components:

(1) *A coupling strategy.* As we discuss below, there are several choices: the domain decomposition method, adaptive model refinement or adaptive model reduction and the heterogeneous multiscale method (HMM);

(2) *The data*, usually in the form of boundary conditions, to be passed between the macro and micro models.

(3) *The implementation of the boundary conditions* for the macro and micro models.

The second and third components constitute the macro–micro interface conditions. This is the main issue for type A problems, namely, how should we formulate the interface conditions in such a way that the overall macro–micro algorithm is stable and accurate? Of particular interest is the consistency between the different models at the interface where the two models meet.

Since the nature of the different models can be quite different, linking them together consistently can be a considerable task. For example, quantum mechanics models describe explicitly the charge density due to the electrons; this information is lost in classical models, however. Consequently charge balance is a very serious issue in coupled QM–MM formulations. As another example, the temperature is treated as a continuous field in continuum mechanics. However, in molecular dynamics the temperature is a measure of the magnitude of the velocity fluctuations of the atoms. These are drastically different representations of the same physical quantity. For this reason, coupling continuum mechanics and molecular dynamics at finite temperatures is a highly nontrivial task.

In what follows, we will first discuss general coupling strategies. We will then discuss how to assess the performance of these coupling strategies in terms of stability and consistency, two most important notions in classical numerical analysis. As expected these classical notions take on a new dimension in the current context, owing to the different natures of the physical models involved.

7.1 Domain decomposition method for type A problems

This is a rather popular strategy, used, for example in the QM–MM method. The basic idea is similar to that of the standard domain decomposition method (see Chapter 3), except that different models are used on different domains. The places where interesting local events occur are covered by domains on which microscale models are used. The sizes of the domains are chosen for maximum accuracy and efficiency. The domains may or may not overlap, and they may change as the computation proceeds.

Specifically, let Ω be the physical domain of interest, and assume that Ω can be decomposed into the union of two domains, Ω_1 and Ω_2: $\Omega = \Omega_1 \cup \Omega_2$. Different models will be used on Ω_1 and Ω_2. Our task is to formulate a model for the whole domain Ω.

Regarding whether Ω_1 and Ω_2 overlap, we may have several different situations:

(1) Ω_1 and Ω_2 overlap;
(2) Ω_1 and Ω_2 do not overlap but are appended with buffer regions to deal with the boundary conditions for the subsystems on Ω_1 and Ω_2;
(3) Ω_1 and Ω_2 intersect only along an interface, and no buffer regions are used.

7.1.1 Energy-based formulation

If models Ω_1 and Ω_2 are both in the form of variational principles (which usually amount to minimizing some energy, perhaps the free energy) then it is natural to look for a variational formulation for the full problem on Ω. The starting point of an energy-based formulation is a total energy in the form

$$E_{\text{tot}} = E_1 + E_2 + E_{\text{int}}, \tag{7.1.1}$$

where E_{tot} is the total energy of the system, E_1 and E_2 are the energies associated with Ω_1 and Ω_2 respectively and E_{int} is the energy due to the interaction between the subsystems on Ω_1 and Ω_2.

The QM–MM method The most well-known example of this class of models is the QM–MM method. In QM–MM models of macromolecules, one uses a QM model on one part of the molecule, say Ω_1, and an MM model on the rest of the molecule, Ω_2. Equation (7.1.1) becomes

$$H_{\text{tot}} = H_1 + H_2 + H_{\text{int}}, \tag{7.1.2}$$

where H_1 is the quantum mechanical Hamiltonian on Ω_1, H_2 is the contribution from Ω_2 computed using a classical mechanics model and H_{int} is the contribution to the energy due to the interaction between the subsystems on Ω_1 and Ω_2.

There are two important issues for this model. The first is how to compute H_{int}. The second is how to impose boundary conditions for the quantum mechanical component of the system. In particular, how should we terminate covalent bonds?

Regarding the first issue, the simplest strategy is to use a classical empirical potential to treat the long-range interactions, i.e. the electrostatic and van der Waals interactions [52]:

$$H_{\text{int}} = H_{\text{elec}} + H_{\text{vdW}}.$$

For example, for the van der Waals interaction one may use

$$H_{\text{vdW}} = \sum_{I,j} 4\varepsilon_{I,j} \left(\left(\frac{\sigma_{I,j}}{|\mathbf{x}_I - \mathbf{x}_j|} \right)^{12} - \left(\frac{\sigma_{I,j}}{|\mathbf{x}_I - \mathbf{x}_j|} \right)^{6} \right),$$

where the subscripts I and j denote atoms in the QM and MM regions respectively. The quantities $\varepsilon_{I,j}$ and $\sigma_{I,j}$ are empirical parameters obtained from standard optimization procedures. The electrostatic interaction is often modeled in the form

$$H_{\text{elec}} = -\sum_{j} \int_{\Omega_1} \frac{Q_j \, \rho(\mathbf{r})}{|\mathbf{r} - \mathbf{x}_j|} \, d\mathbf{r} + \sum_{I,j} \frac{Z_I Q_j}{|\mathbf{x}_I - \mathbf{x}_j|},$$

where Z_I is the nuclear charge of the Ith atom in the QM region, Q_j is the partial charge of the jth atom in the MM region and $\rho(\cdot)$ is the electron density in the QM region.

We now turn to the second issue. When a covalent bond cuts across the interface between the QM and MM regions, the electrons that contribute to the covalent bond are explicitly represented on the QM side but not on the MM side. This gives rise to some difficulties, such as the stabilization of the covalent bonds and the charge balance. In addition, the bond, bond angle and dihedral angle terms in the MM potential need a partner in the QM region on which to act.

Numerous proposals have been made to deal with this issue. Some representative proposals are as follows:

(1) *The use of "link atoms"* [47]. An atom, usually hydrogen, is added to the QM–MM bond. This link atom satisfies the valence of the QM region. Depending on the formulation, it may or may not interact with the atoms in the MM region.

(2) *The use of hybrid orbitals* [52, 19]. An orbital is added to the QM–MM bond. Depending on the formulation, it may or may not enter into the self-consistent iteration for the whole system.

(3) *The pseudo-bond approach* [55]. The boundary atom in the MM region is replaced by an atom with one valence electron and an effective core potential. This atom then forms a pseudo-bond with the boundary atom in the QM region. This pseudo-bond is taken into account only in the QM part of the energy.

At this point in time the QM–MM method is still something of an art. For a thorough review of the subject we refer to [18].

For crystalline solids, similar strategies were used in [1] to construct coupled tight-binding and molecular dynamics models for the simulation of crack propagation in silicon, which is a covalently bonded crystal. The extension to metallic systems was considered in [5], in which methods that couple orbital-free density functional theory and empirical embedded atom models were proposed.

The Arlequin method One strategy for coupling different models is to blend them together using a transition function. One of the earliest examples of this type was the *Arlequin method*, which is based on blending together the energies of the different models in the overlap region, using a transition function, and enforcing consistency between the macro and micro states by explicit constraints [8]. Take for example the case when the macroscale model on Ω_1 is in the form

$$E_1(\mathbf{U}) = \int_{\Omega_1} W(\nabla \mathbf{U}(\mathbf{x})) \, d\mathbf{x},$$

and the microscale model on Ω_2 is an atomistic model in the form

$$E_2(\mathbf{u}) = \sum_{\mathbf{x}_j \in \Omega_2} V_j,$$

where V_j is the energy associated with the jth atom. Let $S = \Omega_1 \cap \Omega_2$; then the total energy for the coupled system has the form

$$E_{\text{tot}} = \int_{\Omega_1} \alpha(\mathbf{x}) W(\nabla \mathbf{U}(\mathbf{x})) \, d\mathbf{x} + \sum_{\mathbf{x}_j \in \Omega_2} \beta(\mathbf{x}_j) V_j + C(\lambda, \mathbf{U}, \{\mathbf{x}_j\}), \quad (7.1.3)$$

where the transition functions α and β satisfy the conditions

$$\alpha + \beta = 1, \quad \text{on } \Omega$$
$$\alpha = 1, \quad \text{on } \Omega_1 \setminus S$$
$$\beta = 1, \quad \text{on } \Omega_2 \setminus S.$$

The last term in (7.1.3) is a Lagrange multiplier term enforcing the constraint that the macro- and microstates are consistent with each other on S.

Other strategies that use similar ideas can be found in [3, 2]. See also [6] for an example of the blending of different differential equations by use of a transition function.

7.1.2 Dynamic atomistic and continuum methods for solids

We now turn to methods for modeling dynamics. We first discuss solids with defects. Many proposals have been made on constructing coupled atomistic and continuum models for simulating the dynamics of such solids. We will mention briefly a few representative ideas.

Energy-based formulation One first constructs a total energy for the coupled system. The dynamics is then modeled simply by Hamilton's equation for with the total energy. This is the strategy followed in [1, 43]. This idea is appealing for its simplicity. However, it has two notable shortcomings: it is very difficult to take into account thermal effects, particularly the effect of thermal gradients, in this setting. Secondly, it is also difficult to improve the coupling conditions along the interface between the atomistic and continuum models.

Force-based formulation So far, this has been limited to the case of zero temperature. The continuum model is elastodynamics and the atomistic model is molecular dynamics. To ensure consistency between the two models, the continuum elastodynamics model is derived from the Cauchy–Born rule applied to the atomic model (see Section 4.2). The main question is how to ensure consistency between the two models at the atomistic–continuum interface. This is a rather complex issue. We will discuss this issue for static problems later in the chapter, but much remains to be done for dynamic problems.

Examples of such an approach can be found in [2, 12, 39].

As we will explain in the notes at the end of this chapter, this approach allows us to improve the coupling condition at the interface. However, it is still limited to the zero temperature case: taking into account thermal effects is rather difficult in this setting.

Formulation based on conservation laws The most general formulation uses conservation laws. As explained in Chapter 4, one can formulate conservation laws of mass, momentum and energy for both the continuum and atomistic models. This provides a natural starting point for formulating coupled models. A notable example is given in [27]. We will return to this in Section 7.3.

7.1.3 Coupled atomistic and continuum methods for fluids

There is a similar story for fluids. The literature on this topic is equally vast; see for example [36, 24, 23, 25, 26, 35, 53, 17, 7, 41]. We will not attempt to review the literature. Instead, we will examine a flow between parallel plates, to illustrate the main ideas. Our presentation follows closely the setup in [40].

The channel is defined by $\{(x, y, z) \in \mathbb{R}^3, \ -\infty < x, z < \infty, \ -L \leq y \leq L\}$. We will consider the simplest situation, when the macroscopic flow is a shear flow, i.e. if we denote the velocity field by $\mathbf{U} = (U, V, W)$ then $U = U(t, y), V = 0, W = 0$.

The continuum equation in this case reduces to

$$\rho \partial_t U - \partial_y \sigma_{12} = 0, \tag{7.1.4}$$

where ρ is the density of the fluid and σ_{12} is the value of the shear stress. We will consider Lennard-Jones fluids, for which σ_{12} is very accurately modeled by a linear constitutive relation for the type of shear rate considered here:

$$\sigma_{12} = \mu \partial_y U, \tag{7.1.5}$$

where μ is the viscosity of the fluid. The value of σ_{12} can be obtained from a MD simulation of shear flows as the ratio between the measured shear stress and the imposed shear rate. Combining (7.1.4) and (7.1.5), we obtain

$$\rho \partial_t U = \mu \partial_y^2 U. \tag{7.1.6}$$

At the boundary $y = \pm L$, we will use the no-slip boundary condition $U = 0$.

At the microscopic level the system is made up of particles interacting with the Lennard-Jones potential (see Section 4.2). Their dynamics obey Newton's equation. A thermostat is applied to maintain the system at the desired temperature T.

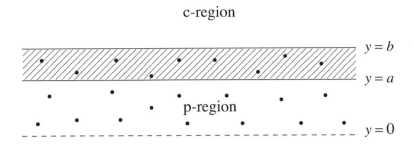

Figure 7.1. Illustration of the domain decomposition method for coupled atomistic–continuum simulation; the shaded region is the overlap between the c- and p-regions. (Courtesy of Weiqing Ren.) Owing to the symmetry of the situation, only half the computational domain is shown.

For the coupled scheme, we define the atomistic region by $-b \leq y \leq b$. The two continuum regions are defined by $-L \leq y \leq -a$ and $a \leq y \leq L$ respectively. Here $0 < a < b$. The two regions defined by $-b \leq y \leq -a$ and $a \leq y \leq b$ are the overlap regions. In general, we should have $L \gg b$, i.e. the continuum regions are much larger than the atomistic regions. Following [40] we will call the continuum and atomistic regions the *c-region* and the *p-region* respectively; see Figure 7.1.

The dynamics in the c-region and in the p-region are coupled through the exchange of velocity or shear stress in the overlap region; this is expressed in the form of boundary conditions. Boundary conditions are required for the continuum model at $y = \pm a$ and for the atomistic model at $y = \pm b$. The required quantities in these boundary conditions are measured from the other model. We have two natural choices. Either we can impose velocity boundary conditions, e.g. we obtain the velocity from MD and use the resulting value in the continuum model, or we can impose flux boundary conditions, e.g. we obtain the momentum flux, here the shear stress, from MD and use the resulting value in the continuum model. This gives us four different combinations (see [40]):

(1) velocity–velocity (VV = VpVc);
(2) flux–velocity (FV = FpVc);
(3) velocity–flux (VF = VpFc);
(4) flux–flux (FF = FpFc).

Here, for example, the notation FV (= FpVc) means that the momentum flux (shear stress) obtained from the continuum model is imposed on the particle dynamics (this is the meaning of F = Fp), and the mean velocity obtained from the particle dynamics is imposed as the boundary condition for the continuum model (this is the meaning of V = Vc). The meanings of the other three notations are similar.

Next we discuss some details of the coupled scheme. In the c-region, say from $y = a$ to $y = L$, we define the grid points $\{y_i = a + i\Delta y, i = 0, 1, \ldots, N\}$, where $\Delta y = (L - a)/N$. The velocity U is defined at $y_{i+1/2}$, the center of each cell. The shear stress σ_{12} is defined at the grid points y_i and calculated using central differences:

$$(\sigma_{12})_i = \mu\,\frac{U_{i+1/2} - U_{i-1/2}}{\Delta y}.$$

A simple way of discretizing (7.1.4) is as follows:

$$\rho\frac{U_{i+1/2}^{n+1} - U_{i+1/2}^{n}}{\Delta t} - \frac{(\sigma_{12})_{i+1}^{n} - (\sigma_{12})_{i}^{n}}{\Delta y} = 0. \qquad (7.1.7)$$

The no-slip boundary condition at $y = L$ is imposed by setting $(1/2)(U_{N-1/2}^{n} + U_{N+1/2}^{n}) = 0$, and similarly for the boundary condition at $y = -L$. The boundary condition at $y = a$ is provided by the particle dynamics in the p-region: depending on the coupling scheme we choose, either the velocity $U_{1/2}^{n}$ at $y = y_{1/2}$ or the shear stress $(\sigma_{12})_0$ at $y = y_0$ is calculated from MD. Specifically, $U_{1/2}^{n}$ is calculated by averaging the particle velocities between $y = y_0$ and $y = y_1$ over some time interval T_c. The shear stress $(\sigma_{12})_0$ is calculated using the modified Irving–Kirkwood formula discussed in Sections 4.2 and 6.6; see (6.6.11). This formula is averaged over the region between $y = y_0 - \Delta y/2$ and $y = y_0 + \Delta y/2$ and also over the time interval T_c.

The imposition of boundary conditions for the MD is a much more difficult task. Since the dynamics is homogeneous in the x and z directions, straightforward periodic boundary conditions can be used in these two directions. The boundary condition in the y direction for the MD is much harder to deal with. At the present time, the situation is still quite unsatisfactory and all existing implementations contain some ad hoc components: there are no quantitative ways of accessing the accuracy of these algorithms other than performing numerical tests. Obviously, accuracy has to be understood in a statistical sense. But, at the present time, a systematic statistical theory for atomistic models of fluids with boundary conditions is yet to be established. We refer to [40, 53] for discussions of these questions.

Now the overall procedure is a generalization of alternating Schwarz iteration in the domain decomposition method [50]. Starting from some initial condition, the continuum equation and the MD are alternatively solved in the c- and p-regions, with time-steps Δt and δt respectively and for a time interval T_c, using some provisional boundary conditons in the overlap region. The results are used to update the boundary conditions in the overlap region and the process is then repeated.

7.2 Adaptive model refinement or model reduction

Adaptive model refinement is a modification of the well-known procedure of adaptive mesh refinement, in which not only the mesh but also the physical models are chosen adaptively. One of the earliest examples of adaptive model refinement was presented in [20], where the authors developed algorithms for gas dynamics simulations using adaptive mesh refinement, but locally, around the shocks, direct simulation Monte Carlo (DSMC) instead of the equations of gas dynamics was used on the finest mesh. The nonlocal quasicontinuum method can also be viewed as adaptive model refinement strategy [49]. Ideas involving adaptive model refinement were also proposed in [37] and applied to the modeling of stress distribution in a composite material. Averaged equations were used over most of the computational domain except at places of stress concentration, where the original microscale model was used instead.

An important aspect of adaptive model refinement is that the models used at the different levels have to be consistent with each other, i.e. away from defects both models should apply and they should produce approximately the same results. This is guaranteed when the macroscopic model is the reduction of the microscopic model in the macroscopic limit. For example, the equations of gas dynamics are often the reduction of the kinetic equation under the local equilibrium approximation. The nonlinear elasticity model obtained using the Cauchy–Born rule is the reduction of the atomistic model for the situation when the displacement of the atoms follows a smooth vector field.

In such circumstances, instead of thinking about adaptively refining the macroscopic model one can also think about adaptively simplifying

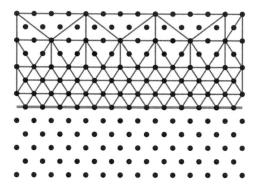

Figure 7.2. Illustration of the nonlocal quasicontinuum method. The top is the local region. The bottom is the nonlocal region. (Courtesy of Pingbing Ming.)

or coarsening the microscopic model. The former is a more top-down strategy. The latter is a more bottom-up strategy. Of course they may produce the same results.

7.2.1 The nonlocal quasicontinuum method

The local quasicontinuum (QC) method was discussed in Chapter 6. Its main purpose is to model the elastic deformation of crystalline solids using atomistic models instead of empirical continuum models. The nonlocal QC method has the additional feature that it can handle crystal defects with atomistic accuracy. This is done by creating a nonlocal region around each defect and using the atomistic model in these nonlocal regions. The main algorithmic components are therefore:

(1) the definition of the local and nonlocal regions;
(2) a summation rule for computing the energy or forces for a given trial function or trial configuration;

For the first component, nonlocal QC uses a finite element setting. The vertices in the finite element triangulation are *representative atoms* (rep-atoms). The rep-atoms are chosen adaptively on the basis of the estimated size of the local deformation gradient, which serves as the local error indicator: regions with smaller deformation gradients contain a lower density of rep-atoms. In the nonlocal region every atom becomes a rep-atom; see Figure 7.2.

As in local QC, the displaced positions of the atoms are represented by continuous, piecewise-linear, trial functions and they define the finite element space. To compute the energy or forces associated with a given trial function, we separate the atoms into three different classes:

(1) atoms in the local region;
(2) atoms in the nonlocal region;
(3) atoms at the interface between the local and nonlocal regions.

In the local region one may use the Cauchy–Born rule, as in local QC. In the nonlocal region one uses the full atomistic model. The atoms at the interface are treated basically in the same way as those in the nonlocal region.

The treatment of the local region can be thought of equivalently as being a finite element discretization of the nonlinear continuum elasticity model in which the stored energy density is obtained from the Cauchy–Born rule. Therefore, effectively the local region plays the role of the continuum region.

To illustrate the details of the treatment at the interface and to prepare for the discussion in Section 7.5 on errors in QC, we consider the following example, which provides the simplest nontrivial setting [34].

We will consider an infinite one-dimensional chain of atoms, interacting by means of a two-body potential V that depends on the distance between the atoms. We will write $V(y) = V_0(y/\varepsilon)$, where ε is the lattice constant, and define $G(y) = \partial_y V(y)$. We will consider only nearst-neighbor and next-nearest-neighbor interactions. The equilibrium positions of the atoms are given by

$$x_j = j\varepsilon, \quad j = \ldots, -2, -1, 0, 1, 2, \ldots \qquad (7.2.1)$$

We will use the more compact notation $\bar{j} = -j$ to number the indices.

Since the main source of error at the local–nonlocal interface comes from the transition between the Cauchy–Born rule, which is naturally an element-based summation rule in the local region, and an atom-based summation rule in the nonlocal region, the change in the element size (or the density of the rep-atoms) is less important even though it is crucial for the efficiency of QC. Therefore we will focus on the case when every atom is a rep-atom.

Given an external force $\mathbf{f} = (\ldots, f_{\bar{2}}, f_{\bar{1}}, f_0, f_1, f_2, \ldots)$, the total energy of the atomistic model can be written as

$$E_{\text{tot}} = \sum_j E_j - \sum_j f_j u_j,$$

where $u_j = y_j - x_j$ and

$$
\begin{aligned}
E_j &= \tfrac{1}{2}(V(y_j - y_{j-1}) + V(y_{j+1} - y_j) + V(y_j - y_{j-2}) + V(y_{j+2} - y_j)) \\
&= \tfrac{1}{2}(V(r_{j-1,j}) + V(r_{j,j+1}) + V(r_{j-2,j}) + V(r_{j,j+2})).
\end{aligned} \tag{7.2.2}
$$

Here, and in what follows, we use the notation $r_{j,k} = y_k - y_j$.

There is an obvious problem: for an infinite system, this energy is usually infinite. In reality, the system of interest is always finite and we should supplement the problem with some boundary conditions. However, we will simply ignore this problem since our main interest is the forces acting on the atoms, not the total energy.

The Euler–Lagrange equation for the minimization of E_{tot} is given by

$$
-(G(y_j^\varepsilon - y_{j-1}^\varepsilon) + G(y_j^\varepsilon - y_{j-2}^\varepsilon) + G(y_j^\varepsilon - y_{j+1}^\varepsilon) + G(y_j^\varepsilon - y_{j+2}^\varepsilon)) = f_j \tag{7.2.3}
$$

or

$$
\begin{aligned}
-\frac{1}{\varepsilon}\left(\partial_y V_0\left(\frac{y_j^\varepsilon - y_{j-1}^\varepsilon}{\varepsilon}\right) + \partial_y V_0\left(\frac{y_j^\varepsilon - y_{j-2}^\varepsilon}{\varepsilon}\right) + \partial_y V_0\left(\frac{y_j^\varepsilon - y_{j+1}^\varepsilon}{\varepsilon}\right)\right. \\
\left. + \partial_y V_0\left(\frac{y_j^\varepsilon - y_{j+2}^\varepsilon}{\varepsilon}\right)\right) = f_j
\end{aligned} \tag{7.2.4}
$$

for $j = \ldots, \bar{2}, \bar{1}, 0, 1, 2, \ldots$ We will write this in a compact form:

$$
\mathcal{L}_{\text{atom}}^\varepsilon(\mathbf{y}^\varepsilon) = \mathbf{f}. \tag{7.2.5}
$$

In nonlocal QC we assume that the local region consists of atoms indexed by the integers $j \geq 0$ and that the nonlocal region consists of atoms indexed by negative integers. The region with $x \geq 0$ will be treated using a piecewise linear finite element method, each atom being a vertex. We will denote the jth element $[x_j, x_{j+1}]$ by I_j. The total energy of the system can be written as

$$
E_{\text{qc}} = E_{\text{local}} + E_{\text{nonlocal}} + E_{\text{int}} - \sum_j f_j u_j,
$$

where E_{nonlocal} is the total potential energy due to the atoms in the nonlocal region,

$$
E_{\text{nonlocal}} = \sum_{j<0} E_j,
$$

and, E_{local} is the contribution from the local region and is a sum over all the elements,

$$E_{\text{local}} = \sum_{j \geq 0} e_j.$$

Here e_j is the potential energy from element I_j, computed using the Cauchy–Born rule,

$$e_j = V(r_{j,j+1}) + V(2r_{j,j+1}), \tag{7.2.6}$$

and E_{int} is the additional contribution from the local–nonlocal interface, i.e. the atom with index 0. Since the interaction between this atom and the atoms in the local region is already taken into account by the Cauchy–Born rule, we have only to consider the interaction with the atoms in the nonlocal region. This gives

$$E_{\text{int}} = \tfrac{1}{2}(V(r_{\bar{1},0}) + V(r_{\bar{2},0})).$$

The interfacial region can be larger than this and depends on the range of the interatomic potential and the details of the QC algorithm.

We can also rewrite the total energy as a sum over contributions from each atom:

$$E_{\text{qc}} = \sum_j \tilde{E}_j - \sum_j f_j u_j,$$

where

$$\tilde{E}_{\bar{j}} = E_{\bar{j}}, \quad \tilde{E}_j = \tfrac{1}{2}(e_{j-1} + e_j) \tag{7.2.7}$$

for $j > 0$ and $\tilde{E}_0 = \tfrac{1}{2}e_0 + E_{\text{int}}$. For example,

$$\tilde{E}_{\bar{3}} = \tfrac{1}{2}\left(V(r_{\bar{5},\bar{3}}) + V(r_{\bar{4},\bar{3}}) + V(r_{\bar{3},\bar{2}}) + V(r_{\bar{3},\bar{1}})\right),$$
$$\tilde{E}_{\bar{2}} = \tfrac{1}{2}\left(V(r_{\bar{4},\bar{2}}) + V(r_{\bar{3},\bar{2}}) + V(r_{\bar{2},\bar{1}}) + V(r_{\bar{2},0})\right),$$
$$\tilde{E}_{\bar{1}} = \tfrac{1}{2}\left(V(r_{\bar{3},\bar{1}}) + V(r_{\bar{2},\bar{1}}) + V(r_{\bar{1},0}) + V(r_{\bar{1},1})\right),$$
$$\tilde{E}_0 = \tfrac{1}{2}\left(V(r_{\bar{2},0}) + V(r_{\bar{1},0}) + V(r_{0,1}) + V(2r_{0,1})\right),$$
$$\tilde{E}_1 = \tfrac{1}{2}\left(V(2r_{0,1}) + V(r_{0,1}) + V(r_{1,2}) + V(2r_{1,2})\right),$$
$$\tilde{E}_2 = \tfrac{1}{2}\left(V(2r_{1,2}) + V(r_{1,2}) + V(r_{2,3}) + V(2r_{2,3})\right).$$

The Euler–Lagrange equations are the same as (7.2.4) for $j \leq -1$. For $j \geq 2$ we have

$$\mathcal{L}_{\text{cb}}^{\varepsilon}(\mathbf{y})_j = f_j,$$

where

$$\mathcal{L}_{cb}^{\varepsilon}(\mathbf{y})_j = -\left(G(r_{j,j-1}) + G(r_{j,j+1}) + 2G(2r_{j,j-1}) + 2G(2r_{j,j+1})\right).$$

For $j = -\bar{1}, 0, 1$, we have

$$-\left(G(r_{\bar{1},\bar{3}}) + G(r_{\bar{1},\bar{2}}) + G(r_{\bar{1},0} + \tfrac{1}{2}G(r_{\bar{1},1})\right) = f_{\bar{1}},$$

$$-\left(G(r_{0,\bar{2}}) + G(r_{0,\bar{1}}) + G(r_{0,1}) + G(2r_{0,1})\right) = f_0,$$

$$-\left(\tfrac{1}{2}G(r_{1,\bar{1}}) + G(r_{1,0}) + G(r_{1,2}) + 2G(2r_{1,0}) + 2G(2r_{1,2})\right) = f_1.$$

As before, we will write these equations in a compact form as

$$\mathcal{L}_{qc}^{\varepsilon}(\mathbf{y}) = \mathbf{f}. \tag{7.2.8}$$

It is quite obvious that nonlocal QC can be thought of both as an adaptive model-refinement procedure and as an adaptive model-reduction or coarsening procedure. As an adaptive model-refinement procedure, one starts with an adaptive finite element method for a Cauchy–Born-based continuum model. Near defects the mesh is refined to the atomic level and, at the same time, one replaces the Cauchy–Born rule by the full atom model. As an adaptive model-reduction procedure, one starts with a full atom model and coarsens the full atom representation to a rep-atom representation. Away from the interface one also makes the approximation that the deformation is locally homogeneous, i.e. one approximates the full atom summation by the Cauchy–Born rule.

In comparison with the example of fluids, discussed in the previous subsection, QC is much simpler owing to the absence of statistical fluctuations. However, consistency errors still exist at the interface. The simplest and most well-known form of consistency error is the "ghost force," i.e. a force acting on atoms when they are in the equilibrium position. These forces are the result of numerical error. We will come back to this problem later in the chapter.

7.2.2 Coupled gas-dynamic–kinetic models

In our next example we consider continuum gas dynamics locally corrected near shocks by the kinetic model. The macroscopic model is the gas dynamics equation, say, Euler's equation. The microscopic model is a kinetic equation such as the Boltzmann equation or the BGK model (see Chapter 4). We will assume that the Euler and the kinetic equations are

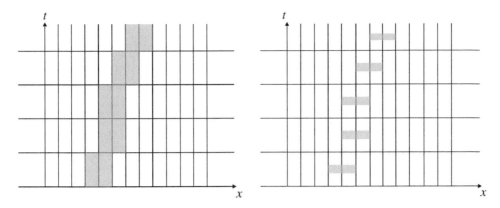

Figure 7.3. Schematics for the coupled kinetic–gas dynamics simulation. A finite volume method is imposed everywhere in the xt domain. The numerical fluxes are computed using the kinetic scheme in the region away from the shocks and directly from the solutions of the kinetic equation around the shocks. The shaded regions indicate where the kinetic equation needs to be solved. The left-hand panel illustrates the case when there is no scale separation between the relaxation time scale inside the shock and the hydrodynamic time scale. The right-hand panel shows the case when there is time scale separation [15].

consistent in the sense that Euler's equation is obtained from the kinetic equation under the local equilibrium approximation. As the macroscale solver it is natural to choose the kinetic scheme for Euler's equation (see Chapter 6). Recall that this is a finite-volume scheme, with numerical fluxes computed using (6.1.12). This helps to guarantee consistency between Euler's equation and the kinetic equation even at the numerical level. See Figure 7.3.

Near the shocks the numerical fluxes are computed by solving the kinetic model directly, using micro time-steps. Consider a one-dimensional example: assume that the interface is located at $x_{k+1/2}$, with the Euler equation to be solved on the left-hand side of $x_{k+1/2}$ and the kinetic equation to be solved on the right-hand side of $x_{k+1/2}$. For the kinetic equation we need a boundary condition when the velocity $v > 0$ (See (6.1.12)). It is natural to choose

$$f(x, v, t) = M(x_{k+1/2}^-, v, t), \quad v > 0, \qquad (7.2.9)$$

where the right-hand side is the local Maxwellian corresponding to the macroscopic state at $x_{k+1/2}$ and the superscript "$-$" means that the left-hand limit is taken. The numerical flux at $x_{k+1/2}$ is computed using

$$\mathbf{F}_{k+1/2} = \frac{1}{\Delta t} \int_{t^n}^{t^{n+1}} dt \left\{ \int_{\mathbb{R}^+} M(x_{k+1/2}^-, v, t) \begin{pmatrix} v \\ v^2 \\ \frac{1}{2}v^3 \end{pmatrix} dv \right.$$

$$\left. + \int_{\mathbb{R}^-} f(x_{k+1/2}^+, v, t) \begin{pmatrix} v \\ v^2 \\ \frac{1}{2}v^3 \end{pmatrix} dv \right\}. \quad (7.2.10)$$

The final component of the method is a criterion for locating the kinetic region. In principle, we would like to have something like the *a posteriori* error estimators discussed in Chapter 3 to help in locating the kinetic regions. Up to the present time, little has been done in this direction. As a working scheme one may simply use the size of the local gradients as the local error indicator. If this size is above a certain threshold in a particular region, we make it a shock region or kinetic region.

Examples of numerical results computed using this method are shown in Figure 7.4 [15]. Here the results from the kinetic equation are taken as the exact solution. Note that the coupled kinetic–Euler equation gives more accurate shock profiles than the Euler equation by itself.

For this one-dimensional example, one may solve the kinetic equation using finite difference or other deterministic grid-based methods. For high-dimensional problems it might be advantageous to use discrete simulation Monte Carlo methods [4], as was done in [20].

Again we may regard the strategy discussed above either as an adaptive model-refinement strategy or as an adaptive model-reduction strategy. Indeed, the macroscopic model used in the region away from the shocks can be regarded as a simplification of the kinetic equation under the local equilibrium approximation.

7.3 The heterogeneous multiscale method

In some cases we are interested only in obtaining the right macroscale behavior, not the microscopic details near the defects or singularities. In this situation, the heterogeneous multiscale method (HMM) might be a good choice. One starts with a macro solver that covers the whole computational domain. The information needed about the structure or dynamics of the defects in order to understand the macroscale behavior

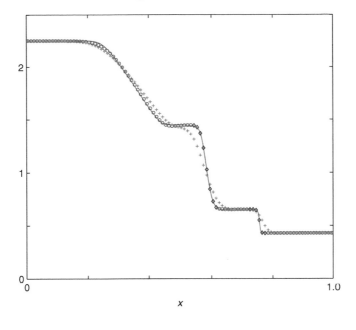

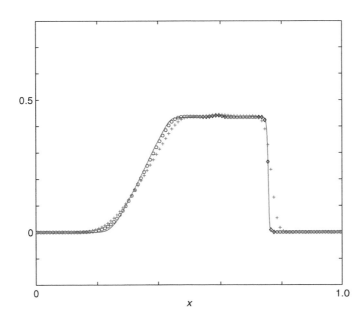

Figure 7.4. Numerical results for the shock tube problem: solid line, results from the kinetic equation; ∘, results from the coupled kinetic–gas-dynamics simulation; +, results from the gas dynamics equation. Upper panel, density; lower panel, velocity. Note that the coupled approach gives shock profiles that are more accurate than that of the gas dynamics equation (courtesy of Xiantao Li).

is extracted from microscopic simulations near the defects. Such a strategy has been applied to several examples, including the dynamics of the moving contact line [41] and the interaction of elastic waves with crack dynamics [28] (see also [27] for a description of the main ideas in the algorithm).

In the moving contact line problem the macroscale behavior of interest is the dynamics of the two-phase flow, modeled by equations of the form

$$\rho(\partial_t \mathbf{u} + \nabla \cdot (\mathbf{u} \otimes \mathbf{u})) + \nabla p = \mu \Delta \mathbf{u} + \nabla \cdot \boldsymbol{\sigma} + \mathbf{f},$$
$$\nabla \cdot \mathbf{u} = 0, \tag{7.3.1}$$
$$v_\mathrm{n} = \mathbf{u} \cdot \hat{\mathbf{n}},$$

where

$$\boldsymbol{\sigma} = -\gamma (I - \hat{\mathbf{n}} \otimes \hat{\mathbf{n}}) \delta_\Gamma \tag{7.3.2}$$

models the surface tension force along the fluid–fluid interface Γ, γ is the surface tension coefficient, $\hat{\mathbf{n}}$ is the unit normal vector along Γ, δ_Γ is the surface delta function associated with Γ and $\mathbf{f}$ is some external forcing. The final equation of (7.3.1) says that the fluid–fluid interface is advected by the velocity field of the fluid, v_n being the normal component of velocity.

The data that need to be estimated from microscopic models such as MD are:

(1) the shear stress near the contact line;
(2) the velocity of the contact line, which provides the boundary condition for the dynamics of Γ.

The schematics are shown in Figure 7.5. See [41] for details.

Compared with adaptive model refinement, this method might provide a more effective way of handling the case when the relaxation time of the local defects τ_ε is much smaller than the time scale t_M for interesting dynamics in the local region. This is possible, even for type A problems, since τ_ε is governed by microscopic processes whereas t_M depends on the external driving force. In such a case, HMM provides a natural setting for capturing the correct local structure around the defects without the need to follow exactly the microscopic dynamics.

In the same spirit, one can also extend the seamless strategy discussed in the previous chapter to the case discussed in the present chapter.

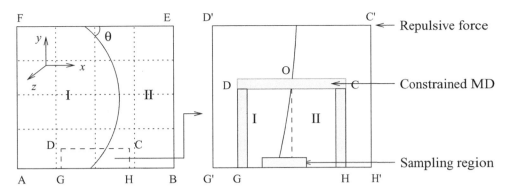

Figure 7.5. Schematic of the multiscale method for the moving contact line problem. The macro grid used in the computation is represented by the dotted lines in the left-hand panel. The molecular dynamics is carried out in a local region around the contact line. The dynamics in the shaded strips GD, DC and CH is constrained by the continuum velocity field. The shear stress and the position of the contact point are computed in the strip along GH. From [41].

Once a general strategy is chosen, the questions to be addressed next are: which data will be exchanged between the macro and micro models; how will the exchanged data be used? The most important concern when addressing these questions is the consistency between the macro and micro models at the interface where they meet. The coupling scheme should not create significant artifacts or numerical errors at the interface. However, this is not the only concern. As we see below, numerical stability might also be an issue.

7.4 Stability issues

We will focus on the domain decomposition strategy, since this is the simplest coupling strategy. We saw in Chapter 3 that the domain decomposition method may not converge if we do not exchange data between the different domains in the right way, since then some components of the error are not damped during the iteration. This has important consequences in a multiscale, multi-physics context.

To illustrate this point we will discuss the work of Ren, [40], who considered coupled atomistic–continuum models of fluids based on the domain decomposition method. Using a simple example, Ren demonstrated that the performance of the coupled scheme depends very much on how data are exchanged between the atomistic and continuum models. Some coupling schemes lead to numerical instabilities.

The general setup of the domain-decomposition-based coupling strategy was introduced in Section 7.1. Our focus is the four coupling schemes discussed, i.e. the VV, VF, FV and FF schemes.

To begin with, let us note that the domain decomposition method considered here is different from that discussed in Chapter 3 in at least two respects:

(1) the p-region is usually much smaller than the c-region, which means that $L \gg b$;

(2) one cannot avoid statistical errors.

We will see later that these differences have important consequences.

Let us first consider the simple case when all the following hold.

(1) The system of interest is in equilibrium. Therefore the mean velocity should be zero. Of course, the velocities of the individual particles in the p-region will not actually, be zero, neither will the averaged velocity obtained from simulations, because of statistical fluctuations.

(2) We assume reflection symmetry with respect to the xz-plane for the coupled scheme. Therefore the mean velocity on the xz-plane vanishes.

(3) $T_c = \infty$. This means that we only exchange data after the results of each simulation (the continuum model or the MD) have reached a (statistical) steady state.

(4) We will also assume that the (spatiotemporal) averaged velocity $\tilde{u}$ in the p-region is a linear function of y plus random fluctuations. Similarly, we will assume that the averaged shear stress in the p-region is a constant plus random fluctuations; this is the case if the effective constitutive relation is linear. As we said earlier, this is a very good approximation in the regime we are considering.

Admittedly this is a very idealized case but, as we will see below, the results obtained are already quite instructive.

Lemma ([40]) *Under these assumptions, the numerical solutions of the coupled scheme have the following form:*

$$u_n(y) = \left(\sum_{i=1}^{n} k^{n-i} \xi_i \right) g(y), \qquad (7.4.1)$$

where $u_n(\cdot)$ is the velocity at the nth iteration in the c-region and ξ_i is the statistical error introduced in the boundary condition at the ith iteration. The amplification (or damping) factor k is given for the different coupling schemes by

$$k_{\mathrm{VV}} = \frac{a}{b}\frac{L-b}{L-a}, \tag{7.4.2}$$

$$k_{\mathrm{FV}} = \frac{a}{a-L}, \tag{7.4.3}$$

$$k_{\mathrm{VF}} = \frac{b-L}{b}, \tag{7.4.4}$$

$$k_{\mathrm{FF}} = 1. \tag{7.4.5}$$

The function $g(\cdot)$ is given by

$$g_{\mathrm{VV}}(y) = g_{\mathrm{FV}}(y) = \frac{L-y}{L-a}, \tag{7.4.6}$$

$$g_{\mathrm{VF}}(y) = g_{\mathrm{FF}}(y) = y - L. \tag{7.4.7}$$

This lemma can be proved by induction. Thanks to the symmetry assumption, it is enough to consider only the upper half-channel. Consider the VV coupling scheme, for example. In the first iteration one starts with an equilibrium MD in the p-region. The average velocity at $y = a$ is denoted by ξ_1 and is used as the Dirichlet boundary condition for the continuum model in the c-region. The steady state solution of the continuum equation (7.1.4), with boundary conditions $u(L) = 0$ and $u(a) = \xi_1$, is given by

$$u_1(y) = \xi_1 \frac{L-y}{L-a} = \xi_1 g_{\mathrm{VV}}(y).$$

This proves (7.4.1) for $n = 1$. Assume that (7.4.1) is valid for the nth iteration. In the $(n+1)$th iteration, the MD in the p-region is constrained so that the mean particle velocity at $y = b$ is equal to $u_n(b) = \sum_{i=1}^{n} k_{\mathrm{VV}}^{n-i} \xi_i g(b)$. Therefore the average velocity in the p-region in the $(n+1)$th iteration is given by

$$\tilde{u}_{n+1}(y) = u_n(b)\frac{y}{b} + \eta_n(y),$$

where $\eta_n(y)$ denotes the random fluctuations in the averaged velocity.

Therefore, if we let $\xi_{n+1} = \eta_n(a)$, we have

$$u_{n+1}(a) = u_n(a)\frac{a}{b} + \xi_{n+1} = \frac{a}{b}g(a)\sum_{i=1}^{n}k_{VV}^{n-i}\xi_i + \xi_{n+1} = \sum_{i=1}^{n+1}k_{VV}^{n+1-i}\xi_i.$$

The solution to the continuum equation with boundary conditions $u(a) = u_{n+1}(a)$, $u(L) = 0$ is then given by

$$u_{n+1}(y) = u_{n+1}(a)g_{VV}(y) = \left(\sum_{i=1}^{n+1}k_{VV}^{n+1-i}\xi_i\right)g_{VV}(y).$$

This proves (7.4.1) for the $(n+1)$th iteration.

By following a similar procedure, one can prove the lemma for the other three coupling schemes.

Note that the amplification factors can also be computed much more simply by replacing the MD by the continuum model, which in this case becomes

$$\partial_y^2 u = 0. \qquad (7.4.8)$$

For example, for the VV coupling scheme this results in the following deterministic domain decomposition method. At the nth iteration:

(1) one first solves the continuum equation (7.4.8) in the c-region with boundary conditions $u_n(a) = \tilde{u}_{n-1}(a)$, $u_n(L) = 0$, to obtain u_n;
(2) one then solves the continuum equation (7.4.8) in the p-region with the boundary conditions $\tilde{u}_n(b) = u_n(b)$, $\tilde{u}_n(0) = 0$, to obtain $\tilde{u}_n$.

To find the amplification factor for such an iterative process, we look for solutions of this problem in the form

$$u_n(y) = k^n U(y), \quad \tilde{u}_n(y) = k^n \tilde{U}(y).$$

It is easy to see that U and $\tilde{U}$ must be of the form

$$U(y) = C\frac{y}{b}, \quad \tilde{U}(y) = \tilde{C}\frac{L-y}{L-a}.$$

From the boundary conditions at $y = a, b$ we get

$$k = \frac{a}{b}\frac{L-b}{L-a}, \quad \tilde{C} = \frac{b}{a}kC.$$

Similarly, one can compute the amplification factors for the VF, FV and FF schemes.

Let us try to understand what these analytical results tell us. First note that, according to the size of the amplification factor, we can put the four coupling schemes into three categories:

(1) *The FF scheme.* In this case, we have $k_{FF} = 1$. If we assume that the $\{\xi_i\}$ are independent then we have

$$\langle u^n(y)^2 \rangle^{1/2} \sim \sqrt{n}\sigma|g(y)|,$$

where $\sigma = \langle \xi_i^2 \rangle^{1/2}$. The important fact is that the error grows as $\sqrt{n}$ due to the random fluctuations.

(2) *The VV and FV scheme.* Since $|k_{VV}|, |k_{FV}| < 1$, for both schemes we have

$$\langle u^n(y)^2 \rangle^{1/2} \leq C\sigma|g(y)|,$$

i.e. the error is bounded. This is the best situation as far as stability is concerned.

(3) *The VF scheme*, for which the amplification factor satisfies $|k_{VF}| > 1$. We will show later that this is an artifact of using $T_c = \infty$. For practical values of T_c, $|k_{VF}|$ is usually less than 1. Therefore this case should be combined with the second case.

Now we consider the case when T_c is finite. We can also define the amplification factor. For example, for the VV coupling scheme one simply has to replace the continuum equation (7.4.8) by a time-dependent version,

$$\rho\partial_t u = \mu\partial_y^2 u,$$

and the boundary conditions by

$$u_n(a, t) = \tilde{u}_{n-1}(a, T_c), \quad \tilde{u}_n(b, t) = u_n(b, T_c).$$

To compute the amplification factor, we look for solutions that satisfy

$$u_n(y, T_c) = k^n U(y), \quad \tilde{u}_n(y, T_c) = k^n \tilde{U}(y).$$

In this case, it is much simpler to compute k numerically (we refer to [40] for the details of how to compute k); the results are shown in Figure 7.6 for all four coupling schemes.

In general, we are interested in the case when $T_c \sim \Delta t$, i.e. when we exchange data after a small number of macro time-steps. In this case the results in Figure 7.6 agree with those discussed earlier for $T_c = \infty$ (except

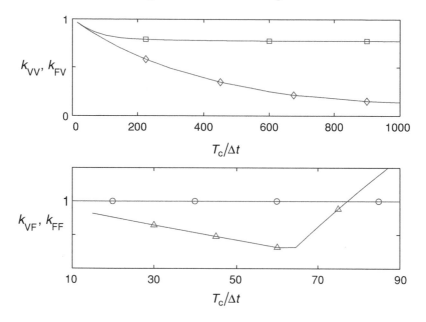

Figure 7.6. The amplification factor k versus $T_c/\Delta t$ for the four schemes: VV (squares), FV (diamonds), VF (triangles) and FF (circles) (courtesy of Weiqing Ren).

in the VF coupling scheme): if T_c is not too large, the VF coupling scheme is quite stable.

These predictions are confirmed by the numerical results presented in [40] for more general situations. In particular, it was found that the FF coupling scheme leads to large errors. Consider an impulsively started shear flow with a constant pressure gradient. A term of the form ρf with $f = 2 \times 10^{-3}$ (in atomic units) is added to the right-hand side of the continuum equation. In the MD region, in addition to the intermolecular forces the fluid particles also feel the external force $\mathbf{f} = (f, 0, 0)$. The numerical results for the case when $T_c = \Delta t$ and $t = 1500$ obtained using the FV and FF coupling schemes are shown in Figure 7.7. We see from the bottom panel that the error eventually grows roughly linearly and the mean velocity profile becomes discontinuous, as shown in the middle panel. Note that the setup has been changed slightly: the particle regions are at the boundary and the continuum region is now in the middle.

7.5 Consistency questions illustrated using nonlocal QC

In general, the question of the consistency in a coupled multi-physics algorithm should be examined at two different levels.

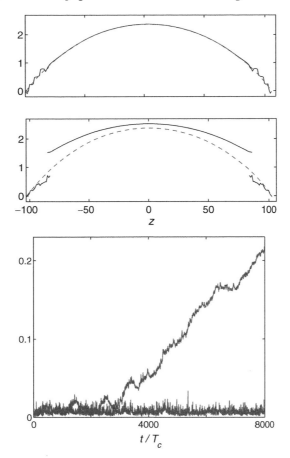

Figure 7.7. Top and middle panels: numerical solutions for the channel flow u driven by the pressure gradient at $t = 1500$, obtained using the FV scheme (top panel) and the FF scheme (middle panel) respectively. The broken curve is the reference solution. The oscillatory curves at the two ends are the numerical solution in the p-region; the smooth solid curve in the middle panel is the numerical solution in the c-region. Bottom panel: the error e_u of the numerical solution versus time. The two curves correspond to the FV scheme (lower curve) and the FF scheme (upper curve) respectively. From [40].

(1) *Consistency between the different physical models* when they are used separately for systems for which they are each supposed to be adequate. For example, a continuum elasticity model and an atomistic model should produce approximately the same result when they are used for analyzing the mechanical response of an elastically deformed single-crystal material.

(2) *Consistency at the interface* where the different physical models are coupled together. In general it is inevitable that coupling introduces extra error. A main task in formulating good coupling schemes is to minimize this error.

An example of such a coupling error is the *ghost force* in the nonlocal quasi-continuum (QC) method, mentioned at the end of subsection 7.2.1. In what follows, we will use this example to illustrate the origin of a coupling error, what its consequences are and how to reduce this error.

A very important question is how to quantify such as error. We will see that in the case of nonlocal QC we can use a simple generalization of the notion of the *local truncation error* to quantify the error due to the mismatch between the different models at the interface.

To illustrate the heart of the matter, we will focus on the one-dimensional model problem discussed in Section 7.2. We assume that there exists a smooth function $f(x)$ such that

$$f(x_j) = f_j, \quad j = \ldots, \bar{2}, \bar{1}, 0, 1, 2, \ldots \tag{7.5.1}$$

where f_j is the external force acting on the jth atom.

7.5.1 The appearance of the ghost force

First, let us consider the simplest case, when there are no external forces: $\mathbf{f} = 0$. The exact solution should be given by $y_j = x_j$ for all j, i.e. the atoms should be in their equilibrium positions. Obviously this solution satisfies the equations (7.2.4). However, this is not true for nonlocal QC. Indeed, it is easy to see that if we substitute $y_j = x_j$ (for all j) into the nonlocal QC equations (7.2.8), we have

$$\mathcal{L}_{\text{qc}}^{\varepsilon}(\mathbf{x})_{\bar{1}} = -G(2), \quad \mathcal{L}_{\text{qc}}^{\varepsilon}(\mathbf{x})_0 = G(2), \quad \mathcal{L}_{\text{qc}}^{\varepsilon}(\mathbf{x})_1 = -G(2).$$

The force components that appear on the right-hand sides constitute the ghost force [49]: They are entirely an artifact of the numerical method. Note that they are $\mathcal{O}(1/\varepsilon)$.

The origin of this ghost force is intuitively quite clear: the influence between the atoms in the local and nonlocal regions is asymmetric. For the example discussed here, the atom at y_1 influences the energy associated with the atom indexed by $\bar{1}$, but not the other way around.

The error induced by the ghost force can be analyzed explicitly in the case when the interaction potential is harmonic, as was done in [34]

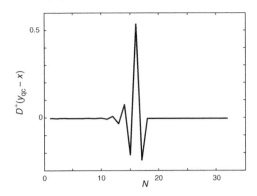

Figure 7.8. The error in the deformation gradient for the original nonlocal QC with a nearest neighbor and next nearest neighbor quadratic interaction potential. From [34]. (Copyright © 2009 Society for Industrial and Applied Mathematics. Reprinted with permission. All rights reserved.)

(see also [9] for a more streamlined argument); see Figure 7.8. This explicit calculation tells us several things:

(1) the deformation gradient has $\mathcal{O}(1)$ error at the interface;
(2) the influence of the ghost force decays exponentially away from the interface;
(3) away from an interfacial region of width $\mathcal{O}(\varepsilon|\log \varepsilon|)$, the error in the deformation gradient is $\mathcal{O}(\varepsilon)$.

The large error at the interface is a concern [34, 32]. In realistic situations, it is conceivable that artificial defects may appear as a result of the ghost force. In any case, it is of interest to design modifications of nonlocal QC that eliminate or reduce the ghost force.

7.5.2 Removing the ghost force

Force-based ghost force removal The simplest idea for removing the ghost force is a force-based method [45]. In this approach one defines

$$(\mathcal{L}^{\varepsilon}_{\text{fqc}}(\mathbf{y}))_j = \begin{cases} (\mathcal{L}^{\varepsilon}_{\text{atom}}(\mathbf{y}))_j & \text{if } j \leq 0, \\ (\mathcal{L}^{\varepsilon}_{\text{cb}}(\mathbf{y}))_j & \text{if } j \geq 1, \end{cases}$$

where $\mathcal{L}^{\varepsilon}_{\text{cb}}$ is the Cauchy–Born operator. The deformed positions of the atoms are found by solving

$$\mathcal{L}^{\varepsilon}_{\text{fqc}}(\mathbf{y}) = \mathbf{f}. \tag{7.5.2}$$

Quasi-nonlocal QC The force-based approach discussed above does not admit a variational formulation, i.e. there is no underlying energy. Shimokawa *et al.* [46] introduced the concept of quasi-nonlocal atoms, which are interfacial atoms. When computing the energy, a quasi-nonlocal atom acts like a nonlocal atom on the nonlocal side of the interface and like a local atom on the local side of the interface. As we will see below, if the interatomic interaction is limited to the nearest and next nearest neighbors, as in the case we are considering, a quasi-nonlocal treatment of the interfacial atoms is sufficient for removing the ghost force.

For the simple example considered here, the quasi-nonlocal atoms are indexed by $\bar{1}$ and 0, and their energies are computed using

$$E_{\bar{1}} = \tfrac{1}{2} \left(V(r_{\bar{1},\bar{2}}) + V(r_{\bar{1},0}) + V(r_{\bar{1},3}) + V(2r_{\bar{1},0}) \right),$$
$$E_0 = \tfrac{1}{2} \left(V(r_{1,\bar{1}}) + V(r_{1,0}) + V(r_{1,2}) + V(2r_{1,2}) \right).$$

The energies for the other atoms are calculated in the same way as before for nonlocal QC (see (7.2.7)). Similarly, the Euler–Lagrange equations are the same as for the original nonlocal QC except for atoms indexed by $\bar{1}, 0$:

$$- \left(G(r_{\bar{1},3}) + G(r_{\bar{1},\bar{2}}) + G(r_{\bar{1},0}) + G(2r_{\bar{1},0}) \right) = f_{\bar{1}},$$
$$- \left(G(r_{0,\bar{1}}) + G(r_{0,\bar{2}}) G(2r_{0,\bar{1}}) + G(r_{0,1}) + 2G(2r_{0,1}) \right) = f_0. \quad (7.5.3)$$

Again following [34], we will write them in a compact form as

$$\mathcal{L}^{\varepsilon}_{\mathrm{qqc}}(\mathbf{y}) = \mathbf{f}. \quad (7.5.4)$$

It is straightforward to check that there are no ghost forces, i.e. that when $\mathbf{f} = 0$ the Euler–Lagrange equations are satisfied by $\mathbf{y} = \mathbf{x}$.

Introducing quasi-nonlocal atoms is only sufficient when the interatomic interaction is limited to the nearest and next nearest neighbors. For the general case, E, Lu and Yang introduced the notion of *geometrically consistent schemes*, which contain the quasi-nonlocal construction as a special case [16].

7.5.3 Truncation error analysis

In classical numerical analysis, the standard way of analyzing the errors in a discretization scheme for a differential equation is to study the truncation error of the scheme, which is defined as the error obtained when an exact solution of the differential equation is substituted into the discretization scheme. If in addition the scheme is stable then we may conclude that the error in the numerical solution is of the same order as the truncation error [42].

In general, it is unclear how to extend this analysis to multi-physics problems, as remarked earlier. However, QC provides an example for which the classical truncation error analysis is just as effective. The main reason is that the models used in the local and nonlocal regions can both be considered as consistent numerical discretizations of the same continuum model, the elasticity model obtained using the Cauchy–Born rule. That this holds for the local region is obvious; that it holds for the atomistic model in the nonlocal region requires some work, and this was carried out in [14]. For this reason we will focus on the truncation error at the interfacial region between the local and nonlocal regions. Note that the ghost force is the truncation error for the special case when $\mathbf{f} = 0$.

Definition 7.1   Let $\mathbf{y}^\varepsilon$ be the solution of the atomistic model (7.2.5). We define the *truncation error* $\mathbf{F} = (\ldots, F_{\bar{2}}, F_{\bar{1}}, F_0, F_1, F_2, \ldots)$ by

$$F_k = (\mathcal{L}^\varepsilon_{\text{atom}} - \mathcal{L}^\varepsilon)(\mathbf{y})_k, \quad k = \ldots, \bar{2}, \bar{1}, 0, 1, 2, \ldots,$$

where $\mathcal{L}^\varepsilon$ stands for the Euler–Lagrange operator for the different versions of nonlocal QC.

It is now a simple matter to compute the truncation error for the various nonlocal QC schemes that we have discussed. The results are as follows [34].

Truncation error of the original nonlocal QC Let $\mathbf{F} = (\mathcal{L}^\varepsilon_{\text{atom}} - \mathcal{L}^\varepsilon_{\text{qc}})(\mathbf{y}^\varepsilon)$. Then we have

$$F_k = 0, \quad k \leq \bar{2},$$
$$F_{\bar{1}} = 2G(2\widehat{D}y_0^\varepsilon),$$
$$F_0 = \left(G(2\widehat{D}y_1^\varepsilon) - 2G(2D^+ y_0^\varepsilon) \right),$$
$$F_1 = \left(-\tfrac{1}{2}G(2\widehat{D}y_0^\varepsilon) + G(2\widehat{D}y_2^\varepsilon) + 2G(2D^- y_1^\varepsilon) - 2G(2D^+ y_1^\varepsilon) \right),$$
$$\begin{aligned} F_k &= (\mathcal{L}^\varepsilon_{\text{atom}} - \mathcal{L}^\varepsilon_{\text{cb}})_k \\ &= \left(G(2\widehat{D}y_{k+1}^\varepsilon) - G(2\widehat{D}y_{k-1}^\varepsilon) - 2G(2D^+ y_k^\varepsilon) + 2G(2D^- y_k^\varepsilon) \right), \quad k \geq 2, \end{aligned}$$

where $\hat{D}$, D^- and D^+ are the central divided-difference, backward divided-difference and forward divided difference operators respectively. Straightforward Taylor expansion gives

$$F_k = \begin{cases} 0, & k \leq \bar{2}, \\ \mathcal{O}(1/\varepsilon), & k = \bar{1}, 0, 1, \\ \mathcal{O}(\varepsilon^2) & k \geq 2. \end{cases}$$

Truncation error of the force-based nonlocal QC Let $\mathbf{F} = (\mathcal{L}_{\text{atom}}^\varepsilon - \mathcal{L}_{\text{fqc}}^\varepsilon)(\mathbf{y}^\varepsilon)$. Then

$$F_k = 0, \quad k < 0,$$
$$F_k = (\mathcal{L}_{\text{atom}}^\varepsilon - \mathcal{L}_{\text{cb}}^\varepsilon)_k, \quad k \geq 0.$$

For $k \geq 0$, we can write F_k as

$$F_k = \begin{cases} 0, & k \leq \bar{1}, \\ D^+ Q_k, & k \geq 0, \end{cases}$$

where

$$Q_k = G(2\hat{D}y_k^\varepsilon) + G(2\hat{D}y_{k-1}^\varepsilon) - 2G(2D^+ y_{k-1}^\varepsilon).$$

By Taylor expansion we have

$$
\begin{aligned}
Q_k &= \varepsilon \left[\int_0^1 \overline{V}''\big((1+t)D^+ y_{k-1}^\varepsilon + (1-t)D^+ y_k^\varepsilon\big)dt \right] (D^+)^2 y_{k-1}^\varepsilon \\
&\quad - \varepsilon \left[\int_0^1 \overline{V}''\big((1+t)D^+ y_{k-1}^\varepsilon + (1-t)D^+ y_{k-2}^\varepsilon\big)dt \right] (D^+)^2 y_{k-2}^\varepsilon \\
&= \varepsilon^2 \left[\int_0^1 \overline{V}''\big((1+t)D^+ y_{k-1}^\varepsilon + (1-t)D^+ y_k^\varepsilon\big)dt \right] (D^+)^3 y_{k-2}^\varepsilon \\
&\quad - \varepsilon^2 \left[\int_0^1 \int_0^1 \overline{V}'''\big((1+t)D^+ y_{k-1}^\varepsilon + (1-t)D^+ \right. \\
&\quad \left. \times (sy_k^\varepsilon + (1-s)y_{k-2}^\varepsilon))ds\,dt \right] (D^+)^2 (y_{k-2}^\varepsilon + y_{k-1}^\varepsilon)(D^+)^2 y_{k-2}^\varepsilon.
\end{aligned}
$$
$$(7.5.5)$$

This gives $Q = \mathcal{O}(\varepsilon^2)$. Furthermore, we also have $D^+ Q_k = \mathcal{O}(\varepsilon^2)$. Therefore,

$$\mathbf{F} = \mathcal{O}(\varepsilon^2). \tag{7.5.6}$$

Truncation error of the quasi-nonlocal method Let $\mathbf{F} = (\mathcal{L}_{\text{atom}}^{\varepsilon} - \mathcal{L}_{\text{qqc}}^{\varepsilon})(\mathbf{y}^{\varepsilon})$. Then

$$F_k = 0, \quad k \leq \bar{2},$$
$$F_{\bar{1}} = \left(G(2\widehat{D}y_0^{\varepsilon}) - G(2D^+y_{\bar{1}}^{\varepsilon}) \right),$$
$$F_0 = \left(G(2\widehat{D}y_{\bar{1}}^{\varepsilon}) + G(2D^-y_0^{\varepsilon}) - 2G(2D^+y_0^{\varepsilon}) \right),$$
$$F_k = (\mathcal{L}_{\text{atom}}^{\varepsilon} - \mathcal{L}_{\text{cb}}^{\varepsilon})_k, \quad k \geq 1.$$

Taylor expansion gives

$$F_k = \begin{cases} 0, & k \leq \bar{2}, \\ \mathcal{O}(1), & k = \bar{1}, 0, \\ \mathcal{O}(\varepsilon^2), & k \geq 1. \end{cases}$$

We see that the truncation error is small except at the interface, where it is $\mathcal{O}(1)$. Using this fact, one can show under sharp stability conditions that the maximum error in the deformation gradient is $\mathcal{O}(\varepsilon)$. The details can be found in [10, 34].

As we said earlier, what makes this kind of analysis possible is that the models used in both the local and the nonlocal regions can be regarded as consistent approximations to the same continuum model. The extension of this kind of analysis to the case when statistical errors are involved is still an open problem.

7.6 Notes

This was the most difficult chapter to write, and perhaps many readers will not find what they expected to see. The difficulty is the following. On the one hand, the idea of developing hybrid formulations or hybrid algorithms that couple different levels of physical models together is extremely popular in many disciplines nowadays. On the other hand, much existing work involves some ad hoc components in one way or another. Except for very few cases, it is not clear how one can assess the accuracy of these hybrid schemes. Consequently, a description of these methodologies will inevitably become a set of recipes rather than a coherent set of ideas with a solid foundation. For this reason, we have opted to discuss the general principles rather than the latest exciting examples. Our selection of the general principles also reflect the desire to put things on a

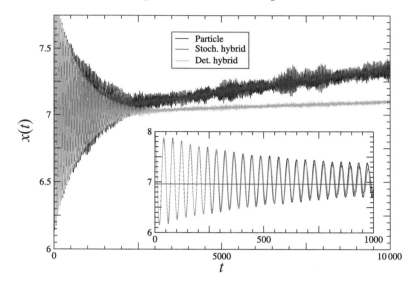

Figure 7.9. Relaxation of a massive rigid piston ($M/m = 4000$), which is initially not in mechanical equilibrium. Shown in the figure is the position of the piston as a function of time. Mechanical equilibrium is established through rapid oscillations. Both the hybrid method with fluctuating hydrodynamics and direct particle simulation are able to predict the very slow relaxation to the thermal equilibrium. The hybrid method with deterministic hydrodynamics fails to do so or at least under-predicts the relaxation rate. The inset highlights the initial oscillations and shows that in the initial regime, where the fluctuations do not matter, the deterministic and stochastic hybrid methods agree well (from [11], courtesy of John Bell).

solid foundation: after discussing the general coupling strategies we chose to focus on two classical issues in numerical analysis, namely, stability and accuracy.

The most unsatisfactory aspect of all this is the lack of understanding of the effects of noise or fluctuations. We have seen from Ren's work that the accumulation of noise can lead to numerical instabilities. But what about accuracy? There are examples which demonstrate that the amplitude of the fluctuations is reduced when kinetic Monte Carlo models are replaced in part of the computational domain by deterministic continuum equations [20, 44]. This suggests that, instead of deterministic continuum models, one should use stochastic continuum models when developing coupled atomistic–continuum simulation methods [20, 44]. Doing this consistently is still a challenge.

Figure 7.9 is an image taken from [11] for the adiabatic piston problem. A long quasi-one-dimensional box with adiabatic walls is divided into two halves by a thermally insulating piston of mass M that can move along the length of the box without friction. The two halves of the box are filled with fluid particles of mass m ($m \ll M$). The pressure difference on the two sides of the piston drives the piston to mechanical equilibrium, in which the pressure difference vanishes. More interestingly, it was shown that the mechanical asymmetric fluctuations of the piston due to its thermal motion eventually drive the system to thermal equilibrium, in which the temperature difference of the two sides also vanishes. This process can in principle be simulated using molecular dynamics or fluctuating hydrodynamics. However, classical (deterministic) hydrodynamics fails to predict the eventual full thermal equilibrium. The consequence of this is that the hybrid method with deterministic hydrodynamics fails to predict the long-term behavior of the system accurately.

There is another difficulty. Take for example QM–MM methods at zero temperature. In this case, we do not have to worry about noise. However, we still do not know how to quantify the error in QM–MM methods systematically, since we do not have a natural small parameter to measure the size of the error. In classical numerical analysis, we measure the numerical error in terms of the size of the grid. In homogenization theory, we measure the error of the homogenized model in terms of the small parameter ε (see Section 2.3). Here, there is no such small parameter, at least, not one that is obvious.

References for Chapter 7

[1] F.F. Abraham, J.Q. Broughton, N. Bernstein, and E. Kaxiras, "Spanning the continuum to quantum length scales in a dynamic simulation of brittle fracture," *Europhys. Lett.*, vol. 44, no. 6, pp. 783–787, 1998.

[2] S. Badia, P. Bochev, J. Fish *et al.* "A force-based blending model for atomistic-to-continuum coupling," *Int. J. Multiscale Comp. Engrg*, vol. 5, pp. 387–406, 2007.

[3] T. Belytschko, and S.P. Xiao, "Coupling methods for continuum model with molecular model," *Int. J. Multiscale Comput. Engrg*, vol 1, pp. 115–126, 2003.

[4] G.A. Bird, *Molecular Gas Dynamics*, Oxford University Press, 1976.

[5] N. Choly, G. Lu, W. E and E. Kaxiras, "Multiscale simulations in simple metals: a density-functional based methodology," *Phys. Rev. B*, vol. 71, no. 9, pp. 094101-1–094101-16, 2005.

[6] P. Degond, J.-G. Liu and L. Mieussens, "Macroscopic fluid modes with localized kinetic upscaling effects," *Multiscale Model. Simu.*, vol. 5, pp. 940–979, 2006.

[7] R. Delgado-Buscalioni and P.V. Coveney, "Continuum–particle hybrid coupling for mass, momentum, and energy transfers in unsteady fluid flow," *Phys. Rev. E*, vol. 67, pp. 046 704-1–046 704-13, 2003.

[8] H. Ben Dhia, "Multiscale mechanical problems: the Arlequin method," *C. R. Acad. Sci. IIB*, vol. 326, pp. 899–904, 1998.

[9] M. Dobson and M. Luskin "An analysis of the effect of ghost force oscillation on quasicontinuum error," *Math. Model. Numer. Anal.*, vol. 43, pp. 591–604, 2009.

[10] M. Dobson, M. Luskin and C. Ortner, "Stability, instability and error for the force-based quasicontinuum approximation," *Arch. Rat. Mech. Anal.*, vol 197, pp. 179–202, 2010.

[11] A. Donev, J.B. Bell, A.L. Garcia, and B.J. Alder, "A hybrid particle–continuum method for hydrodynamics of complex fluids," *Multiscale Modeling and Simulation*, vol. 8, no. 3, pp. 871–911, 2010.

[12] W. E, and Z. Huang, "Matching conditions in atomistic–continuum modeling of materials," *Phys. Rev. Lett.*, vol. 87, no. 13, pp. 135 501-1–135 501-4, 2001.

[13] W. E, and J. Lu, "The continuum limit and the QM-continuum approximation of quantum mechanical models of solids," *Comm. Math. Sci.*, vol. 5, pp. 679–696, 2007.

[14] W. E and P.B. Ming, "Cauchy–Born rule and the stability of the crystalline solids: static problems," *Arch. Rat. Mech. Anal.*, vol. 183, pp. 241–297, 2007.

[15] W. E, B. Engquist, X. Li, W. Ren and E. Vanden-Eijnden, "Heterogeneous multiscale methods: a review," *Commun. Comput. Phys.*, vol. 2, pp. 367–450, 2007.

[16] W. E, J. Lu and J.Z. Yang, "Uniform accuracy of the quasicontinuum method," *Phys. Rev B*, vol. 74, pp. 214 115-1–214 115-12, 2006.

[17] E.G. Flekkoy, G. Wagner, and J. Feder, "Hybrid model for combined particle and continuum dynamics," *Europhys. Lett*, vol. 52, pp. 271–276, 2000.

[18] J. Gao, "Methods and applications of combined quantum mechanical and molecular mechanical potentials," in *Reviews in Computational Chemistry*, vol. 7, K.B. Lipkowitz and D.B. Boyd, eds., pp. 119–185, VCH Publishers, 1995.

[19] J. Gao, P. Amara, C. Alhambra and J.J. Field, "A generalized hybrid orbital (GHO) method for the treatment of boundary atoms in combined QM/MM calculations," *J. Phys. Chem. A*, vol. 102, pp. 4714–4721, 1998.

[20] A.L. Garcia, J.B. Bell, W.Y. Crutchfield, and B.J. Alder "Adaptive mesh and algorithm refinement using direct simulation Monte Carlo," *J. Comput. Phys.*, vol. 154, pp. 134–155, 1999.

[21] C. Garcia-Cervera, J. Lu and W. E, "A sublinear scaling algorithm for computing the electronic structure of materials," *Comm. Math. Sci.*, vol. 5, no. 4, pp. 990–1026, 2007.

[22] V. Gavini, K. Bhattacharya and M. Ortiz, "Quasi-continuum orbital-free density functional theory: a route to multi-million atom non-periodic DFT calculation," *J. Mech. Phys. solids*, vol. 55, pp. 697–718, 2007.

[23] N. Hadjiconstantinou "Hybrid atomistic–continuum formulations and the moving contact-line problems," *J. Comp. Phys*, vol. 154, pp. 245–265, 1999.

[24] N. Hadjiconstantinou and A.T. Patera, "Heterogeneous atomistic–continuum representation for dense fluid systems," *Int. J. Mod. Phys. C*, vol. 8, pp. 967–976, 1997.

[25] J. Li, D. Liao and S. Yip, "Coupling continuum to molecular-dynamics simulation: reflecting particle method and the field estimator," *Phys. Rev. E*, vol. 57, pp. 7259–7267, 1998.

[26] J. Li, D. Liao and S. Yip, "Nearly exact solution for coupled continuum/MD fluid simulation," *J. Computer-Aided Material Design*, vol. 6, pp. 95–102, 1999.

[27] X. Li and W. E, "Multiscale modeling of the dynamics of solids at finite temperature," *J. Mech. Phys. Solids*, vol. 53, pp. 1650–1685, 2005.

[28] X. Li, J.Z. Yang and W. E., "A multiscale coupling method for the modeling of dynamics of solids with application to brittle cracks," *J. Comput. Phys.*, vol. 229, no. 10, pp. 3970–3987, 2010.

[29] P. Lin, "Theoretical and numerical analysis for the quasi-continuum approximation of a material particle model," *Math. Comp.*, vol. 72, pp. 657–675, 2003.

[30] P. Lin, "Convergence analysis of a quasi-continuum approximation for a two-dimensional material without defects," *SIAM J. Numer. Anal.*, vol. 45, pp. 313–332, 2007.

[31] G. Lu and E. Kaxiras, "Overview of multiscale simulations of materials," in *Handbook of Theoretical and Computational Nanotechnology*, M. Rieth and W. Schommers, eds., vol. X, pp. 1–33, American Scientific Publishers, 2005.

[32] R. Miller and E. Tadmor, "A unified framework and performance benchmark of fourteen multiscale atomistic/continuum coupling methods," *Modelling and Simulation in Mater Sci. Engrg*, vol. 17, pp. 053 001-1–053 001-51, 2009.

[33] P.B. Ming "Error estimates of forced-based quasicontinuum method," *Comm. Math. Sci.*, vol. 5, pp. 1089–1095, 2008.

[34] P.B. Ming and J.Z. Yang "Analysis of the quasicontinuum method," *Multiscale Model Simul.*, vol. 7, pp. 1838–1875, 2009.

[35] X. Nie, S. Chen, W. E and M.O. Robbins, "A continuum and molecular dynamics hybrid method for micro- and nano-fluid flow," *J. Fluid Mech.*, vol. 500, pp. 55–64, 2004.

[36] S.T. O'Connell and P.A. Thompson, "Molecular dynamics–continuum hybrid computations: a tool for studying complex fluid flows," *Phys. Rev. E*, vol. 52, pp. 5792–5795, 1995.

[37] J.T. Oden and K.S. Vemaganti, "Estimation of local modeling error and goal-oriented adaptive modeling of heterogeneous materials. Part I: Error estimates and adaptive algorithms," *J. Comput. Phys.*, vol. 164, pp. 22–47, 2000.

[38] C. Ortner and E. Süli, "Analysis of a quasicontinuum method in one dimension," *ESAIM M2AN*, vol. 42, pp. 57–92, 2008.

[39] H.S. Park, E.G. Karpov, W.K. Liu and P.A. Klein, "The bridging scale for two-dimensional atomistic/continuum coupling," *Phil. Mag. A*, vol. 85, pp. 79–113, 2005.

[40] W. Ren "Analytical and numerical study of coupled atomistic-continuum methods for fluids," *J. Compututut. Phys.*, vol. 227, pp. 1353–1371, 2007.

[41] W. Ren and W. E, "Heterogeneous multiscale method for the modeling of complex fluids and micro-fluidics," *J. Comput. Phys.*, vol. 204, pp. 1–26, 2005.

[42] R.D. Richtmyer and K.W. Morton, *Difference Schemes for Initial Value Problems*, Interscience, 1967.

[43] R.E. Rudd and J.Q. Broughton, "Coarse-grained molecular dynamics and the atomic limit of finite elements," *Phys. Rev. B*, vol. 58, no. 10, pp. R5893–R5896, 1998.

[44] T. Schulze, P. Smereka and W. E, "Coupling kinetic Monte-Carlo and continuum models with application to epitaxial growth," *J. Comput. Phys.*, vol. 189, no. 1, pp. 197–211, 2003.

[45] V.B. Shenoy, R. Miller, E.B. Tadmor, R. Phillips and M. Ortiz, "An adaptive finite element approach to atomic scale mechanics – the quasicontinuum method," *J. Mech. Phys. Solids*, vol. 47, pp. 611–642, 1999.

[46] T. Shimokawa, J.J. Mortensen, J. Schiøz and K.W. Jacobsen, "Matching conditions in the quasicontinuum method: Removal of the error introduced in the interface between the coarse-grained and fully atomistic region," *Phys. Rev. B*, vol. 69, pp. 214 104-1–214 104-10, 2004.

[47] U.C. Singh and P.A. Kollman, "A combined ab initio quantum-mechanical and molecular mechanical method for carrying out simulations on complex molecular systems applications to the CH3Cl + Cl-exchange reaction and gas-phase protonation of polyethers," *J. Comput. Chem.*, vol. 7, pp. 718–730, 1986.

[48] G. Strang, "Accurate partial difference methods. Part II: Non-linear problems," *Numer. Math*, vol. 6, pp. 37–46, 1964.

[49] E.B. Tadmor, M. Ortiz and R. Phillips, "Quasicontinuum analysis of defects in crystals," *Phil. Mag. A*, vol. 73, pp. 1529–1563, 1996.

[50] A. Toselli and O. Widlund, *Domain Decomposition Methods*, Springer-Verlag, 2004.

[51] A. Warshel and M. Karplus, "Calculation of ground and excited state potential surfaces of conjugated molecules. Part I: Formulation and parametrization," *J. Amer. Chem. Soc.*, vol. 94, pp. 5612–5625, 1972.

[52] A. Warshel and M. Levitt, "Theoretical studies of enzymic reactions," *J. Mol. Biol.*, vol. 103, pp. 227–249, 1976.

[53] T. Werder, J.H. Walther, J. Asikainen and P. Koumoutsakos, "Continuum-particle hybrid methods for dense fluids," in *Multiscale Modeling and Simulation*, S. Attinger and P. Koumoutsakos, eds., *Lecture Notes in Computational Science and Engineering*, vol. 39, pp. 35–68, Springer-Verlag, 2004.

[54] T. Werder, J.H. Walther and P. Koumoutsakos, "Hybrid atomistic-continuum method for the simulation of dense fluid flows," *J. Comput. Phys.*, vol. 205, pp. 373–390, 2005.

[55] Y. Zhang, T.-S. Lee and W. Yang "A pseudobond approach to combining quantum mechanical and molecular mechanical methods," *J. Chem. Phys.*, vol. 110, pp. 46–54, 1999.

8

Elliptic equations with multiscale coefficients

8.1 Introduction

So far our discussions have been rather general. We have emphasized the issues that are common to different multiscale, or multi-physics problems. In this and the next two chapters, we will be more concrete and focus on two representative classes of multiscale problems. In this chapter we will discuss a typical class of problems with multiple spatial scales: elliptic equations with multiscale solutions. In the following two chapters we will discuss two typical classes of problems with multiple time scales: systems with slow and fast dynamics and systems with rare events.

We have in mind two kinds of elliptic equation with multiscale solutions:

$$-\nabla \cdot (\mathbf{a}^\varepsilon(\mathbf{x})\nabla)u^\varepsilon(\mathbf{x}) = f(\mathbf{x}), \tag{8.1.1}$$

$$\Delta u_\omega(\mathbf{x}) + \omega^2 u_\omega(\mathbf{x}) = f(\mathbf{x}), \quad \omega \gg 1. \tag{8.1.2}$$

Here $\mathbf{a}^\varepsilon$ is a multiscale-coefficient tensor. For (8.1.1) the multiscale nature may come in a variety of ways: the coefficients may have discontinuities or other types of singularities or they may have oscillations at disparate scales. Similar ideas have been developed for these two kinds of problem. Therefore we will discuss them on the same footing and will use (8.1.1) as our primary example.

Another closely related example is

$$-\varepsilon\Delta u(\mathbf{x}) + (\mathbf{b}(\mathbf{x}) \cdot \nabla)u(\mathbf{x}) = f(\mathbf{x}). \tag{8.1.3}$$

As we saw in Section 2.2, in general solutions to this problem have large gradients; this is also a type of multiscale behavior. Some technique

334

discussed below, such as residual-free bubble finite element methods, can also be applied to this class of problem.

The behavior of solutions to (8.1.1) and (8.1.2) was analyzed in Chapter 2 for various special cases. Roughly speaking, in the case when $\mathbf{a}^\varepsilon$ is oscillatory, say $\mathbf{a}^\varepsilon(\mathbf{x}) = \mathbf{a}(\mathbf{x}, \mathbf{x}/\varepsilon)$, one expects the solutions to (8.1.1) to behave as $u^\varepsilon(\mathbf{x}) \sim u_0(\mathbf{x}) + \varepsilon u_1(\mathbf{x}, \mathbf{x}/\varepsilon) + \cdots$. Consequently, to leading order, u^ε does not exhibit small-scale oscillations. Note, however, that $\nabla u^\varepsilon(\mathbf{x})$ is not accurately approximated by $\nabla u_0(\mathbf{x})$.

In the case of (8.1.2) one expects the solutions to be oscillatory with frequencies $\mathcal{O}(\omega)$. This is easy to see from the one-dimensional examples discussed in Section 2.2.

From a numerical viewpoint, multiscale methods have been developed in the settings of finite element, finite difference and finite volume methods. Again, the central issues in these different settings are quite similar. Therefore, for the sake of argument, we will focus on finite element methods.

Babuska pioneered the study of finite element methods for elliptic equations with multiscale coefficients. He recognized, back in the 1970s, the importance of developing numerical algorithms that specifically target this class of problems rather than using traditional numerical algorithms or relying solely on analytical homogenization theory. Among other things, Babuska made the following observations [7].

(1) Traditional finite element methods have to resolve the small scales. Otherwise the numerical solution may converge to the wrong solution.

(2) Analytical homogenization theory might not be accurate enough, or it might be difficult to use the analytical theory owing to the complexity of the multiscale structure in the coefficients or to complications from the boundary conditions, or to the fact that the small parameters in the problem may not be sufficiently small.

In 1983, Babuska and Osborn developed the so-called *generalized finite element method* for problems with rough coefficients, including those that have discontinuous as well as those that have multiple scales [10]. The main idea was to modify the finite element space to take into account explicitly the microstructure of the problem. Although the generalized finite element method has since evolved considerably [12], the original idea of modifying the finite element space to include functions with the correct microstructure has been further developed by many people, in two main directions:

(1) increasing finite element spaces by adding functions with the right microstructure; the residual-free bubble method is a primary example of this kind [18, 19, 17];

(2) changing the finite element space by modifying its basis functions [37, 42, 36].

The *variational multiscale method* (VMS) developed by Hughes and his co-workers is based on a somewhat different idea. Its starting point is a decomposition of the finite element space into the sum of coarse- and fine-scale components [40, 41]. Hughes recognized explicitly the importance of localization in dealing with the fine-scale components [41]. Indeed, as we have seen before and will see again below, localization is a central issue in developing multiscale methods.

These methods are all fine-scale solvers: their purpose is to resolve efficiently the details, i.e. components at all scales, of the solution. In many cases we are only interested in obtaining an effective macroscale model. Homogenization theory is very effective when it applies, but it is also very restrictive. To deal with more general problems, various algorithms have been developed to compute numerically the effective macroscopic model. This procedure is generally referred to as *upscaling*. The simplest example of an upscaling procedure is the technique that uses the representative averaging volume (RAV), discussed in Chapter 1. We will discuss another systematic upscaling procedure based on successively eliminating the fine-scale components.

The algorithms discussed above all have a complexity that scales linearly or worse with respect to the number of degrees of freedom needed to represent the fine-scale details of the problem. In many cases it is desirable to have sublinear scaling algorithms. The heterogeneous multiscale method (HMM) provides a framework for developing such algorithms. Unlike linear scaling algorithms, which probe the microstructure over the entire physical domain, HMM probes the microstructure in subsets of the physical domain. The idea is to capture the macroscale behavior of the solution by locally sampling the microstructure of the problem.

In this chapter we will make use of some standard function spaces such as L^2 and H^1. The reader may consult standard textbooks such as [21] for the definitions of these spaces.

8.2 Multiscale finite element methods

8.2.1 The generalized finite element method

We start our discussion with the generalized finite element method (GFEM) proposed in 1983 by Babuska and Osborn [10]. This work focused on one-dimensional examples, but in it the important idea was put forward of adapting the finite element space to the particular fine-scale features of the problem. Traces of such ideas can be found earlier, for example in the context of enriching the finite element space using special functions that resolve the corner singularities better [53]. However, the setup of Babuska and Osborn is more general and has led to many other developments.

Consider the one-dimensional example

$$(Lu)(x) \equiv -\partial_x(a(x)\partial_x u(x)) + b(x)\partial_x u(x) + c(x)u(x) = f(x), \quad 0 < x < 1, \tag{8.2.1}$$

with boundary conditions

$$u(0) = u(1) = 0. \tag{8.2.2}$$

Define the bilinear form

$$B(u,v) = \int_0^1 \left(a(x)\partial_x u(x)\partial_x v(x) + b(x)\partial_x u(x)v(x) + c(x)u(x)v(x) \right) dx$$

for $u, v \in H^1(I)$, $I = [0,1]$. We will also use the standard inner product for L^2 functions defined on $[0,1]$:

$$(f,g) = \int_0^1 f(x)g(x)\, dx.$$

With this notation we can write down the weak form of the differential equation (8.2.1) with boundary conditions (8.2.2) as follows. Find $u \in H_0^1(I)$ such that

$$B(u,v) = (f,v)$$

for every $v \in H_0^1(I)$.

The key to the generalized finite element method proposed in [10] is to replace the standard finite element spaces of piecewise polynomial functions by a more general finite element space over a macro mesh. Denote by $\mathcal{T}_H$ such a mesh with mesh size H: $\mathcal{T}_H = \{I_j, j = 1, \ldots, n\}$. The standard finite element space takes the form

$$V_H^k = \{v \in H_0^1(I) : v|_{I_j} \in \mathcal{P}^k(I_j), j = 1, \ldots, n\},$$

where $\mathcal{P}^k(I_j)$ denotes the space of polynomials of degree k on I_j. It is instructive to write

$$V_H^k = V_H^1 \oplus \hat{V}_H^k,$$

where $\hat{V}_H^k$ is the subspace of functions in V_H^k which vanish at the nodes. One can then regard $\hat{V}_H^k$ as the space of internal degrees of freedom (or bubble space, to make an analogy with what will be discussed below). However, we do not have to use polynomials to represent the internal degrees of freedom. Other functions can also be used, and this is the main idea proposed by Babuska and Osborn.

One choice they suggested is to use

$$\hat{V}_H^k = \{v \in H_0^1(I) : L_0 v|_{I_j} \in \mathcal{P}^{k-2}(I_j), j = 1, \ldots, n\}, \quad k \geq 2,$$

where $L_0 v(x) = -\partial_x(a(x)\partial_x v(x))$. When $k = 1$, this is replaced by

$$\hat{V}_H^1 = \{v \in H_0^1(I) : L_0 v|_{I_j} = 0, j = 1, \ldots, n\}.$$

The purpose of choosing this finite element space is to select functions that have the same fine-scale structure as the expected solution of the differential equation (8.1.1). Babuska and Osborn proved that if the problem is uniformly elliptic, i.e. if there exist constants α and β such that

$$0 < \alpha \leq a(x) \leq \beta < \infty,$$

then the error of the finite element solution depends on the size of H and the regularity of f, not on the regularity of a, e.g.

$$\|u - u_H\|_{H^1(I)} \leq CH\|f\|_{L^2(I)}, \tag{8.2.3}$$

where C depends only on α and β and u and u_H are respectively the exact solution to the original PDE (8.2.1) and the numerical solution using the generalized finite element method. Note that the right-hand side of (8.2.3) depends only on f, not on u.

The generalization of this idea to higher dimensions is a nontrivial matter and different strategies have been proposed. We will discuss a few representative ideas.

8.2.2 *Residual-free bubbles*

For simplicity we will consider only PDEs in the form of (8.1.1). Let $\mathcal{T}_H$ be a regular triangulation of Ω: $\bar{\Omega} = \bigcup_{K \in \mathcal{T}_H} K$. As before, we use H

instead of h to denote the mesh size in order to emphasize the fact that the mesh $\mathcal{T}_H$ should be regarded as a macro mesh. Let $V_{\mathrm{p}} \subset H_0^1(\Omega)$ be the standard conforming piecewise-linear finite element space on $\mathcal{T}_H$. Define the bubble space by

$$V_{\mathrm{b}} = \{v \in H_0^1(\Omega), v|_{\partial K} = 0, \quad \text{for any } K \in \mathcal{T}_H\}$$

and let $V_H = V_{\mathrm{p}} \oplus V_{\mathrm{b}}$. The *residual-free bubble* (RFB) finite element method proposed by Brezzi and co-workers can be formulated as follows [18, 19, 17]: find $u_H = u_{\mathrm{p}} + u_{\mathrm{b}} \in V_H$, such that

$$a(u_H, v_H)$$
$$\equiv \int_\Omega a^\varepsilon(\mathbf{x}) \nabla u_H(\mathbf{x}) \cdot \nabla v_H(\mathbf{x}) dx = (f, v_H), \quad \text{for all } v_H = v_{\mathrm{p}} + v_b \in V_H.$$

This scheme can be split naturally into two parts: a coarse-scale part

$$a(u_{\mathrm{p}} + u_{\mathrm{b}}, v_{\mathrm{p}}) = (f, v_{\mathrm{p}}), \quad \text{for all } v_{\mathrm{p}} \in V_{\mathrm{p}}; \tag{8.2.4}$$

and a fine-scale part

$$a(u_{\mathrm{b}}, v_{\mathrm{b}}) = (f - Lu_{\mathrm{p}}, v_{\mathrm{b}}), \quad \text{for all } v_{\mathrm{b}} \in V_{\mathrm{b}}, \tag{8.2.5}$$

where L is the differential operator defined by the left-hand side of (8.1.1).

The RFB finite element method can also be regarded as an upscaling procedure for computing the coarse-scale part u_{p} [51]: we just have to solve the fine-scale equation (8.2.5) for the bubble component u_{b} and substitute the latter into the coarse-scale equation (8.2.4). The remaining problem for u_{p} is the upscaled problem.

An important feature is that the bubble components are localized to each element: for every $K \in \mathcal{T}_H$ the fine-scale equation (8.2.5) is the following local problem for u_{b}:

$$L(u_{\mathrm{p}} + u_{\mathrm{b}}) = f \quad \text{in } K, \tag{8.2.6}$$
$$u_{\mathrm{b}} = 0 \quad \text{on } \partial K. \tag{8.2.7}$$

Hence $u_H = u_{\mathrm{p}} + u_{\mathrm{b}}$ is *residual-free* inside each element: $Lu_H - f = 0$ on K for all K.

Consider the one-dimensional example

$$-\partial_x(a^\varepsilon(x)\partial_x u(x)) = f(x), \quad 0 < x < 1,$$

with boundary conditions $u(0) = u(1) = 0$. Let V_H be the standard piecewise-linear finite element space, and let u_{p} be a function in V_H over the partition $\mathcal{T}_H$. Let $I_j = [x_j, x_{j+1}]$ be an element in $\mathcal{T}_H$. Then on I_j the RFB function u_b that corresponds to u_p is given by the solution to the following problem:

$$-\partial_x(a^\varepsilon(x)\partial_x u_b(x)) = f(x) + c_j\partial_x a^\varepsilon(x), \quad x_j < x < x_{j+1},$$

with boundary conditions $u_b(x_j) = u_b(x_{j+1}) = 0$, where c_j is the value of $\partial_x u_p(x)$ on I_j.

To gain some insight into the performance of the RFB method for multiscale problems, let us look at the case of (8.1.1). The following result was proved in [51].

Theorem *Let $a^\varepsilon(\mathbf{x}) = a(\mathbf{x}/\varepsilon)$, where a is a smooth periodic function with period $I = [0, 1]^d$. Denote by u^ε the exact solution of (8.1.1) and by u_H the numerical solution with grid size H. Then there exists a constant C, independent of ε and H, such that*

$$\|u^\varepsilon - u_H\|_{H^1(\Omega)} \leq C\left(\varepsilon + H + \sqrt{\frac{\varepsilon}{H}}\right)\|f\|_{L^2(\Omega)}. \tag{8.2.8}$$

8.2.3 Variational multiscale method

The variational multiscale method (VMS), proposed by Hughes and co-workers [40, 41], is another general framework for designing multiscale methods in a variational setting. The basic idea in VMS is to decompose the numerical solution u into a sum of coarse- and fine-scale parts, i.e. setting

$$u = U + \tilde{u},$$

and to use a coarse mesh to compute the coarse-scale part U and a nested fine mesh (or higher-order polynomials as in the p-version finite element method or spectral methods) to compute the fine-scale part $\tilde{u}$. The key to VMS is to apply various localization procedures for the computation of $\tilde{u}$. One can also think of VMS as an approach for subgrid modeling and $\tilde{u}$ as the subgrid component. In this regard, VMS can be viewed as an upscaling procedure.

In the context of finite element methods, the computational domain Ω is triangulated into a macro finite element mesh with mesh size H. The coarse-scale component of the finite element solution is sought in a standard finite element space, say the space of piecewise-linear functions over this triangulation $\mathcal{T}_H$. We will denote this coarse-scale finite element space by V_H. Each element in $\mathcal{T}_H$ is further triangulated into a fine-scale mesh. The union of these fine-scale meshes constitutes a fine-scale mesh over the whole domain Ω, which we denote by $\mathcal{T}_h$.

Let V_h be the standard piecewise-linear finite element space over $\mathcal{T}_h$ and let

$$\tilde{V}_h = \{v_h \in V_h, \text{such that } v_h \text{ vanishes at the nodes of } \mathcal{T}_H\}.$$

Then

$$V_h = V_H \oplus \tilde{V}_h.$$

Let $\{\Phi_j\}$ and $\{\varphi_k\}$ be the standard nodal bases of V_H and $\tilde{V}_h$ respectively. We can express any function in V_h in the form

$$u_h = \sum_j U_j \Phi_j + \sum_k u_k \varphi_k.$$

Let $U = (U_1, \ldots, U_n)^T$, $u = (u_1, \ldots, u_N)^T$; then the finite element solution in V_h satisfies a linear system of equations of the form

$$\begin{pmatrix} A_{11} & A_{12} \\ A_{21} & A_{22} \end{pmatrix} \begin{pmatrix} U \\ u \end{pmatrix} = \begin{pmatrix} B \\ b \end{pmatrix}.$$

Eliminating u, we obtain

$$(A_{11} - A_{12} A_{22}^{-1} A_{21}) U = B - A_{12} A_{22}^{-1} b. \tag{8.2.9}$$

The term $-A_{12} A_{22}^{-1} A_{12}$ expresses the effect of the fine-scale components on the coarse-scale components.

In general A_{22}^{-1} is a nonlocal operator. In order to obtain an efficient algorithm, the authors of [41] localized this operator. The simplest localization procedure is to decouple the fine-scale component over the different macro elements. One way of realizing this is simply to set the boundary value of $\tilde{u}$ over each macro element to 0. This eliminates the degrees of freedom of u_h over the edges of the macro elements and decouples the microscale component on different macro elements. Other localization procedures are discussed in [49].

The assumption that the fine scale components vanish at the macro element boundaries is a strong assumption. Nolen *et al.* [46] explored the possibility of extending the over-sampling technique proposed in [37] (also discussed below) to this setting.

As was pointed out in [40, 41], the fine-scale component over each macro element is closely related to the bubble functions in the RFB method [18]. In fact, for the one-dimensional example discussed at the end of the last subsection, the bubble functions and the finite-scale components are the same. This is true in high dimensions also as long as we require that the fine-scale components vanish at the macro element boundaries.

8.2.4 *Multiscale basis functions*

The idea in RFB and VMS is to add fine-scale components to a coarse finite element solution. An alternative idea is to modify the basis functions so that they contain the same fine-scale components as the solution to the microscale problem. This was pursued in [36, 42] in the context of solving the Helmholtz equation (8.1.2) and in [37] in the context of solving elliptic equations with multiscale coefficients such as (8.1.1). As usual, the difficulty is in the boundary conditions for the modified basis functions. There is a major difference between one dimension and higher dimensions. In one dimension, we simply replace the standard nodal basis functions by functions which satisfy the homogeneous equation over each element, with the same values at the nodes as the nodal basis functions. The situation is not as simple in high dimensions. Below we discuss the approach proposed in [37] for handling this problem, in the context of (8.1.1).

Let $\mathcal{T}_H$ be a triangulation of the domain Ω with mesh size H. For any given element $K \in \mathcal{T}_H$, denote by $\{\mathbf{x}_{K,j}, j = 1, \ldots, d\}$ the vertices of K. We will construct basis functions $\{\phi_{K,j}, j = 1, \ldots, d\}$ such that

$$\phi_{K,j}(\mathbf{x}_{K,i}) = \delta_{ij}. \qquad (8.2.10)$$

This way of indexing the basis functions introduces some redundancy: neighboring elements may share vertices and therefore basis functions. Nevertheless, there is not going to be any inconsistency since we will focus on a single element.

As in the one-dimensional case, $\phi_{K,j}$ is required to solve the homogeneous version of the original microscale problem:

$$(L\phi_{K,j})(\mathbf{x}) = -\nabla \cdot (\mathbf{a}^\varepsilon(\mathbf{x})\nabla)\phi_{K,j}(\mathbf{x}) = 0, \quad \mathbf{x} \in K. \qquad (8.2.11)$$

The main question is how to impose the boundary condition on ∂K. The simplest proposal is to require that

$$\phi_{K,j}|_{\partial K} = \varphi_{K,j}|_{\partial K}, \qquad (8.2.12)$$

where $\varphi_{K,j}$ is the nodal basis function associated with the jth vertex of the element K. We will see later that the basis functions resulting from this choice of boundary condition is very closely related to the RFB method: $\phi_{K,j}$ is equal to $\varphi_{K,j}$ plus its RFB correction.

Of course there is no guarantee that such a procedure captures the correct microstructure at the boundaries of the elements, and this can also affect the accuracy in the interior of the elements. Motivated by the fact that, when the microscale is much smaller than the macroscale, the influence of the boundary condition is sometimes limited to a thin boundary layer, it was proposed in [37] that the domain on which (8.2.11) is solved should be extended. Then the basis functions should be formed as linear combinations of the extended functions, restricted to the original elements. The intuition is that the precise boundary condition will have less effect on the domain of interest. Specifically, one may proceed as follows.

(1) Solve (8.2.11) on a domain S which contains, and which is a few times larger than, K as a subdomain. The boundary condition at ∂S can be quite arbitrary. The minimum requirement is that the matrix Ψ defined below be nonsingular. Denote these temporary basis functions by $\{\psi_{K,j}\}$.

(2) Let

$$\phi_{K,i} = \sum_{j,K} c_K^{ij} \psi_{K,j}, \qquad (8.2.13)$$

where the coefficients $\{c_K^{ij}\}$ are chosen in such a way that the condition (8.2.10) is satisfied by $\{\phi_{K,i}\}$. Let $\Psi = (\psi_{K,i}(x_{K,j}))$ and $C = (c_K^{ij})$. It is easy to see that $C = \Psi^{-1}$ if the coefficients are ordered consistently.

This procedure allows us to construct basis functions for each element K. The overall basis functions are constructed by patching together the basis functions on each element. The basis functions constructed in this way are in general nonconforming, owing to the discontinuity across the element boundaries. However, as we see below, convergence can still be established in some interesting cases. We will denote the numerical solution by $\tilde{u}_H$. This version of the multiscale finite element method is often denoted as MsFEM.

The above procedure is called "oversampling," even though "overlapping" might be a more suitable term. It has been observed numerically that this change improves the accuracy of the algorithm [37]. Analytical results for homogenization problems also support this numerical observation [27].

For the case when $\mathbf{a}^\varepsilon(\mathbf{x}) = \mathbf{a}(\mathbf{x}, \mathbf{x}/\varepsilon)$, where $\mathbf{a}$ is a smooth periodic tensor function, the following error estimates have been proved [38] when linear boundary conditions are used at the element boundaries to construct basis functions:

$$\|u^\varepsilon - \tilde{u}_H\|_{H^1(\Omega)} \leq C_1 H \|f\|_{L^2(\Omega)} + C_2 \sqrt{\frac{\varepsilon}{H}}, \qquad (8.2.14)$$

where C_1 and C_2 are constants independent of ε and H. When the oversampling technique is used, (8.2.14) can be improved to [27]

$$\|u^\varepsilon - \tilde{u}_H\|_{H^1(\Omega)} \leq C_1 H \|f\|_{L^2(\Omega)} + C_2 \frac{\varepsilon}{H} + C_3 \sqrt{\varepsilon}. \qquad (8.2.15)$$

8.2.5 *Relations between the various methods*

It is quite obvious that the RFB finite element method is closely related to VMS. As was pointed out in [40, 41], the RFB finite element method can be viewed as a special case of VMS in which the bubble functions represent the fine-scale component of the solution.

The relation between the RFB finite element method and MsFEM has also been investigated [19, 51, 55]. The most thorough discussion is found in [55]. We now summarize its main conclusions.

We split the bubble component u_{b} into two parts, so that $u_{\mathrm{b}} = M_0(u_{\mathrm{p}}) + M(f)$ where $M_0(u_{\mathrm{p}}), M(f) \in V_{\mathrm{b}}$ and

$$a(u_{\mathrm{p}} + M_0(u_{\mathrm{p}}), v) = 0 \quad \text{for all } v \in V_{\mathrm{b}}, \qquad (8.2.16)$$
$$a(M(f), v) = (f, v) \quad \text{for all } v \in V_{\mathrm{b}}. \qquad (8.2.17)$$

Roughly speaking, $M_0(u_{\mathrm{p}})$ is the intrinsic part of the bubble component i.e. the part due to the multiscale structure of the coefficients in the PDE (8.1.1), and $M(f)$ is the part due to the forcing term.

Let $K \in \mathcal{T}_H$. In (8.2.6) let $f = 0$ and $u_{\mathrm{p}} = \varphi_{K,j}$. Then $u_{\mathrm{b}} = M_0(\varphi_{K,j})$, and (8.2.6) becomes

$$L(\varphi_{K,j} + M_0(\varphi_{K,j})) = 0 \quad \text{on } K$$

with the boundary condition $M_0(\varphi_{K,j}) = 0$ on ∂K. Comparing with (8.2.11), we see that if $\phi_{K,j}$ satisfies the boundary condition (8.2.12) then we must have

$$\phi_{K,j} = \varphi_{K,j} + M_0(\varphi_{K,j}). \qquad (8.2.18)$$

A similar statement also holds for the numerical solutions, namely

$$\tilde{u}_H = u_{\mathrm{p}} + M_0(u_{\mathrm{p}}) \qquad (8.2.19)$$

if the bilinear form a is symmetric. Here u_{p} and $\tilde{u}_H$ are respectively the coarse component of the numerical solution of the RFB method and the solution of MsFEM. This can be seen as follows. With the above notation, we can reformulate the RFB finite element method as

$$a(u_{\mathrm{p}} + M_0(u_{\mathrm{p}}) + M(f), v_{\mathrm{p}}) = (f, v_{\mathrm{p}}), \quad \text{for all } v_{\mathrm{p}} \in V_{\mathrm{p}},$$
$$a(u_{\mathrm{p}} + M_0(u_{\mathrm{p}}) + M(f), v_{\mathrm{b}}) = (f, v_{\mathrm{b}}), \quad \text{for all } v_{\mathrm{b}} \in V_{\mathrm{b}}.$$

Letting $v_{\mathrm{b}} = M_0(v_{\mathrm{p}})$ in the second equation and adding the two equations, we obtain

$$a(u_{\mathrm{p}} + M_0(u_{\mathrm{p}}) + M(f), v_{\mathrm{p}} + M_0(v_{\mathrm{p}})) = (f, v_{\mathrm{p}} + M_0(v_{\mathrm{p}})), \quad \text{for all } v_{\mathrm{p}} \in V_{\mathrm{p}}.$$

If the bilinear form $a(\cdot, \cdot)$ is symmetric then, from (8.2.16),

$$a(M(f), v_{\mathrm{p}} + M_0(v_{\mathrm{p}})) = a(v_{\mathrm{p}} + M_0(v_{\mathrm{p}}), M(f)) = 0.$$

Therefore

$$a(u_{\mathrm{p}} + M_0(u_{\mathrm{p}}), v_{\mathrm{p}} + M_0(v_{\mathrm{p}})) = (f, v_{\mathrm{p}} + M_0(v_{\mathrm{p}})) \text{ for all } v_{\mathrm{p}} \in V_{\mathrm{p}}. \quad (8.2.20)$$

Using (8.2.18) we obtain (8.2.19).

It should be noted that both the RFB finite element method and MsFEM have been extended to situations when $a(\cdot, \cdot)$ is asymmetric. In that case (8.2.19) may not hold.

For a discussion on the connection between the variational multiscale method and MsFEM, see [6].

8.3 Upscaling via successive elimination of fine-scale components

In many situations we are interested in finding an effective model for the coarse-scale component of the solution. In some cases such models can be obtained analytically, using, for example, the techniques discussed in Chapter 2. However, analytical approaches are often quite limited. They rely on very special features such as the periodicity of the fine-scale structure of the problem. Therefore, it is of interest to develop numerical techniques for deriving such effective models, i.e. for the purpose of upscaling. The objective then is to eliminate the fine-scale components. At an abstract level, such a procedure was discussed at the end of Chapter 2. However, many practical issues have to be resolved in order to implement this procedure efficiently.

To begin with, we can either eliminate the fine scales in one step or do it successively. Using the notation of Chapter 3, if we write

$$u = \sum_{k=1}^{N} u_k$$

and choose to keep only the components with $k \leq K_0$, we can use one of the following two approaches.

(1) Eliminate the components in $\sum_{k=K_0+1}^{N} u_k$ in one step. Using the notation of Chapter 2, this can be done by choosing H_0 to be the space that contains the components we would like to keep and performing the procedure discussed at the end of Chapter 2.

(2) Eliminate the fine-scale components successively, i.e. we first eliminate u_N, then u_{N-1}, u_{N-2} and so on.

Experience with renormalization group analysis suggests that the second approach is to be preferred in general. The main reason is that, with the second approach, each step is a perturbation of the previous step. Therefore it is more likely that the effective operator only changes slightly at each step. This is important when making approximations to the effective operators. For this reason we will focus on the second approach, which was initially suggested by Beylkin and Brewster [15] and has been pursued systematically by Dorobantu, Engquist, Runborg *et al.* [23, 4]. A useful review can be found in [29].

The basic idea is to implement the scale elimination procedure outlined at the end of Chapter 2 in a multi-resolution framework. For this purpose we need a sequence of multi-resolution subspaces of L^2, as discussed in Chapter 3. The sequence of subspaces generated by wavelets is a natural choice, even though other choices are also possible. In renormalization group analysis, for example, one often chooses the Fourier spaces.

Recall the basic setup for multi-resolution analysis:

$$L^2 = \bigcup_j V_j,$$

where

$$V_{j+1} = V_j \oplus W_j, \quad W_j = V_j^\perp.$$

To start the upscaling procedure, one first chooses a V_{j_0} and discretizes the original problem (8.1.1) in V_{j_0}, which needs to be large enough to resolve the fine-scale details of the solution of (8.1.1). After discretization, one obtains a linear system of equations:

$$A_h u_h = f_h, \tag{8.3.1}$$

where the subscript h refers to the size of the grid. Let $u_{j_0} = u_h$, $A_{j_0} = A_h$, $f_{j_0} = f_h$. In V_{j_0}, define P_{j_0} as the projection operator onto the subspace V_{j_0-1} and put $Q_{j_0} = I - P_{j_0}$. Write $u_{j_0} = u_{j_0-1} + w_{j_0-1}$, where $u_{j_0-1} = P_{j_0} u_{j_0}$ and $w_{j_0-1} = Q_{j_0} u_{j_0}$. Using the basis functions in V_{j_0-1} and W_{j_0-1}, one can rewrite (8.3.1) as

$$\begin{pmatrix} A_{j_0}^{11} & A_{j_0}^{12} \\ A_{j_0}^{21} & A_{j_0}^{22} \end{pmatrix} \begin{pmatrix} u_{j_0-1} \\ w_{j_0-1} \end{pmatrix} = \begin{pmatrix} f_{j_0}^1 \\ f_{j_0}^2 \end{pmatrix}.$$

Eliminating w_{j_0-1} one obtains

$$\left(A_{j_0}^{11} - A_{j_0}^{12}(A_{j_0}^{22})^{-1} A_{j_0}^{21} \right) u_{j_0-1} = f_{j_0}^1 - A_{j_0}^{12}(A_{j_0}^{22})^{-1} f_{j_0}^2. \tag{8.3.2}$$

Letting $A_{j_0-1} = A_{j_0}^{11} - A_{j_0}^{12}(A_{j_0}^{22})^{-1} A_{j_0}^{21}$ and $f_{j_0-1} = f_{j_0}^1 - A_{j_0}^{12}(A_{j_0}^{22})^{-1} f_{j_0}^2$, one can then repeat the above procedure.

An obvious problem with the procedure described so far is the following. Our original operator was a differential operator, and therefore one would expect the matrix A_h obtained from finite difference or finite element discretizations to be sparse. This property is lost after the elimination step, however, since $(A_{j_0}^{22})^{-1}$ is in general a full matrix and so is A_{j_0-1}. Fortunately, even though A_{j_0-1} is full, many of its components are very close to zero owing to their fast decay property, as shown in Figure 8.1. The decay rate in the wavelet representation was analyzed systematically in [16]; see also [29]. Therefore one should be able to approximate the matrix A_{j_0-1} by a sparse matrix and still retain good accuracy. Two ways of approximating this matrix have been proposed and tested. The first is a direct truncation to a banded or block-banded matrix. For one-dimensional problems, this is done as follows. Given a matrix A, define a

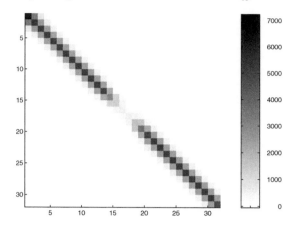

Figure 8.1. The structure of the matrices obtained during the elimination process (courtesy of Olof Runborg). The axis numbers label the rows and columns of the matrices. The gray scale indicates the sizes of the elements. Note that they are very small in the middle of a matrix.

new banded matrix trunc (A, ν) of bandwidth ν by

$$
\operatorname{trunc}(A, \nu)_{ij} = \begin{cases} A_{ij} & \text{if } 2|i - j| \leq \nu - 1, \\ 0 & \text{otherwise.} \end{cases}
$$

For higher-dimensional problems, one should consider block-banded matrices (see [29]).

In the second approach the matrix A_{j_0-1} is projected onto the space of banded (or block-banded) matrices, with the constraint that the action of the original matrix and the projected matrix be the same on some special subspaces, e.g. subspaces representing smooth functions. This is called *band projection*.

In principle there is a third possibility, namely, thresholding: if a matrix element is less than a prescribed threshold value, it is set to 0. This is less practical, since one does not have *a priori* control of the sparsity pattern in such an approach.

An example of the numerical solutions obtained using this successive scale-elimination procedure is shown in Figure 8.2.

The procedure described above is an example of numerical homogenization techniques. It has been applied to several different classes of problems, including elliptic equations with multiscale coefficients and Helmholtz equations as well as some hyperbolic and nonlinear problems

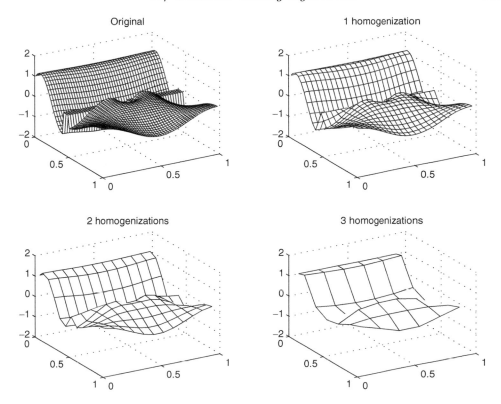

Figure 8.2. An example of the numerical solutions of the Helmholtz equation obtained after successively eliminating the small scales; see [29]. (Used with kind permission of Springer Science + Business Media.)

(see [29]). It has a close resemblance to the renormalization group procedure. Its main advantage is its generality: there are no special requirements on either the PDE or the multiscale structure. Nevertheless, it requires one to start from a fully resolved numerical discretization. Therefore, the best one can hope for is a linear scaling algorithm. In addition, information about the fine-scale components of the solution is lost during the process. For example, in the context of the stress analysis of composite materials, special efforts have to be made to recover information about the stress distribution.

8.4 Sublinear scaling algorithms

The approaches discussed above are all linear or superlinear scaling algorithms. Take the example of the homogenization problem with periodic coefficients (8.1.1): $\mathbf{a}^\varepsilon(\mathbf{x}) = \mathbf{a}(\mathbf{x}, \mathbf{x}/\varepsilon)$. In their current formulation, the overheads for forming the finite element spaces in GFEM, in RFB finite element, in VMS and in MsFEM all scale as $1/\varepsilon^d$ or more, where d is the spatial dimension of the problem. The cost for forming the fine-scale discretization operator in the numerical homogenization procedure just discussed also scales as $1/\varepsilon^d$. In some applications algorithms with such a cost are not feasible and one has to look for more efficient techniques, i.e. sublinear scaling algorithms.

As explained in the first chapter, we cannot expect to have sublinear scaling algorithms for completely general situations. We have to rely on special features of the problem, such as the separation of scales or similarity relations between different scales. In other words, the problem has to have a special structure in the space of scales. This special structure allows us to make predictions on coarser scales using information gathered from fine-scale simulations, therefore achieving sublinear scaling. For the problems treated here, the heterogeneous multiscale method (HMM) discussed in Chapter 6 provides a possible framework for developing such algorithms. We will discuss two cases for which sublinear scaling algorithms can be developed. The first is the case with scale separation. In the second case, discussed in the notes at the end of the chapter, the fine-scale features exhibit statistical self-similarity.

There is another reason for designing algorithms such as HMM which focus on small windows. For practical problems, we often do not have full information about the coefficients everywhere in the physical domain of interest. In the context of subsurface porous-medium modeling, for example, we often have rather precise information locally near the wells and some coarse and indirect information away from them. The permeability and porosity fields used in porous-medium simulations are obtained through geostatistical modeling. This is also a source of significant error. Therefore it might be of interest to consider the reverse approach: instead of first performing geostatistical modeling on the coefficients and solving the resulting PDE, one could develop algorithms for predicting the solution by using only the data obtained directly from measurements. The heterogeneous multiscale method can be viewed as an example of such a procedure.

8.4.1 Finite element HMM

To set up HMM we need to have some knowledge of the macroscale model. In the case of (8.1.1), abstract homogenization theory tells us that the macroscale component of the solution satisfies an effective equation of the form [14]

$$-\nabla \cdot (\mathbf{A}(\mathbf{x})\nabla U(\mathbf{x})) = f(\mathbf{x}), \qquad \mathbf{x} \in D, \tag{8.4.1}$$

where $\mathbf{A}(\mathbf{x})$ is the effective coefficient at the macroscale. These coefficients are not explicitly known. It is possible to precompute them but, if we are interested in a strategy that works equally effectively for more complicated situations such as nonlinear or time-dependent problems, we should compute them "on-the-fly."

As the macro solver it is natural to use the standard finite element method. The simplest choice is standard C^0 piecewise-linear elements, on a triangulation $\mathcal{T}_H$ where H denotes the element size. We will denote by V_H the finite element space. The size of the elements should resolve the macroscale computational domain D, but they do not have to resolve the fine scales.

The needed data is the stiffness matrix for the finite element method:

$$\mathcal{A} = (A_{ij}),$$

where

$$A_{ij} = \int_D (\nabla \Phi_i(\mathbf{x}))^{\mathrm{T}} \mathbf{A}(\mathbf{x}) \nabla \Phi_j(\mathbf{x}) \, d\mathbf{x},$$

$\mathbf{A}(\mathbf{x})$ is the effective coefficient (say, the conductivity) at the scale H and $\{\Phi_i(\mathbf{x})\}$ are the standard nodal basis functions of V_H; if $\mathbf{A}(\mathbf{x})$ were known we could have evaluated (A_{ij}) by numerical quadrature. Let $f_{ij}(\mathbf{x}) = (\nabla \Phi_i(\mathbf{x}))^{\mathrm{T}} \mathbf{A}(\mathbf{x}) \cdot \nabla \Phi_j(\mathbf{x})$; then

$$A_{ij} = \int_D f_{ij}(\mathbf{x}) \, d\mathbf{x} \simeq \sum_{K \in \mathcal{T}_H} |K| \sum_{\mathbf{x}_\ell \in K} w_\ell f_{ij}(\mathbf{x}_\ell), \tag{8.4.2}$$

where $\{\mathbf{x}_\ell\}$ and $\{w_\ell\}$ are the quadrature points and weights respectively and $|K|$ is the volume of the element K. Therefore the data that we need to estimate are the values of $\{f_{ij}(\mathbf{x}_\ell)\}$. This estimation is done by solving the original microscopic model, properly reformulated, locally around each quadrature point $\{\mathbf{x}_\ell\}$ on the domain $I_\delta(\mathbf{x}_\ell)$, as shown in Figure 8.3.

The key component of this method is the boundary condition for the microscale problem (8.1.1) on $I_\delta(\mathbf{x}_\ell)$. This will be discussed in the next subsection.

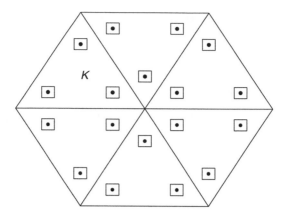

Figure 8.3. Illustration of HMM for solving (8.1.1). A typical triangular element is denoted by K. The dots are the quadrature points. The little squares are the microcell domains $I_\delta(\mathbf{x}_\ell)$.

From the solution to the microscale problem, we estimate the needed data $f_{ij}(\mathbf{x}_\ell)$ by

$$f_{ij}(\mathbf{x}_\ell) \simeq \frac{1}{\delta^d} \int_{I_\delta(\mathbf{x}_\ell)} (\nabla \varphi_i^\varepsilon(\mathbf{x}))^{\mathrm{T}} \mathbf{a}^\varepsilon(\mathbf{x}) \nabla \varphi_j^\varepsilon(\mathbf{x}) \, d\mathbf{x}, \qquad (8.4.3)$$

where φ_j^ε is the solution to the microscale problem on $I_\delta(\mathbf{x}_\ell)$ that corresponds to Φ_j (a more precise formulation will be discussed below):

$$-\nabla(\mathbf{a}^\varepsilon(\mathbf{x}) \cdot \nabla)\varphi_j^\varepsilon(\mathbf{x}) = 0, \qquad (8.4.4)$$

$$\langle \nabla \varphi_j^\varepsilon \rangle_{I_\delta} = \nabla \Phi_j(\mathbf{x}_\ell), \qquad (8.4.5)$$

and similarly for φ_i^ε. Knowing $\{f_{ij}(\mathbf{x}_\ell)\}$, we can obtain the stiffness matrix $\mathcal{A}$ from (8.4.2).

8.4.2 The local microscale problem

We now come to the boundary conditions for the local microscale problem. Ideally, one would like to formulate the boundary conditions as if the boundary was not present, i.e. we would like to capture exactly the microscale solution on I_δ (see Chapter 7), which is the solution of the microscale problem over the whole macroscale domain restricted to I_δ. This is not possible, owing to the intrinsic nonlocality of the elliptic equations. Still, we would like to minimize the error due to the artificial boundary conditions.

To formulate the local microscale problem on I_δ, we ask the question: if the average gradient is $\mathbf{G}$, what is the corresponding microscale behavior? To answer this question we solve the microscale problem (8.4.4) subject to the constraint

$$\langle \nabla \varphi^\varepsilon \rangle_{I_\delta} = \mathbf{G}. \tag{8.4.6}$$

We will denote the solution to the microscale problem with the constraint (8.4.6) by $\varphi^\varepsilon_{\mathbf{G}}$. This problem was investigated in [56]. Below is a summary of the findings in [56]. Three different types of boundary conditions were considered.

Dirichlet formulation In this case, the Dirichlet boundary conditions are used for the local microscale problem:

$$\begin{cases} -\nabla \cdot (\mathbf{a}^\varepsilon(\mathbf{x})\nabla\varphi) = 0 & \text{in } I_\delta, \\ \varphi(\mathbf{x}) \qquad\qquad\ = \mathbf{G} \cdot \mathbf{x}, & \text{on } \partial I_\delta. \end{cases} \tag{8.4.7}$$

One can check easily that

$$\langle \nabla\varphi(\mathbf{x}) \rangle \equiv \fint_{I_\delta} \nabla\varphi(\mathbf{x})\, dx \equiv \frac{1}{|I_\delta|}\int_{I_\delta} \nabla\varphi(\mathbf{x})\, dx = \mathbf{G},$$

where the second identity symbol defines the previous expression.

In the Dirichlet formulation the effective conductivity tensor $\mathbf{A}^*_{\mathrm{D}}$ is defined by the relation

$$\langle \mathbf{a}^\varepsilon(\mathbf{x})\nabla\varphi \rangle = \mathbf{A}^*_{\mathrm{D}}\langle \nabla\varphi \rangle = \mathbf{A}^*_{\mathrm{D}}\mathbf{G}. \tag{8.4.8}$$

Periodic formulation In this case, the microscale problem is given by

$$\begin{cases} -\nabla \cdot (\mathbf{a}^\varepsilon(\mathbf{x})\nabla)\varphi = 0 & \text{in } I_\delta, \\ \varphi(\mathbf{x}) - \mathbf{G} \cdot \mathbf{x} & \text{is periodic with period } I_\delta. \end{cases} \tag{8.4.9}$$

It can be easily checked that the constraint (8.4.6) is satisfied. The effective conductivity tensor $\mathbf{A}^*_{\mathrm{P}}$ is defined through

$$\langle \mathbf{a}^\varepsilon(\mathbf{x})\nabla\varphi \rangle = \mathbf{A}^*_{\mathrm{P}}\langle \nabla\varphi \rangle = \mathbf{A}^*_{\mathrm{P}}\mathbf{G}. \tag{8.4.10}$$

It is obvious that when the microstructure of $\mathbf{a}^\varepsilon$ is periodic, $\mathbf{A}^*_{\mathrm{P}}$ will be the same as the homogenized coefficient tensor obtained in the homogenization theory ([57], see also Section 2.4) if I_δ is chosen to be an integer multiple of the period.

Neumann formulation In this case, the microscale problem is formulated as

$$\begin{cases} -\nabla \cdot (\mathbf{a}^\varepsilon(\mathbf{x})\nabla)\varphi = 0 & \text{in } I_\delta, \\ \mathbf{a}^\varepsilon(\mathbf{x})\nabla\varphi(\mathbf{x}) \cdot \mathbf{n} = \boldsymbol{\lambda} \cdot \mathbf{n} & \text{on } \partial I_\delta, \end{cases} \tag{8.4.11}$$

where the constant vector $\boldsymbol{\lambda} \in \mathbb{R}^d$ is the Lagrange multiplier for the constraints

$$\langle \nabla\varphi \rangle = \mathbf{G}. \tag{8.4.12}$$

The effective conductivity tensor is then given by

$$\langle \mathbf{a}^\varepsilon(\mathbf{x})\nabla\varphi \rangle = \mathbf{A}_N^* \langle \nabla\varphi \rangle = \mathbf{A}_N^* \mathbf{G}. \tag{8.4.13}$$

Take for example $d = 2$; to solve problem (8.4.11) with the constraint (8.4.12), we first solve for φ_1 and φ_2 from

$$\begin{cases} -\nabla \cdot (\mathbf{a}^\varepsilon(\mathbf{x})\nabla\varphi_i) = 0 & \text{in } I_\delta, \\ \mathbf{a}^\varepsilon(\mathbf{x})\nabla\varphi_i(\mathbf{x}) \cdot \mathbf{n} = \mu_i \cdot \mathbf{n} & \text{on } \partial I_\delta, \end{cases}$$

for $i = 1, 2$, where $\boldsymbol{\mu}_1 = (1 \quad 0)^T$, $\boldsymbol{\mu}_2 = (0 \quad 1)^T$. Given an arbitrary $\mathbf{G}$, the Lagrange multiplier vector $\boldsymbol{\lambda} = (\lambda_1 \quad \lambda_2)^T$ is determined by the linear equation

$$\lambda_1 \langle \nabla\varphi_1 \rangle + \lambda_2 \langle \nabla\varphi_2 \rangle = \mathbf{G} \tag{8.4.14}$$

and the solution of (8.4.11), (8.4.12) is given by $\varphi = \lambda_1\varphi_1 + \lambda_2\varphi_2$.

It is easy to see that for the one-dimensional case the three quantities $\mathbf{A}_D^*$, $\mathbf{A}_P^*$, $\mathbf{A}_N^*$ are equal.

Theorem *In high dimensions, we have ([56])*

$$\mathbf{A}_N^* \le \mathbf{A}_P^* \le \mathbf{A}_D^*. \tag{8.4.15}$$

To see the influence of the cell size and the particular boundary conditions on the accuracy of the effective coefficient tensor, Yue and E [56] performed a systematic numerical study on various examples. In particular, the two-dimensional random checker-board problem was used as a test case. This is a random two-phase composite material: the plane is covered by a square lattice, and each square is randomly assigned one of the two phases with equal probability. Both phases are assumed to be isotropic. Interest in this problem stems from the fact that the effective conductivity for this case can be computed analytically using some duality

Table 8.1. *Random checker-board: ensemble-averaged effective conductivities $\bar{K}^*$ computed using different boundary conditions. Also shown are results for the sample variances σ*

Cell size	4	6	8	10	16
Number of realizations	1000	800	800	400	100
$\bar{K}_D^*$	4.354	4.253	4.182	4.164	4.109
$\bar{K}_P^*$	4.105	4.061	4.044	4.021	4.016
$\bar{K}_N^*$	3.790	3.838	3.848	3.883	3.925
σ_D^2	0.619	0.295	0.153	0.090	0.035
σ_P^2	0.645	0.267	0.153	0.093	0.038
σ_N^2	0.502	0.204	0.134	0.080	0.030

relations and is found to be the geometric mean of the conductivities of the two phases [54]. The main conclusions of this study were as follows.

(1) The periodic boundary condition performs better for both the random and periodic problems tested.
(2) In general the Neumann formulation underestimates the effective conductivity tensor and the Dirichlet formulation overestimates it; this is consistent with the above theorem.
(3) When computing the effective tensor by averaging over the cell, the accuracy can be improved if weighted or truncated averaging is used. In general, the results of weighted averaging, with smaller weights near the boundary, are more robust.
(4) The variance of the estimated effective conductivity tensor behaves as $\sigma^2 \sim L^{-2}$ for the random checker-board problem.

Table 8.1 gives a sample of the results for the random checker-board problem. The exact effective conductivity is 4.

8.4.3 Error estimates

To get an idea about how the method works, we follow the framework discussed in Chapter 6 for the error analysis of HMM. Define

$$e(\text{HMM}) = \max_{\mathbf{x}_\ell \in K, K \in \mathcal{T}_H} \|\mathbf{A}(\mathbf{x}_\ell) - \tilde{\mathbf{A}}(\mathbf{x}_\ell)\|,$$

where $\tilde{\mathbf{A}}(\mathbf{x}_\ell)$ is the estimated coefficient at $\mathbf{x}_\ell$ using (8.4.2) and (8.4.3) and $\mathbf{A}(\mathbf{x}_\ell)$ is the coefficient of the effective model. Following the general strategies discussed for HMM in Chapter 6, one can prove the following [26].

Theorem *Assume that* (8.1.1) *and* (8.4.1) *are uniformly elliptic. Denote by* U_0 *and* U_{HMM} *the solution of* (8.4.1) *and the* HMM *solution respectively. If* U_0 *is sufficiently smooth then there exists a constant* C, *independent of* ε, δ *and* H, *such that*

$$\|U_0 - U_{\mathrm{HMM}}\|_1 \leq C\left(H + e(\mathrm{HMM})\right), \qquad (8.4.16)$$

$$\|U_0 - U_{\mathrm{HMM}}\|_0 \leq C\left(H^2 + e(\mathrm{HMM})\right). \qquad (8.4.17)$$

This result is completely general. However, to estimate $e(\mathrm{HMM})$ quantitatively we need to make specific assumptions on the structure of the coefficients in (8.1.1). In general, if we assume that $\mathbf{a}^\varepsilon(\mathbf{x})$ is of the form $\mathbf{a}(\mathbf{x}, \mathbf{x}/\varepsilon)$ then we have

$$|e(\mathrm{HMM})| \leq C\left(\frac{\varepsilon}{\delta}\right)^\alpha, \qquad (8.4.18)$$

where δ is the size of the computational domain for the microscale problem. The value of the exponent α depends on the rate of convergence of the relevant homogenization problem as well as on the boundary conditions used in the microscopic problem. Some interesting results are given in [34].

An error analysis of the full discretization that includes the effect of discretization of the microscale problem is considered in [1].

8.4.4 Information about the gradients

There is another important point to be made. So far in this section, we have emphasized capturing the macroscale behavior of the system. We have paid little attention to what happens at the microscale. In fact, we have tried to get away with knowing as little as possible about the microscale behavior, so long as we were able to capture the macroscale behavior. For some applications involving (8.1.1) this is not what is needed. For example, in the context of composite materials, we are often interested in the stress distribution, which is not captured by the effective macroscale model or the homogenized equation: information about stress (or the gradients) is part of the microscale details, to which we did not pay much attention. The same is true in the context of modeling multiphase flow in a porous medium. There we are also interested in the velocity field for the purpose of modeling the transport, and information about the velocity field is also part of the microscale details to which we have not paid much attention.

It is possible to recover some information about the gradients from the HMM solutions. Let u_H be an HMM solution. Clearly, ∇u_H is not going to be close to ∇u^ε, where u^ε is the solution to the original microscale problem (8.1.1), since u_H contains only the macroscopic component. To recover information about the gradients note that, on each element $K \in \mathcal{T}_\mathrm{H}$, ∇u_H is a constant vector $\mathbf{G}_K = \nabla u_\mathrm{H}|_K$. On K, we expect $\nabla \phi^\varepsilon_{\mathbf{G}_K}$ to contain some information about ∇u^ε. For example, we expect that their probability distribution functions are close. At this point in time, though, this remains speculative since no systematic quantitative study has been carried out on how much information can be recovered about ∇u^ε from this procedure.

For the general case, knowing u_H one can also obtain locally the microstructural information using an idea in [47]. Assume that we are interested in recovering u^ε and ∇u^ε only in the subdomain $D \subset \Omega$. Let D_η be a domain with the properties $D \subset D_\eta \subset \Omega$ and $\mathrm{dist}\,(\partial D, \partial D_\eta) = \eta$. Consider the following auxiliary problem:

$$-\nabla \cdot (\mathbf{a}^\varepsilon(\mathbf{x})\nabla)\tilde{u}(\mathbf{x}) = f(\mathbf{x}) \qquad \mathbf{x} \in D_\eta, \qquad (8.4.19)$$
$$\tilde{u}(\mathbf{x}) = u_\mathrm{H}(\mathbf{x}) \qquad \mathbf{x} \in \partial D_\eta. \qquad (8.4.20)$$

We then have [26]

$$\left(\frac{1}{|D|}\int_D |\nabla(u^\varepsilon - \tilde{u})|^2 \, d\mathbf{x}\right)^{1/2} \le \frac{C}{\eta}\|u^\varepsilon - u_\mathrm{H}\|_{L^\infty(D_\eta)}. \qquad (8.4.21)$$

See [2] for a review of other relevant issues concerning finite element HMM.

8.5 Notes

Other approaches There are many other related contributions that we ought to have described but could not owing to lack of space. These include:

- *Efficient use of the homogenization approach* [22, 52]. In particular, Schwab and co-workers have studied the use of sparse representation to solve homogenized models efficiently.
- *Mixed formulation* [5, 20]. This is preferred in some applications since it tends to have better conservation properties.

Adaptivity This is obviously important. See [3, 47] for some initial results in this direction. We will come back to this point at the end of the book.

Generality versus efficiency The main issue is still the conflict between generality and efficiency. Here, by generality we mean the class of problems that can be handled. The most general methods are those such as the classical multigrid solver for the original microscale problem. However, they are still too costly for many practical problems. The most efficient algorithms are HMM or approaches based on precomputing the effective homogenized equation, such as the RAV approach.

Central questions are whether one can make the former class of methods more efficient by taking into account the special features that a problem might have and whether we can make the HMM type of algorithm more general.

It should be noted that, when the coefficients have multiscale features, the simple version of the multigrid algorithm discussed in Chapter 3 is no longer adequate since the choice of the coarse-grid operator discussed there does not necessarily reflect the effective properties on the coarse scale. How one should choose the effective operators in this case is still somewhat of an open question for general elliptic PDEs with multiscale coefficients. Some work has been reported in [28, 45].

We have discussed cases of spatial scales separation in which sublinear scaling algorithms can be constructed. Other situations can also be contemplated. For example, if we are solving a dynamic problem and the microstructure is frozen in time, or more generally when the microstructure changes much more slowly than the macro dynamics of interest, then the kind of algorithms discussed in Section 8.2, such as the RFB finite element method and MsFEM, might also be made sublinear since the overhead involved in formulating the finite element space (such as finding the basis functions) can be greatly reduced. Yet another case for which sublinear scaling algorithms can be constructed is discussed next.

Exploring statistical self-similarity In the example discussed above, sublinear scaling was possible owing to the disparity between the macro- and microscales. Another situation for which it is possible to develop sublinear scaling algorithms and sample the microscale behavior on small domains to predict the behavior on larger scales, was explored in [24]; this is the case when the small-scale components at different scales are statistically self-similar. This work is still quite far from being mature. Nevertheless, we will summarize it here since it gives some interesting suggestions on how one may develop sublinear scaling algorithms for problems without scale separation.

General strategy for data estimation In the problems discused above and in Section 6.3, scale separation was very important for the efficiency of HMM: it allows us to work with small simulation boxes for the microscale problem and still retain reasonable accuracy for the estimated data. Working with larger simulation boxes gives roughly the same estimates for the data. In other words, if we denote the value of the estimated data on a box of size L as $f = f(L)$, then when L is above some critical size, f is approximately independent of L:

$$f(L) \approx \text{constant}. \tag{8.5.1}$$

Here the critical size is determined by the characteristic scale of the microscopic model, which might simply be the correlation length. This idea can be generalized in an obvious way. As long as the scale dependence of $f = f(L)$ is of a simple form, with a few parameters, we can make use of this simple relationship by performing a few fine-scale simulations (not just one, as for problems with scale separation) and using the results to establish an accurate approximation for the dependence of f on L. Once we have $f = f(L)$, we can use it to predict f at a much coarser scale.

One example of such a situation is when the system exhibits local self-similarity. In this case the dependence is of the form $f(L) = C_0 L^\beta$ with only two parameters C_0 and β. Therefore we can use the results of microscopic simulations at two different values of L to predict the result at a much coarser scale, namely that of the macroscale mesh H.

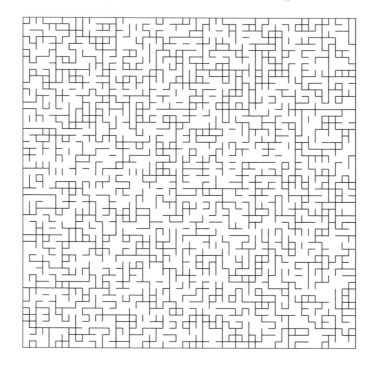

Figure 8.4. Bond percolation network on a square lattice at $p = 0.5$ (courtesy of Xingye Yue).

Transport on a percolation network at criticality This idea was demonstrated in [24] for the example of the effective transport on a two-dimensional bond percolation network at the percolation threshold.

In the standard two-dimensional bond percolation model with parameter p, $0 \leq p \leq 1$, we start with a square lattice, each bond of which is either kept with probability p or deleted with probability $1 - p$. Of particular interest is the size of the percolation clusters formed by the remaining bonds in the network (see Figure 8.4 for an example of bond percolation).

The critical value p^* for this model is 0.5. For an infinite lattice, if $p < p^*$ then the probability of having infinite-sized clusters is zero. For $p > p^*$ then the probability of having infinite-sized clusters is 1. Given the parameter value p, the network has a characteristic length scale: the correlation length, denoted by ξ_p. As $p \to p^*$, ξ_p diverges as follows:

$$\xi_p \sim |p - p^*|^\alpha,$$

where $\alpha = -4/3$ (see [50]). At $p = p^*$, $\xi_p = \infty$. In this case the system has a continuum distribution of scales, i.e. it has clusters of all sizes. In the following we will consider the case $p = p^*$.

We are interested in the macroscopic transport on such a network. To study the transport of, say, some pollutants whose concentration density is denoted by c, we embed this percolation model into a domain $\Omega \in \mathbb{R}^2$. We denote by ε the bond length of the percolation network and by L the size of Ω. We will consider the case when ε/L is very small.

The basic microscopic model is that of mass conservation. Denote by $S_{i,j}$, $i,j = 1,\ldots,N$ the ijth site of the percolation network and by $c_{i,j}$ the concentration at that site. Define the fluxes as follows:

- the flux from the right, $f_{i,j}^{\mathrm{r}} = B_{i,j}^x(c_{i+1,j} - c_{i,j})$;
- the flux from the left, $f_{i,j}^{\mathrm{l}} = B_{i-1,j}^x(c_{i-1,j} - c_{i,j})$;
- the flux from the top, $f_{i,j}^{\mathrm{t}} = B_{i,j}^y(c_{i,j+1} - c_{i,j})$;
- the flux from the bottom, $f_{i,j}^{\mathrm{b}} = B_{i,j-1}^y(c_{i,j-1} - c_{i,j})$.

Here the various B's are the bond conductivities for the specified bonds. The bond conductivity is 0 if the corresponding bond is deleted and 1 if the bond is kept. At each site $S_{i,j}$, from mass conservation we have

$$f_{i,j}^{\mathrm{t}} + f_{i,j}^{\mathrm{b}} + f_{i,j}^{\mathrm{r}} + f_{i,j}^{\mathrm{l}} = 0, \tag{8.5.2}$$

i.e. the total flux to this site is zero. This is our microscale model. It can be viewed as a discrete analog of (8.1.1).

In the HMM framework it is natural to choose a finite volume scheme over a macroscale grid Ω_H, where H is the size of the finite volume cell. The data that we need to estimate from the microscale model, here the percolation model, are the fluxes at the mid-points of the boundaries of the finite volume cells. Since the present problem is linear, we need to estimate the effective conductivity only for a network of size H. We note that, at $p = p^*$, the effective conductivity is strongly size-dependent. In fact there is strong evidence suggesting that [39]

$$\kappa(L) \sim C_0 L^\beta \tag{8.5.3}$$

for some value of β, where $\kappa(L)$ is the mean effective conductivity at size L.

Assuming that (8.5.3) holds, we will estimate numerically the values of β and C_0. For this purpose, we perform a series of microscopic simulations on systems of sizes L_1 and L_2, where $\varepsilon \ll L_1\varepsilon < L_2\varepsilon \ll H$. From the

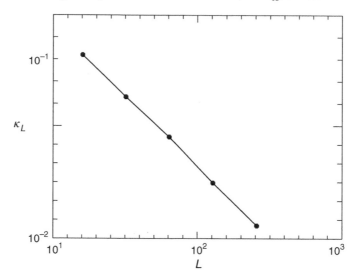

Figure 8.5. The effective conductivity κ_L at several different scales, $L = (16, 32, 64, 128, 256)\varepsilon$, for a realization of the percolation network with $p = p^* = 0.5$ (courtesy of Xingye Yue).

results we estimate $\kappa(L_1)$ and $\kappa(L_2)$. We then use these results to estimate the parameter values C_0 and β in (8.5.3). Once we have these parameter values, we can use (8.5.3) to predict $\kappa(H)$; see Figure 8.5.

One interesting question is the effect of statistical fluctuations. To see this we plot in Figure 8.6 the actual and predicted values (using self-similarity) of the conductivity on a lattice of size $L = 128$ for a number of realizations of the percolation network. The predicted values are computed using the simulated values from lattices of sizes $L = 16$ and $L = 32$. We see that the predicted values have larger fluctuations than the actual values, suggesting that the fluctuations are amplified in the process of predicting the mean values. The latter are predicted with reasonable accuracy considering the relatively small size of the network.

For more details see [24].

The general case In the general case when there are no special features of which use can be made use, one possibility is to use the viewpoint of model reduction, i.e. given the operator $L = -\nabla \cdot (\mathbf{a}^\varepsilon(\mathbf{x})\nabla)$ or its discretization on a fine grid, A_N (where N is the number of grid points), and given a number M ($M \ll N$), one asks: what is its best $M \times M$ approximation $\bar{A}_M$, in the sense that the solutions to the equations $A_N u = f$

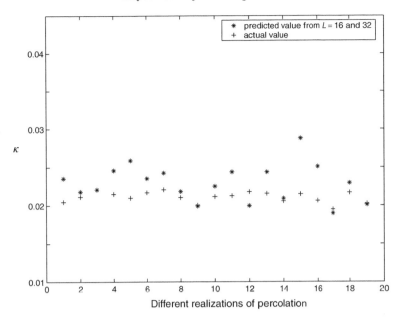

Figure 8.6. Effect of fluctuations: the actual and predicted effective conductivities at scale $L = 128\varepsilon$ for different realizations of the percolation network with $p = 0.5$. The predicted values are computed from $L = 16\varepsilon$ and 32ε lattices (courtesy of Xingye Yue).

and $\bar{A}_M u_M = f$ are the closest, among all $M \times M$ matrices, for a general right-hand side f? We will return to this problem in Chapter 11.

References for Chapter 8

[1] A. Abdulle, "On a priori error analysis of fully discrete heterogeneous multiscale finite element method," *Multiscale Model Simul.*, vol. 4, no 2, pp. 447–459, 2005.

[2] A. Abdulle, "The finite element heterogeneous multiscale method: a computational strategy for multiscale PDEs," *GAKUTO Int. Ser. Math. Sci. Appl.*, vol. 31, pp. 135–184, 2009.

[3] A. Abdulle and A. Nonnenmacher, "Adaptive finite element heterogeneous multiscale method for homogenization problems," *Comput. Meth. Appl. Mech. Engrg*, in press.

[4] U. Andersson, B. Engquist, G. Ledfelt and O. Runborg, "A contribution to wavelet-based subgrid modeling," *Appl. Comput. Harmon. Anal.*, vol. 7, pp. 151–164, 1999.

[5] T. Arbogast, "Numerical subgrid upscaling of two-phase flow in porous media," in *Numerical Treatment of Multiphase Flows in Porous Media,*

Z. Chen *et al.*, eds., *Lecture Notes in Physics*, vol. 552, pp. 35–49, Springer-Verlag, 2000.

[6] T. Arbogast and K.J. Boyd, "Subgrid upscaling and mixed multiscale finite elements," *SIAM J. Numer. Anal.*, vol. 44, no. 3, pp. 1150–1171, 2006.

[7] I. Babuska, "Homogenization and its applications, mathematical and computational problems," in *Numerical Solutions of Partial Differential Equations III*, B. Hubbard, ed., pp. 89–116, Academic Press, 1976.

[8] I. Babuska, "Solution of problems with interface and singularities," in *Mathematical Aspects of Finite Elements in Partial Differential Equations*, Carl de Boor, ed., pp. 213–277, Academic Press, 1974.

[9] I. Babuska, "Solution of interface by homogenization, I, II, III," *SIAM J. Math. Anal.*, vol. 7, pp. 603–634, 635–645, 1976; vol. 8, pp. 923–937, 1977.

[10] I. Babuska and J.E. Osborn, "Generalized finite element methods: their performance and their relation to mixed methods," *SIAM J. Numer. Anal.*, vol. 20, pp. 510–536, 1983.

[11] I. Babuska and J.E. Osborn, "Finite element methods for the solution of problems with rough data," in *Singularities and Constructive Methods and Their Treatment*, P. Grisvard, W. Wendland and J.R. Whiteman, eds., Lecture Notes in Mathematics, vol. 1121, pp. 1–18, Springer-Verlag, 1985.

[12] I. Babuska, U. Banerjee and J. Osborn, "Survey of meshless and generalized finite element methods: a unified approach," *Acta Numerica*, pp. 1–125, Cambridge University Press, 2003.

[13] I. Babuska, G. Caloz and J. Osborn, "Special finite element methods for a class of second order elliptics with rough coefficients," *SIAM J. Numer. Anal.*, vol. 31, pp. 945–981, 1994.

[14] A. Bensoussan, J.-L. Lions and G.C. Papanicolaou, "Asymptotic analysis for periodic structures," *Stud. Math. Appl.*, vol. 5, 1978.

[15] G. Beylkin and M. Brewster, "A multiresolution strategy for numerical homogenization," *Appl. Comput. Harmon. Anal.*, vol. 2, pp. 327–349, 1995.

[16] G. Beylkin, R. Coifman and V. Rokhlin, "Fast wavelet transforms and numerical analysis," *Comm. Pure Appl. Math.*, vol. 44, no. 2, pp. 141–183, 1991.

[17] F. Brezzi and L.D. Marini, "Augmented spaces, two-level methods, and stabilizing subgrids," *Int. J. Numer. Meth. Fluids*, vol. 40, pp. 31–46, 2002.

[18] F. Brezzi and A. Russo, "Choosing bubbles for advection–diffusion problems," *Math. Models Meth. Appl. Sci.*, vol. 4, pp. 571–587, 1994.

[19] F. Brezzi, L.D. Marini and E. Süli, "Residual-free bubbles for advection–diffusion problems: the general error analysis," *Numer. Math.*, vol. 85, pp. 31–47, 2000.

[20] Z.M. Chen and T. Y. Hou, "A mixed multiscale finite element method for elliptic problems with oscillating coefficients," *Math. Comp.*, vol. 72, pp. 541–576, 2003.

[21] P.G. Ciarlet, *The Finite Element Method for Elliptic Problems*, SIAM, 2002.

[22] J.Z. Cui and L.Q. Cao, "The two-scale analysis method for woven composite materials," in *Engineering Computation and Computer Simulation I*, Zhi-Hua Zhong, ed., pp. 203–212, Hunan University Press, 1995.

[23] M. Dorobantu and B. Engquist, "Wavelet-based numerical homogenization," *SIAM J. Numer. Anal.*, vol. 35, no. 2, pp. 540–559, 1998.

[24] W. E and X.-Y. Yue, "Heterogeneous multiscale method for locally self-similar problems," *Comm. Math. Sci.*, vol. 2, pp. 137–144, 2004.

[25] W. E, T. Li and J. Lu, "Localized basis of eigen-subspaces and operator compression," *Proc. Nat. Acad. Sci. USA*, vol. 107, pp. 1273–1278, 2010.

[26] W. E, P.B. Ming and P.W. Zhang, "Analysis of the heterogeneous multiscale method for elliptic homogenization problems," *J. Amer. Math. Soc.*, vol. 18, pp. 121–156, 2005.

[27] Y.R. Efendiev, T.Y. Hou and X.-H. Wu, "Convergence of a nonconforming multiscale finite element method," *SIAM J. Numer. Anal.*, vol. 37, pp. 888–910, 2000.

[28] B. Engquist and E. Luo, "Convergence of a multigrid method for elliptic equations with highly oscillatory coefficients," *SIAM J. Numer. Anal.*, vol. 34, no. 6, pp. 2254–2273, 1997.

[29] B. Engquist and O. Runborg, "Wavelet-based numerical homogenization with applications," in *Multiscale and Multiresolution Methods: Theory and Applications*, T.J. Barth, T.F. Chan and R. Haimes, eds., *Lecture Notes in Computational Science and Engineering*, vol. 20, pp. 97–148, Springer-Verlag, 2001.

[30] J.M. Farmer, "Upscaling: a review," *Int. J. Numer. Meth. Fluids*, vol. 40, pp. 63–78, 2002.

[31] J. Fish, "The *s*-version finite element method," *Comp. Structure*, vol. 43, pp. 539–547, 1992.

[32] J. Fish and A. Wagiman, "Multiscale finite element method for a locally nonperiodic heterogeneous medium," *Comput. Mech.*, vol. 12, pp. 164–180, 1993.

[33] M.J. Gander, "Optimized Schwarz methods," *SIAM J. Numer. Anal.*, vol. 44, pp. 699–731, 2006.

[34] A. Gloria and F. Otto, "An optimal variance estimate in stochastic homogenization of discrete elliptic equations," preprint.

[35] G.H. Golub and C. Van Loan, *Matrix Computations*, 3rd edition, Johns Hopkins University Press, 1996.

[36] H. Han, Z. Huang and B. Kellogg, "A tailored finite point method for a singular perturbation problem on an unbounded domain," *J. Sci. Comput.*, vol. 36, pp. 243–261, 2008.

[37] T.Y. Hou and X.-H. Wu, "A multiscale finite element method for elliptic problems in composite materials and porous media," *J. Comput. Phys.*, vol. 134, pp. 169–189, 1997.

[38] T.Y. Hou, X.-H. Wu and Z. Cai, "Convergence of a multiscale finite element method for elliptic problems with rapidly oscillating coefficients," *Math. Comp.*, vol. 68, pp. 913–943, 1999.

[39] B.D. Hughes, *Random Walks and Random Environments, Vol. 2*, Oxford University Press, 1996.

[40] T.J.R. Hughes, "Multiscale phenomena: Green's functions, the Dirichlet-to-Neumann formulation, subgrid scale models, bubbles and the origin of stablized methods," *Comput. Meth. Appl. Mech. Engrg*, vol. 127, pp. 387–401, 1995.

[41] T.J.R. Hughes, G. Feijóo, L. Mazzei and J.-B. Quincy, "The variational multiscale method – a paradigm for computational mechanics," *Comput. Meth. Appl. Mech. Engrg*, vol. 166, pp. 3–24, 1998.

[42] W.K. Liu, Y.F. Zhang and M.R. Ramirez, "Multiple scale finite element methods," *Int. J. Numer. Meth. Engrg*, vol. 32, pp. 969–990, 1991.

[43] P.B. Ming and X.-Y. Yue, "Numerical methods for multiscale elliptic problems," *J. Comput. Phys.*, vol. 214, pp. 421–445, 2006.

[44] J.D. Moulton, J.E. Dendy and J.M. Hyman, "The black box multigrid numerical homogenization algorithm," *J. Comput. Phys.*, vol. 141, pp. 1–29, 1998.

[45] N. Neuss, W. Jäger and G. Wittum, "Homogenization and multigrid," *Computing*, vol. 66, no. 1, pp. 1–26, 2001.

[46] J. Nolen, G. Papanicolaou and O. Pironneau, "A framework for adaptive multiscale methods for elliptic problems," *Multiscale Model. Simul.*, vol. 7, pp. 171–196, 2008.

[47] J.T. Oden and K.S. Vemaganti, "Estimation of local modeling error and goal-oriented adaptive modeling of heterogeneous materials. Part I: Error estimates and adaptive algorithms," *J. Comput. Phys.*, vol. 164, pp. 22–47, 2000.

[48] A. Quarteroni and A. Valli, *Domain Decomposition Methods for Partial Differential Equations*, Oxford University Press, 1999.

[49] E. Ramm, A. Hund and T. Hettich, "A variational multiscale model for composites with special emphasis on the X-FEM and level sets," in *Proc. EURO-C 2006*, G. Meschke, R. de Borst, H. Mang and N. Bicanic, eds., 2006.

[50] M. Sahimi, *Flow and Transport in Porous Media and Fractured Rock*, Wiley, 1995.

[51] G. Sangalli, "Capturing small scales in elliptic problems using a residual-free bubbles finite element method," *Multiscale Model. Simul.*, vol. 1, pp. 485–503, 2003.

[52] C. Schwab and A.-M. Matache, "Two-scale FEM for homogenization problems," in *Proc. Conf. on Mathematical Modelling and Numerical Simulation in Continuum Mechanics*, Yamaguchi, 2000; *Lecture Notes in Computational Science and Engineering*, I. Babuska *et al.*, eds., Springer-Verlag, 2002.

[53] G. Strang and G.J. Fix, *An Analysis of the Finite Element Method*, Wellesley-Cambridge, 2nd edition, 2008.

[54] S. Torquato, *Random Heterogeneous Materials: Microstructure and Macroscopic Properties*, Springer-Verlag, 2002.

[55] X.-Y. Yue, "Residual-free bubble methods for numerical homogenization of elliptic problems," preprint, 2008.

[56] X.-Y. Yue and W. E, "The local microscale problem in the multiscale modelling of strongly heterogeneous media: effect of boundary conditions and cell size," *J. Comput. Phys.*, vol. 222, pp. 556–572, 2007.

[57] V.V. Zhikov, S.M. Kozlov and O.A. Oleinik, *Homogenization of Differential Operators and Integral Functionals*, Springer-Verlag, 1994.

9

Problems that have multiple time scales

The previous chapter was devoted to an example with disparate spatial scales. In this and the following chapters, we will continue to focus on problems with disparate time scales. This chapter is concerned with problems which are "stiff" in a broad sense, i.e. problems for which the fast processes relax in a short time scale and the slow processes obey an effective dynamics obtained by averaging the slow component of the original dynamics over the quasi-equilibrium distribution of the fast processes. Classical stiff ODEs will be considered as a special case. The next chapter is concerned with the other important class of problems with disparate time scales, namely, rare events.

9.1 ODEs with disparate time scales

9.1.1 General setup for limit theorems

Consider the differential equation

$$\dot{\mathbf{x}} = \mathbf{f}(\mathbf{x}, \varepsilon). \qquad (9.1.1)$$

We have used an explicit dependence on ε to indicate that the system has disparate time scales. For example, $\mathbf{f}$ may take the form

$$\mathbf{f}(\mathbf{x}, \varepsilon) = \mathbf{f}_0(\mathbf{x}) + \frac{1}{\varepsilon}\mathbf{f}_1(\mathbf{x}) \qquad (9.1.2)$$

or, more generally,

$$\mathbf{f}(\mathbf{x}, \varepsilon) = \mathbf{f}_0(\mathbf{x}, \varepsilon) + \frac{1}{\varepsilon}\mathbf{f}_1(\mathbf{x}, \varepsilon). \qquad (9.1.3)$$

where $\mathbf{f}_0$ and $\mathbf{f}_1$ are bounded in ε. Roughly speaking, we have in mind the following picture: the fast component of the dynamics brings the system to some local equilibrium states and this relaxation process is faster than the time scale on which the slow variables change. These local equilibrium states are often called *quasi-equilibrium* states . They are parametrized by the local values of the slow variables and they influence the dynamics of the slow variables. We are interested in capturing those dynamics without resolving the details of the fast component of the dynamics in (9.1.1).

Let $\mathbf{z} = \varphi(\mathbf{x})$. If $\mathbf{f}$ takes the form of (9.1.3) then

$$\dot{\mathbf{z}} = \nabla\varphi \cdot \dot{\mathbf{x}} = \nabla\varphi \cdot \left(\mathbf{f}_0 + \frac{1}{\varepsilon}\mathbf{f}_1 \right). \qquad (9.1.4)$$

Therefore, to be a slow variable φ should satisfy

$$\nabla\varphi \cdot \mathbf{f}_1 = 0. \qquad (9.1.5)$$

In other words, if we define a "virtual fast dynamics" [7]

$$\dot{\mathbf{y}} = \frac{1}{\varepsilon}\mathbf{f}_1(\mathbf{y}, \varepsilon) \qquad (9.1.6)$$

then the slow variables are conserved quantities for the virtual fast dynamics. Assume that $\mathbf{z} = (z_1, \ldots, z_J)$ is a complete set of slow variables, i.e. the ergodic components for the virtual fast dynamics (9.1.6) are parametrized completely by the values of $\mathbf{z}$. Denote by $\mu_{\bar{\mathbf{z}}}(d\mathbf{x})$ the equilibrium distribution of the virtual fast dynamics when the slow variable $\mathbf{z}$ is fixed at the value $\bar{\mathbf{z}}$. We have

$$\dot{\mathbf{z}} = \nabla\varphi(\mathbf{x}) \cdot \mathbf{f}_0(\mathbf{x}, \varepsilon). \qquad (9.1.7)$$

The averaging principle discussed in Chapter 2 suggests that the effective dynamics for the slow variable $\mathbf{z}$ should be described by

$$\dot{\mathbf{z}} = \int \nabla\varphi(\mathbf{x}) \cdot \mathbf{f}_0(\mathbf{x}, \varepsilon)\mu_{\bar{\mathbf{z}}}(d\mathbf{x}). \qquad (9.1.8)$$

We note that the set of slow variables has to be sufficiently large such that the quasi-equilibrium distribution of the fast dynamics is completely parametrized by $\mathbf{z}$. See [17] for a more thorough discussion of this and related issues.

Example Consider

$$\frac{dx}{dt} = f(x, y), \tag{9.1.9}$$

$$\frac{dy}{dt} = -\frac{1}{\varepsilon}(y - \phi(x)). \tag{9.1.10}$$

In this case, x is a slow variable. Fix x; the quasi-equilibrium distribution for the fast variable y is given by a delta function $\mu_x(\,dy) = \delta(y - \phi(x))\,dy$. The effective equation for the slow variable x is given by

$$\frac{dx}{dt} = \int f(x, y)\mu_x\,(dy) = f(x, \phi(x)).$$

Example Now consider

$$\frac{dx}{dt} = 1 + y, \tag{9.1.11}$$

$$\frac{dy}{dt} = \frac{1}{\varepsilon}z, \tag{9.1.12}$$

$$\frac{dz}{dt} = -\frac{1}{\varepsilon}y. \tag{9.1.13}$$

It is easy to see that the dynamics of y and z exhibit fast oscillations. One choice of slow variables is $(x, r = \sqrt{y^2 + z^2})$. Given the variables (x, r), the quasi-equilibrium distribution is a uniform distribution concentrated on the circle of radius r in the yz-plane and centered at the origin. The effective equations for the slow variables are simply

$$\frac{dx}{dt} = 1,$$

$$\frac{dr}{dt} = 0.$$

Example Consider

$$\frac{dx}{dt} = f(x, y), \tag{9.1.14}$$

$$\frac{dy}{dt} = -\frac{1}{\varepsilon}(y - \phi(x)) + \sqrt{\frac{2}{\varepsilon}}\dot{w}, \tag{9.1.15}$$

where $\dot{w}$ is a standard Gaussian white noise. Again, x is one possible choice for the slow variable. Given the value of x, the quasi-equilibrium distribution is a normal distribution in y with mean $\phi(x)$ and variance 1.

The effective dynamics for x is obtained by averaging $f(x, y)$ with respect to a standard normal distribution in y.

9.1.2 Implicit methods

For a problem of the type (9.1.9), a standard explicit scheme such as the forward Euler method would require a time-step size of $O(\varepsilon)$. However, if we are only interested in capturing accurately the dynamics of the slow variable rather than the details of the relaxation process for the fast variable, as is often the case in applications, then there are more efficient approaches. The most well-known examples are the implicit methods.

It is more convenient to consider the more general problem

$$\frac{dz}{dt} = F(z, \varepsilon). \tag{9.1.16}$$

The simplest implicit method is the backward Euler scheme:

$$\frac{z^{n+1} - z^n}{\Delta t} = F(z^{n+1}, \varepsilon). \tag{9.1.17}$$

The reason that this is better than the forward Euler scheme

$$\frac{z^{n+1} - z^n}{\Delta t} = F(z^n, \varepsilon) \tag{9.1.18}$$

is that backward Euler has much better stability properties. In order to see this difference, let us introduce the notion of the *stability region* [21].

Given an ODE solver, consider its application to the linear problem

$$\frac{dy}{dt} = \lambda y, \tag{9.1.19}$$

where λ is a complex constant. The stability region of the ODE solver is defined as the subset of complex numbers $z = \lambda \Delta t$ such that the numerical solution $\{y^n\}$ produced by the ODE solver is bounded in n, i.e. there exists a constant C which is independent of n such that

$$|y^n| \leq C. \tag{9.1.20}$$

When the forward Euler method is applied to (9.1.19), we may deduce that $y_{n+1} = (1 + z)y_n$. Hence (9.1.20) is satisfied if

$$|1 + z| \leq 1. \tag{9.1.21}$$

This is a unit disk of radius 1 on the complex z-plane centered at $z = -1$.

For backward Euler, we have $y_{n+1} = (1 - z)^{-1}y_n$. Hence (9.1.21) is satisfied when

$$|1 - z| \geq 1. \tag{9.1.22}$$

This is the complement of the unit disk centered at $z = 1$.

Note that the stability region for the backward Euler method contains the negative real axis. This is an important feature since the fast component of the dynamics in (9.1.9) is of the form of (9.1.19) with $\lambda = -1/\varepsilon$. This is not just a feature of this simple model problem: it is rather common for ODE systems that arise from chemical kinetic models, the discretization of dissipative PDEs etc.

So far we have only discussed the simplest implicit scheme, the backward Euler method. There are other implicit methods such as the Adams–Moulton methods and the backward differentiation formula. We refer to standard ODE textbooks such as [21] for details.

From a practical viewpoint, one disadvantage of implicit schemes is the very fact that they are implicit: at every time-step one has to solve a system of algebraic equations in order to obtain the numerical solution at that time-step. Such algebraic equations are usually solved by some iterative method. While there are very efficient techniques available for solving them, this does represent one additional complication compared with explicit methods.

It is quite remarkable that one can capture the large-scale behavior of solutions to stiff ODEs without the need to resolve the small-scale transients, even though we are so used to this and tend to take it for granted. In fact, this seems to be special to systems whose quasi-equilibrium distributions are delta functions: the same kind of philosophy does not work for systems whose quasi-equilibrium distributions are nontrivial (i.e. not delta functions), for example stochastic ODEs or highly oscillatory systems [31], and it is unclear how useful implicit methods are for these kinds of problem.

9.1.3 Stabilized Runge–Kutta methods

Now we turn to explicit methods. As we said earlier, simple explicit schemes such the forward Euler method are not very effective for the kind of problem discussed here. However, composite schemes, which are sometimes formed as a composition of a large number of forward Euler steps with different stepsizes, can be quite effective. Here we will show how one can design such schemes in the framework of explicit Runge–Kutta methods.

Consider the mid-point rule for the ODE $\dot{x} = f(x)$:

$$x^{n+1/2} = x^n + \frac{\Delta t}{2} f(x^n),$$
$$x^{n+1} = x^n + \Delta t f(x^{n+1/2}). \tag{9.1.23}$$

When this is applied to (9.1.19) we have

$$x_{n+1} = R(z)x_n,$$

where $z = \lambda \Delta t$ and

$$R(z) = 1 + z + \frac{z^2}{2}$$

is the *stability polynomial* of the mid-point rule. Since the exact solution of (9.1.19) satisfies $x(t_{n+1}) = e^{\lambda z} x(t_n)$ and since

$$e^{\lambda z} - R(z) = \mathcal{O}(z^3),$$

we see that the local error is $\mathcal{O}(\Delta t^3)$ and the global error is $\mathcal{O}(\Delta t^2)$. Similarly, for the classical fourth-order Runge–Kutta method, its stability polynomial is

$$R(z) = 1 + z + \frac{z^2}{2} + \frac{z^3}{3!} + \frac{z^4}{4!}.$$

The local error is $\mathcal{O}(\Delta t^5)$ and the global error is $\mathcal{O}(\Delta t^4)$.

The two examples given above are examples of Runge–Kutta methods that have optimal accuracy, in the sense that their order of accuracy is the maximally achievable given the number of stages (which is the same as the number of function evaluations). The stability region for these methods is, however, quite limited.

It is possible to design Runge–Kutta methods with lower accuracy but a much larger stability region. This class of methods is generally referred to as *stabilized Runge–Kutta methods*. We will discuss a particularly interesting class of stabilized Runge–Kutta methods, namely, Chebychev methods [29].

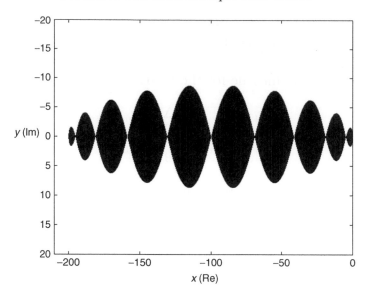

Figure 9.1. Stability region of the Chebychev method with $s = 10$ (courtesy of Assyr Abdulle); x and y are the real and imaginary parts of z.

Consider an s-stage Runge–Kutta method. For consistency, its stability polynomial has to take the form

$$R_s(z) = 1 + z + \sum_{k=2}^{s} C_k z^k, \tag{9.1.24}$$

i.e. the first two terms have to match the Taylor series for e^z. We would like to choose the coefficients $\{C_k\}$ to maximize the stability region of the corresponding Runge–Kutta method. Note that our interest is mainly in the case when the spectrum of the linearized ODE is located near the negative part of the real axis. Therefore our problem can be formulated as: maximize l_s, subject to the condition that

$$(-l_s, 0) \subset \{z : |R_s(z)| < 1\}. \tag{9.1.25}$$

It is easy to see that the solution to this problem is given by

$$R_s(z) = T_s\left(1 + \frac{z}{s^2}\right), \tag{9.1.26}$$

where T_s is the Chebychev polynomial of degree s. In this case, we have

$$l_s = 2s^2. \tag{9.1.27}$$

The stability region is shown in Figure 9.1

How do we turn this into a Runge–Kutta method? There are two ways to proceed. The first is to write

$$R_s(z) = \prod_{j=1}^{s} (1 + c_j z), \tag{9.1.28}$$

where $\{-1/c_j\}$ are the roots of $R_s(z)$ and are all real and negative. Given x^n, we then form

$$
\begin{aligned}
x^{n,0} &= x^n, \\
x^{n,1} &= x^{n,0} + \Delta t \, c_1 f(x^{n,0}), \\
x^{n,2} &= x^{n,1} + \Delta t \, c_2 f(x^{n,1}), \\
&\;\;\vdots \\
x^{n,s} &= x^{n,s-1} + \Delta t \, c_s f(x^{n,s-1}), \\
x^{n+1} &= x^{n,s}.
\end{aligned}
\tag{9.1.29}
$$

It is easy to see that the stability polynomial of this Runge–Kutta method is given by (9.1.26). In fact the stability polynomial up to the kth stage is

$$R_{s,k}(z) = \prod_{j=1}^{k} (1 + c_j z). \tag{9.1.30}$$

The above algorithm can be thought of as being a composition of forward Euler methods with different step sizes $\{c_j \Delta t\}$. As such it can be regarded as a special case of the variable time-step methods. However, we prefer not to do so, partly because when formulated in this way Chebychev methods may suffer from internal instabilities. Even if $|R_s(z)| < 1$, there is no guarantee that $|R_{s,k}(z)| < 1$ for all k. For large s, the numerical errors might be amplified considerably in the intermediate stages before they eventually have a chance to be damped out. This problem of internal instabilities can be alleviated by ordering $\{c_j\}$ in a better way [30].

An easy way to avoid such internal instabilities is to make use of the three-term recurrence relation for Chebychev polynomials,

$$T_{n+1}(x) = 2x T_n(x) - T_{n-1}(x). \tag{9.1.31}$$

Using this, we can reformulate (9.1.29) as

$$x^{n,0} = x^n,$$

$$x^{n,1} = x^{n,0} + \frac{\Delta t}{s^2} f(x^{n,0}),$$

$$\vdots \tag{9.1.32}$$

$$x^{n,j+1} = 2x^{n,j} - x^{n,j-1} + \frac{2\Delta t}{s^2} f(x^{n,j}),$$

$$x^{n+1} = x^{n,s}.$$

It is easy to check that if $f(x) = \lambda x$, $z = \lambda \Delta t$, then

$$x^{n,j} = \tilde{T}_j(z)x^n, \tag{9.1.33}$$

where $\tilde{T}_j(z) = T_j\left(1 + z/s^2\right)$ and T_j is the Chebychev polynomial of degree j. This ensures stability for the intermediate stages even though the scheme is no longer in the form of a composition of forward Euler steps [24, 44].

Another point to be noted is that the width of the stability region vanishes at the points where $T_s = \pm 1$. This is an undesirable feature. It can be avoided by replacing the domain of consideration from $\{z : |R_s(z)| < 1\}$ to $\{z : |R_s(z)| < q\}$, where q is a positive number less than 1. In this case R_s is replaced by

$$\tilde{R}_s(z) = \frac{1}{T_s(\omega_0)} T_s(\omega_0 + \omega_1 z), \tag{9.1.34}$$

where

$$\omega_0 = 1 + \frac{q}{s^2}, \quad \omega_1 = \frac{T_s(\omega_0)}{T_s'(\omega_0)}. \tag{9.1.35}$$

The stability region is shown in Figure 9.2. Note that l_s is no longer equal to $2s^2$, but it is still $\mathcal{O}(s^2)$.

So far, we have only discussed first-order methods. For higher-order Chebychev methods we refer to [5, 2]; for extensions to stochastic ODEs we refer to [3, 4].

Returning to the system (9.1.9) discussed earlier, we have $|\lambda| \sim 1/\varepsilon$. If we require that

$$\Delta t |\lambda| \leq \mathcal{O}(s^2)$$

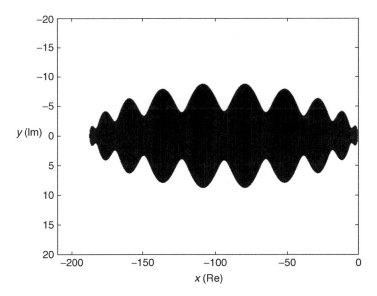

Figure 9.2. Stability region of the improved Chebychev method with $s = 10$ and $q = 0.9$ (courtesy of Assyr Abdulle).

then we must have

$$\Delta t \leq \mathcal{O}(s^2 \varepsilon).$$

Since the cost per step is $\mathcal{O}(s)$, the total cost required to advance to $\mathcal{O}(1)$ time is given by

$$\text{cost} = \mathcal{O}\left(\frac{s}{s^2 \varepsilon}\right) = \mathcal{O}\left(\frac{1}{s\varepsilon}\right).$$

The cost is $\mathcal{O}(1/\varepsilon)$ if we choose s to be $\mathcal{O}(1)$, and $\mathcal{O}(1)$ if we choose s to be $\mathcal{O}(1/\varepsilon)$. In the latter case, however, the accuracy also deteriorates. Therefore the optimal choice depends on considerations of both accuracy and efficiency.

If the spectrum of the problem has a different structure, one can design stabilized Runge–Kutta methods accordingly. For example, if the spectrum is clustered in two regions separated by a gap, one can design Runge–Kutta methods whose stability regions cover these clusters [21, 29]. It can be shown that such stabilized Runge–Kutta methods can be considered as compositions of forward Euler steps whose step sizes cluster into two regions [1, 29]. In other words, they are compositions of the forward Euler method with small and large time-steps.

9.1.4 Heterogeneous multiscale method

To illustrate the idea behind this application of HMM, let us consider the simple example (6.2.1). Here the macroscale variable of interest is $\mathbf{x}$. We will assume that its dynamics is approximated accurately by an ODE, which can be written abstractly as

$$\dot{\bar{\mathbf{x}}} = \mathbf{F}(\bar{\mathbf{x}}, \varepsilon), \quad \bar{\mathbf{x}}(0) = \mathbf{x}_0, \tag{9.1.36}$$

with unknown right-hand side $\mathbf{F}$.

Since the macroscale equation takes the form of an ODE, we will select stable ODE solvers as the macro solver. This suggests an HMM that consists of the following:

(1) *The macro solver*, say forward Euler:

$$\hat{\mathbf{x}}^{n+1} = \hat{\mathbf{x}}^n + \hat{\mathbf{F}}_n \Delta t, \quad \hat{\mathbf{x}}^0 = \mathbf{x}_0,$$

where Δt is the macro time-step and $\hat{\mathbf{F}}_n$ denotes the approximate value of $\mathbf{F}(\hat{\mathbf{x}}^n, \varepsilon)$, which needs to be estimated at each macro time-step.

(2) *The micro solver.* In this case, the constrained microscale model is simply the second equation in (6.2.1) with $\mathbf{x} = \mathbf{x}^n$. In the simplest case, a forward Euler scheme with micro time-step δt, we have

$$\hat{\mathbf{z}}^{n,m+1} = \hat{\mathbf{z}}^{n,m} + \frac{1}{\varepsilon} \mathbf{g}(\hat{\mathbf{x}}^n, \hat{\mathbf{z}}^{n,m}, t^n + m\delta t)\delta t,$$

$m = 0, 1, \ldots$

(3) *A force estimator.* For example, we may use simple time-averaging to extract $\hat{F}_n$:

$$\hat{\mathbf{F}}_n = \frac{1}{M} \sum_{m=0}^{M-1} \mathbf{f}(\hat{\mathbf{x}}^n, \hat{\mathbf{z}}^{n,m}, t^n + m\delta t).$$

The macroscale solver can be quite general. For example, we could also choose higher-order Runge–Kutta methods or linear multi-step methods. If we use the mid-point rule as the macro solver, we have

$$\hat{\mathbf{x}}^{n+1/2} = \hat{\mathbf{x}}^n + \frac{\Delta t}{2} \hat{\mathbf{F}}_n,$$

$$\hat{\mathbf{x}}^{n+1} = \hat{\mathbf{x}}^n + \Delta t \, \hat{\mathbf{F}}_{n+1/2}.$$

Here both $\hat{\mathbf{F}}_n$ and $\hat{\mathbf{F}}_{n+1/2}$ are unknown forces, which have to be estimated from the microscale model.

One should distinguish these methods from multi-time-step methods [42]. In multi-time-step methods, different time-step sizes are used for different terms in the forcing. However, the detailed behavior of the model is computed accurately over the whole time interval of interest. An illustrative example is as follows. Consider

$$\dot{\mathbf{x}} = \mathbf{f}_1(\mathbf{x}) + \mathbf{f}_2(\mathbf{x}).$$

Assume that for some reason, we believe $\mathbf{f}_1$ varies more slowly than $\mathbf{f}_2$. Then we might use a multi-time-step integrator of the type

$$\mathbf{x}^{n,0} = \mathbf{x}^n + \Delta t\, \mathbf{f}_1(\mathbf{x}^n),$$
$$\mathbf{x}^{n,m+1} = \mathbf{x}^{n,m} + \delta t\, \mathbf{f}_2(\mathbf{x}^{n,m}), \quad m = 0, 1, \dots, M-1,$$
$$\mathbf{x}^{n+1} = \mathbf{x}^{n,M},$$

with $M = \Delta t/\delta t$. In contrast, if we are using HMM (or projective integrators for that matter), M is determined by the relaxation time of the fast component of the dynamics.

If the ODE has a stochastic component then the macroscale model might be in the form of stochastic ODEs. In this case one should use a stochastic ODE solver as the macro solver in HMM. The simplest example is the Euler–Maruyama scheme:

$$\mathbf{x}^{n+1} = \mathbf{x}^n + \Delta t\, \hat{\mathbf{B}}_n + \hat{\boldsymbol{\sigma}}_n \sqrt{\Delta t}\, \mathbf{w}^n.$$

Now both the approximate drift $\hat{\mathbf{B}}_n$ and the approximate variance $\hat{\boldsymbol{\sigma}}_n$ need to be estimated. This can be done in a similar way to that described above. The details can be found in [12].

In the same spirit, seamless HMM can also be applied to this class of problems.

9.2 Application of HMM to stochastic simulation algorithms

First, let us review briefly the standard stochastic simulation algorithm (SSA) for chemical kinetic systems proposed in [19, 20], also known as the Gillespie algorithm. This is closely related to the Bort, Kalos and Lebowitz (BKL) algorithm in statistical physics [6]. The general setup for the stochastic chemical kinetic system was discussed in subsection 2.3.3.

Suppose that we are given a chemical kinetic system with reaction channels $\mathbf{R}_j = (a_j, \mathbf{v}_j)$, $j = 1, \ldots, M_R$. Let

$$a(\mathbf{x}) = \sum_{j=1}^{M_R} a_j(\mathbf{x}). \tag{9.2.1}$$

Assume that the current time is t_n and that the system is at state $\mathbf{X}_n$, and perform the following steps.

(1) Generate independent random numbers r_1 and r_2 with uniform distribution on the unit interval $(0, 1]$. Let

$$\delta t_{n+1} = -\frac{\ln r_1}{a(\mathbf{X}_n)}$$

and let k_{n+1} be the natural number such that

$$\frac{1}{a(\mathbf{X}_n)} \sum_{j=0}^{k_{n+1}-1} a_j(\mathbf{X}_n) < r_2 \leq \frac{1}{a(\mathbf{X}_n)} \sum_{j=0}^{k_{n+1}} a_j(\mathbf{X}_n),$$

where $a_0 = 0$ by convention.
(2) Update the time and the state of the system by

$$t_{n+1} = t_n + \delta t_{n+1}, \qquad \mathbf{X}_{n+1} = \mathbf{X}_n + \mathbf{v}_{k_{n+1}}.$$

Then repeat the procedure.

In this algorithm, r_1 is used to update the clock and r_2 is used to select the particular reaction to be executed.

As explained in subsection 2.3.3, in practice reaction rates often have very disparate magnitudes. This can be appreciated from the reaction rate formula from transition state theory (see the next chapter): the rate depends exponentially on the energy barrier. Therefore a small difference in the energy barrier can result in a large difference in the rate. In such a case, we are often interested more in the slow reactions since they are the bottlenecks for the overall dynamics. A theory for the effective dynamics was described in subsection 2.3.3. However, it might be quite difficult to compute the effective reaction rates in the effective dynamics either analytically or by precomputing. Therefore various strategies have been proposed to capture the effective dynamics of the slow variables "on-the-fly." We will discuss an algorithm, proposed in [13], which is a simple modification of the original SSA in which a nested structure is added according to the time scale of the rates. The process at each level of the time scale is simulated with an SSA, using effective rates. Results from simulations on fast time scales are used to compute the rates for the SSA at slower time scales. For simple systems with only two time scales, the nested SSA consists of two SSAs organized so that one is nested in the other: the outer SSA is for the slow reactions only but with modified slow rates computed in an inner SSA modeling the fast reactions only.

Let t_n, $\mathbf{X}_n$ be the current time and state of the system respectively. The two-level nested SSA proceeds as follows.

(1) *Inner SSA.* Pick an integer N. Run N independent replicas of SSA with the fast reactions $\mathbf{R}^f = \{(\varepsilon^{-1}a^f, \mathbf{v}^f)\}$ only, for a time interval $T_0 + T_f$. During this calculation, compute the modified slow rates for $j = 1, \ldots, M_s$:

$$\tilde{a}_j^s = \frac{1}{N} \sum_{k=1}^{N} \frac{1}{T_f} \int_{T_0}^{T_f + T_0} a_j^s(\mathbf{X}_\tau^k) \, d\tau, \qquad (9.2.2)$$

where $\mathbf{X}_\tau^k$ is the result of the kth replica of this auxiliary virtual fast process at virtual time τ, whose initial value is $\mathbf{X}_{t=0}^k = \mathbf{X}_n$, and T_0 is a parameter chosen to minimize the effect of the transients to equilibrium in the virtual fast process.

(2) *Outer SSA.* Run one step of SSA for the modified slow reactions $\tilde{\mathbf{R}}^s = (\tilde{a}^s, \mathbf{v}^s)$ to generate $(t_{n+1}, \mathbf{X}_{n+1})$ from $(t_n, \mathbf{X}_n)$.

Then repeat the procedure.

In HMM language, the macroscale solver is the outer SSA and the data that need to be estimated are the effective rates for the slow reactions. These data are obtained by simulating the virtual fast process which plays the role of the microscale solvers here. However, unlike the standard examples of HMM, here we do not need to know what the slow and fast variables are in order to carry out the computation; the algorithm is formulated in terms of the original variables. This is to be expected, since the effective dynamics can also be formulated in terms of the original variables (see subsection 2.3.3).

Convergence and efficiency of the nested SSA The original SSA is an exact realization of the stochastic chemical kinetic system. The nested SSA, however, is an approximation. The errors in the nested SSA can be analyzed using the general strategy for analyzing HMM (see Section 6.7). The details can be found in [14].

Assume, as in subsection 2.3.3, that there is a complete set of slow variables of the form $\{z_j = \mathbf{b}_j \cdot \mathbf{x}, j = 1, \ldots, J\}$, where $\{\mathbf{b}_j, j = 1, \ldots, J\}$ is a set of basis vectors in the subspace of vectors that are orthogonal to all the vectors $\{\mathbf{v}_k^{\mathrm{f}}\}$. Let f be a smooth function. Denote by $\tilde{\mathbf{X}}_t$ the solution of the nested SSA. Consider the observable $v(\mathbf{x}, t) = \mathbb{E}_{\mathbf{x}} f(\mathbf{b} \cdot \tilde{\mathbf{X}}_t)$, where the expectation is taken with respect to the randomness in the outer SSA only. Let $u(\mathbf{x}, t)$ be the solution of the effective equation (2.3.47) with $u(\mathbf{x}, 0) = f(\mathbf{b} \cdot \mathbf{x})$. The following result is proved in [13, 14].

Theorem *For any $T > 0$, there exist constants C and α, independent of (N, T_0, T_{f}), such that*

$$\sup_{0 \leq t \leq T, \mathbf{x} \in \mathcal{X}} \mathbb{E}\, |v(\mathbf{x}, t) - u(\mathbf{x}, t)| \leq C \left(\varepsilon + \frac{e^{-\alpha T_0 / \varepsilon}}{1 + T_{\mathrm{f}}/\varepsilon} + \frac{1}{\sqrt{N(1 + T_{\mathrm{f}}/\varepsilon)}} \right).$$

$$(9.2.3)$$

This result can be used to analyze the efficiency of the nested SSA. Given a chemical kinetic system with $\mathbf{R} = \{(a_j, \mathbf{v}_j)\}$, we assume that the total rate $a(\mathbf{x}) = \sum a_j(\mathbf{x})$ does not fluctuate much in time: $a(\mathbf{x}) \sim \mathcal{O}(\varepsilon^{-1})$. Given an error tolerance λ, we choose the parameters in the nested SSA such that each term in (9.2.3) is less than $\mathcal{O}(\lambda)$. One possible choice of the parameters is

$$T_0 = 0, \quad N = 1 + \frac{T_{\mathrm{f}}}{\varepsilon} = \frac{1}{\lambda}.$$

The total cost for the nested SSA over a time interval of $\mathcal{O}(1)$ is given by

$$\mathrm{cost} = \mathcal{O}\left(N \left(1 + \frac{T_0}{\varepsilon} + \frac{T_{\mathrm{f}}}{\varepsilon} \right) \right) = \mathcal{O}\left(\frac{1}{\lambda^2} \right).$$

for comparison, the cost for the direct SSA is given by

$$\mathrm{cost} = \mathcal{O}\left(\frac{1}{\varepsilon} \right),$$

$$(9.2.4)$$

since the time-step size is of order ε. When $\varepsilon \ll \lambda^2$, the nested SSA is much more efficient than the direct SSA.

Next we discuss the influence of the other numerical parameters on the efficiency. The parameter T_0 which plays the role of the numerical relaxation time does not influence the efficiency much. Given the same error tolerance λ, for the last term in the error estimate (9.2.3) to be less than $\mathcal{O}(\lambda)$ we need to have

$$N\left(1 + \frac{\varepsilon}{T_{\mathrm{f}}}\right) \geq \mathcal{O}\left(\frac{1}{\lambda^2}\right).$$

Therefore

$$\mathrm{cost} \geq \mathcal{O}\left(N\left(1 + \frac{\varepsilon}{T_{\mathrm{f}}}\right)\right) = \mathcal{O}\left(\frac{1}{\lambda^2}\right), \qquad (9.2.5)$$

which is the same as (9.2.4) regardless of the value of T_0. The above argument also implies that the optimal cost for the nested SSA to achieve an error tolerance of λ is $\mathcal{O}\left(1/\lambda^2\right)$.

Turning now to the effect of the parameter N, the number of realizations for inner SSA, let us see what happens when we take $N = 1$. For the error estimate (9.2.3) to satisfy the same error tolerance λ, we have to choose

$$1 + \frac{\varepsilon}{T_{\mathrm{f}}} = \frac{1}{\lambda^2}.$$

The cost of the nested SSA is given by

$$\mathrm{cost} = \mathcal{O}\left(N\left(1 + \frac{\varepsilon}{T_{\mathrm{f}}}\right)\right) = \mathcal{O}\left(\frac{1}{\lambda^2}\right), \qquad (9.2.6)$$

which is the same as the cost using multiple realizations. This means that using multiple realizations in the inner SSA does not increase the efficiency of the overall scheme either. Obviously, however, using multiple realizations is advantageous for implementation on parallel computers.

Other versions of the nested SSA are discussed in [40, 39]. Although they appear to be quite different, it can be shown that they are essentially the same as the nested SSA discussed here.

A numerical example: a virus infection model As a concrete example we discuss the virus infection model studied in [13, 23]. The model was originally proposed in [41] as an example of the failure of modeling reaction networks with deterministic dynamics. The reactions considered in this model are listed in Table 9.1 for the number of reaction channels $M_R = 6$. The species that are to be modeled are *genome, struct, template* and *virus* ($N_{\mathrm{s}} = 4$). The species *genome* is the vehicle of the viral genetic information, which can take the form of DNA, positive-strand RNA,

Table 9.1. *Reaction channels of the virus infection model*

nucleotides	$\xrightarrow{a_1 = 1.0 \times \textit{template}}$	*genome*
nucleotides + *genome*	$\xrightarrow{a_2 = 0.25 \times \textit{genome}}$	*template*
nucleotides + amino acids	$\xrightarrow{a_3 = 1000 \times \textit{template}}$	*struct* *
template	$\xrightarrow{a_4 = 0.25 \times \textit{template}}$	degraded
struct	$\xrightarrow{a_5 = 1.9985 \times \textit{struct}}$	degraded/secreted *
genome + *struct*	$\xrightarrow{a_6 = 7.5d-6 \times \textit{genome} \times \textit{struct}}$	*virus*

negative-strand RNA or some other variants. The species *struct* represents the structural proteins that make up the virus. The species *template* refers to the form of the nucleic acid that is transcribed and involved in catalytically synthesizing every viral component. The nucleotides and amino acids are assumed to be available at constant concentrations.

When *template* > 0, the production and degradation of *struct*, which respectively comprise the third and fifth reactions marked with * in Table 9.1, are much faster than the other reactions. From the reaction rates we can see that the ratio of time scales ε is about 10^{-3}. In a system that consists only of the fast reactions, *struct* has an equilibrium distribution which is Poisson, with parameter $\lambda = 500 \times \textit{template}$, such that

$$\mathbb{P}_{template}(\textit{struct} = n) = \frac{(500 \times \textit{template})^n}{n!} \exp(-500 \times \textit{template}).$$

Notice that *struct* shows up only in the final slow reaction. A reduced dynamics in the form of the slow reactions ($a_{1,2,4,6}$) with rates averaged with respect to the quasi-equilibrium of the fast reactions ($a_{3,5}$) can be given as a system with four reactions, shown in Table 9.2. The initial condition is chosen to be

$$(\textit{struct, genome, template, virus}) = (0, 0, 10, 0).$$

The mean value and the variance of *template* at time $T = 20$ are used as a benchmark. A computation of these values by a direct SSA using $N_0 = 10^6$ realizations led to

$$\overline{\textit{template}} = 3.7170 \pm 0.005, \quad \texttt{var}(\textit{template}) = 4.9777 \pm 0.005.$$

Table 9.2. *The reduced virus infection model*

nucleotides	$\xrightarrow{a_1 = 1.0 \times template}$	genome
nucleotides + *genome*	$\xrightarrow{a_2 = 0.25 \times genome}$	template
template	$\xrightarrow{a_4 = 0.25 \times template}$	degraded
genome + struct	$\xrightarrow{a_6 = 3.75d-3 \times genome^2 \times struct}$	virus

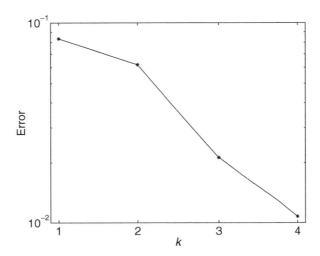

Figure 9.3. Relative error for $\overline{template}$ using the nested SSA for the virus infection model (courtesy of Di Liu).

For the nested SSA, we made a series of simulations in which we chose the size of the ensemble and the simulation time of the inner SSA according to

$$(N, T_0, T/\varepsilon) = (1, 0, 2^{2k}), \tag{9.2.7}$$

for different values of $k = 0, 1, \ldots$ The error estimate in (9.2.3) then implies that the error δ should decay with k at a rate given by

$$\delta = \mathcal{O}(2^{-k}), \tag{9.2.8}$$

which is consistent with the results in Figure 9.3. Table 9.3 gives the total CPU time and the obtained values of $\overline{template}$ and $\mathbf{var}(template)$ when the parameters of inner SSA were chosen according to (9.2.7), using $N_0 = 10^6$ realizations of the outer SSA (as for the direct SSA computation).

Table 9.3. *Efficiency of the nested SSA for the virus infection model*

T_f/ε	1	4	16	64	"exact"
CPU time	154.8	461.3	2068.2	9190.9	34806.4
$\overline{template}$	4.027	3.947	3.796	3.757	3.717 ± 0.005
var($template$)	5.401	5.254	5.007	4.882	4.978 ± 0.005

9.3 Coarse-grained molecular dynamics

Coarse-grained molecular dynamics (CGMD) is a concept that has been with us, maybe implicitly, for a long time. When we develop bead–spring models in polymer dynamics, we have in mind not a full atom model but rather a coarse-grained model, even though we do not necessarily call it such. In the last decade or so this notion has attracted more attention and has been used quite extensively in several contexts:

(1) *Multiscale simulations in material science*, in which CGMD is used as a bridge between full atomistic simulations and finite element methods based on continuum models [37].

(2) *Coarse-grained models of macromolecules.* This is an extension of the bead–spring type of model for polymers but with more attention paid to obtaining a quantitatively correct effective potential for the coarse-grained model. Techniques such as Boltzmann inversion have been used to extract such potentials from the long-time trajectories of full MD simulation [45].

(3) *Free energy exploration for macromolecules.* Free energy is an old concept. It is defined as a function of a set of coarse-grained variables. Until recently it was limited to a very small number of such variables. Recent techniques such as metadynamics and temperature-accelerated molecular dynamics have made it possible to explore the space of many coarse-grained variables [28, 34].

(4) *The application of the Mori–Zwanzig formalism to molecular dynamics* [10, 32].

First, what would be a good choice for the coarse-grained variables? Well, the answer depends on the purpose of the simulation. If the purpose is to understand the mechanism of some chemical reaction then the coarse-grained variables should be able to serve as reaction coordinates. This issue will be discussed in Chapter 10. If our interest is the macroscopic behavior of the system then the coarse-grained variables should be a complete set of slow variables of the system, since in that case the time scale separation between the coarse-grained (slow) variables and the eliminated variables allows us to make satisfactory approximations without the need to keep memory terms. If our interest is the behavior of a specific set of variables, such as the variables that describe the dynamics around a local defect, then we have no choice but to include these variables in the set of coarse-grained variables and to model the memory effects (see for example [32]).

As an example, we will discuss the version of the CGMD proposed by Rudd and Broughton [37]. Starting with a Hamiltonian dynamics,

$$\frac{d\mathbf{x}_j}{dt} = \nabla_{\mathbf{p}_j} H, \quad \frac{d\mathbf{p}_j}{dt} = -\nabla_{\mathbf{x}_j} H,$$

where $j = 1, \ldots, N$ and

$$H = \sum_{j=1}^{N} \frac{1}{2m} |\mathbf{p}_j|^2 + V(\mathbf{x}_1, \ldots, \mathbf{x}_N),$$

we define a set of coarse-grained variables

$$\mathbf{X}_\alpha = \sum_j f_{\alpha,j} \mathbf{x}_j, \quad \mathbf{P}_\alpha = \sum_j f_{\alpha,j} \mathbf{p}_j$$

for some $\{f_{\alpha,j}\}$ where $\alpha = 1, \ldots, K$. In the Rudd–Broughton version of the CGMD, the coarse-grained variables obey the Hamiltonian dynamics

$$\frac{d\mathbf{X}_\alpha}{dt} = \nabla_{\mathbf{P}_\alpha} \mathcal{H}, \quad \frac{d\mathbf{P}_\alpha}{dt} = -\nabla_{\mathbf{X}_\alpha} \mathcal{H}, \tag{9.3.1}$$

with coarse-grained Hamiltonian

$$\mathcal{H}(\mathbf{X}_1, \ldots, \mathbf{X}_K; \mathbf{P}_1, \ldots, \mathbf{P}_K)$$

$$= \frac{1}{Z} \int\!\!\int H e^{-\beta H} \prod_\alpha \delta \left(\mathbf{X}_\alpha - \sum_j f_{\alpha,j} \mathbf{x}_j \right) \delta \left(\mathbf{P}_\alpha - \sum_j f_{\alpha,j} \mathbf{p}_j \right) \prod_{j=1}^{N} d\mathbf{x}_j \, d\mathbf{p}_j,$$

$$\tag{9.3.2}$$

where $\beta = 1/(k_B T)$ and Z is the partition function $Z = \int\int e^{-\beta H} \Pi_{j=1}^{N} d\mathbf{x}_j \times d\mathbf{p}_j$. Equation (9.3.2) is the equilibrium distribution projected onto the submanifold defined by the coarse-grained variables.

When V is a quadratic function,

$$V(\mathbf{x}_1, \ldots, \mathbf{x}_N) = \tfrac{1}{2} \sum_{j,k} \mathbf{x}_j^{\mathrm{T}} D_{jk} \mathbf{x}_k,$$

then a direct computation gives

$$\mathcal{H} = \tfrac{1}{2} \sum_{\alpha,\beta} (M_{\alpha,\beta} \dot{\mathbf{X}}_\alpha \cdot \dot{\mathbf{X}}_\beta + \mathbf{X}_\alpha^{\mathrm{T}} B_{\alpha,\beta} \mathbf{X}_\beta + 3(N-K)k_B T,$$

where

$$M_{\alpha,\beta} = m \left(\sum_k f_{\alpha,k} f_{\beta,k} \right)^{-1},$$

$$B_{\alpha,\beta} = \left(\sum_{j,k} f_{\alpha,j} D_{j,k}^{-1} f_{\beta,k} \right)^{-1}$$

and $\{\dot{\mathbf{X}}_\alpha\}$ and $\{\mathbf{P}_\alpha\}$ are related by

$$\mathbf{P}_\alpha = \sum_\beta M_{\alpha,\beta} \dot{\mathbf{X}}_\beta.$$

This formalism can obviously be generalized to the case when the coarse-grained variables are defined by nonlinear functions. However, there are two main difficulties with this approach.

(1) The question arises whether Hamiltonian dynamics with (9.3.2) as the Hamiltonian is a good approximation to the dynamics of the coarse-grained variables. In the expressions (9.3.1) and (9.3.2) it is assumed that there is a separation of the time scales between the dynamics for the coarse-grained variables and the relaxation dynamics of the remaining degrees of freedom. The accuracy of this approximation depends on the difference between these two time scales.

(2) On the practical side, it is not easy to evaluate the coarse-grained Hamiltonian $\mathcal{H}$ and its derivatives. Analytical evaluation seems only feasible for the case when V is quadratic. Adding cubic terms to V makes the calculation much harder [38].

To illustrate the importance of the first difficulty, let us consider the following example [22]:

$$H = \tfrac{1}{2}(x_1^2 + x_2^2 + x_1^2 x_2^2 + p_1^2 + p_2^2).$$ (9.3.3)

Hamilton's equations are

$$\frac{dx_1}{dt} = p_1, \quad \frac{dp_1}{dt} = -x_1 - x_1 x_2^2,$$

$$\frac{dx_2}{dt} = p_2, \quad \frac{dp_2}{dt} = -x_2 - x_2 x_1^2.$$

If we regard (x_2, p_2) as variables for some thermal bath and use CGMD to eliminate (x_2, p_2), we obtain a system for the remaining variables (x_1, p_1):

$$\frac{dx_1}{dt} = p_1, \quad \frac{dp_1}{dt} = -x_1 - x_1 \, \mathbb{E}\left(x_2^2 \middle| p_1, x_1\right),$$

where the conditional expectation is given by

$$\mathbb{E}\left(x_2^2 \middle| p_1, x_1\right) = \frac{\int x_2^2 e^{-H} \, dx_2 \, dp_2}{\int e^{-H} \, dx_2 \, dp_2} = \frac{1}{1 + x_1^2}.$$

Therefore the CGMD system is

$$\frac{dq_1}{dt} = p_1, \quad \frac{dp_1}{dt} = -q_1 - \frac{q_1}{1 + q_1^2}.$$ (9.3.4)

However, the behavior of this system is quite different from the behavior of the original system, even after averaging (using the equilibrium distribution, the conditioning on (x_1, p_1) being fixed) over different initial conditions of x_2 and p_2, as can be seen from Figure 9.4. In particular, the dissipative effect in the original system is lost in the new system (9.3.4) owing to the fact that the energy in (x_1, p_1) is transferred to the bath variables (x_2, p_2).

As was pointed out in [22], these discrepancies are caused by the drastic approximation used in (9.3.2), which becomes quite inaccurate in nonequilibrium situations. To reduce this kind of error one has to take into account the memory and noise effect, as can be seen in the Mori–Zwanzig formalism (see Section 2.7). This is a general issue, to which we will return to Chapter 11.

We now turn to the second difficulty. One way of overcoming it is to formulate CGMD using an extended Lagrangian formulation [35]. This was done by Iannuzzi, Laio and Parrinello [25]. The idea is to extend the

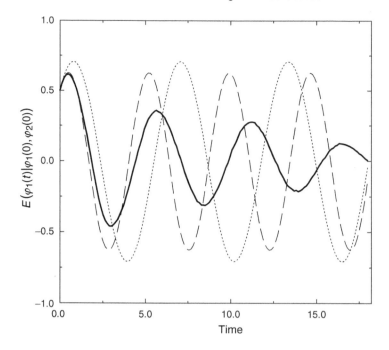

Figure 9.4. For comparison, the ensemble-averaged solution (solid line) of the original Hamiltonian system and the coarse-grained Hamiltonian system for the model problem (9.3.3) (courtesy of Alexandre Chorin) [9]. The broken line is the solution to the coarse-grained molecular dynamics. The dotted line is the solution to a truncated version of (9.3.3) in which the unresolved component (x_2, p_2) is simply set to 0. These two solutions are just about equally bad.

phase space to include the coarse-grained variables as additional dynamic variables. For simplicity, we will follow [11] and consider the case of an overdamped dynamics,

$$\dot{\mathbf{x}} = -\nabla V(\mathbf{x}) + \sqrt{2\beta^{-1}} \dot{\mathbf{w}},$$

in the configuration space instead of MD in the phase space.

Assume that the coarse-grained variables are defined by

$$\mathbf{X} = \mathbf{q}(\mathbf{x}) = (Q_1(\mathbf{x}), \ldots, Q_K(\mathbf{x})).$$

The equilibrium density for $\mathbf{X}$ is given by

$$\bar{\rho}(\mathbf{X}) = \tilde{Z}^{-1} \int_{\mathbb{R}^{3N}} e^{-\beta V(\mathbf{x})} \delta(\mathbf{X} - \mathbf{q}(\mathbf{x})) \, d\mathbf{x}.$$

The question is how to compute this function efficiently.

The idea in [25] is to consider the energy in the extended space of variables $(\mathbf{x}, \mathbf{X})$:

$$V(\mathbf{x}, \mathbf{X}) = V(\mathbf{x}) + \tfrac{1}{2}\mu|\mathbf{X} - \mathbf{q}(\mathbf{x})|^2, \qquad (9.3.5)$$

where $\mu > 0$ is a parameter to be prescribed later. Introducing another friction parameter γ, we can formulate the overdamped dynamics associated with (9.3.5):

$$\begin{cases} \gamma\dot{\mathbf{x}} = -\nabla V(\mathbf{x}) + \mu(\mathbf{X} - \mathbf{q}(\mathbf{x}))\nabla\mathbf{q}(\mathbf{x}) + \sqrt{2\gamma\beta^{-1}}\,\dot{\mathbf{w}}, \\ \dot{\mathbf{X}} = -\mu(\mathbf{X} - \mathbf{q}(\mathbf{x})) + \sqrt{2\beta^{-1}}\,\dot{\mathbf{W}}, \end{cases} \qquad (9.3.6)$$

where $\dot{\mathbf{w}}, \dot{\mathbf{W}}$ are independent white noises.

For all $\gamma > 0$, the equilibrium density for (9.3.6) is

$$\rho_\mu(\mathbf{x}, \mathbf{X}) = Z^{-1}e^{-\beta V(\mathbf{x}) - \mu\beta|\mathbf{X} - \mathbf{q}(\mathbf{x})|^2/2}, \qquad (9.3.7)$$

where Z is a normalization factor such that the total integral of ρ_μ is 1. The corresponding marginal density of $\mathbf{X}$ is

$$\bar{\rho}_\mu(\mathbf{X}) = Z^{-1}\int_{\mathbb{R}^{3N}} e^{-\beta V(\mathbf{x}) - \mu\beta|\mathbf{X} - \mathbf{q}(\mathbf{x})|^2/2}\,d\mathbf{x}. \qquad (9.3.8)$$

As $\mu \to \infty$,

$$\bar{\rho}_\mu \to \bar{\rho}. \qquad (9.3.9)$$

To capture the marginal density of $\mathbf{X}$ we create an artificial separation of the time scales associated with $\mathbf{x}$ and $\mathbf{X}$ by making γ small. In this case, the variables $\mathbf{X}(\cdot)$ evolve much more slowly than $\mathbf{x}(\cdot)$ and only feel the average effect of the latter. This puts us in a situation where averaging theorems for stochastic ODEs can be applied (see the beginning of this chapter, or Section 2.3). The effective equation governing the dynamics of $\mathbf{X}$ is obtained by averaging the right-hand side of the second equation in (9.3.6) with respect to the conditional probability density:

$$\rho_\mu(\mathbf{x}|\mathbf{X}) = Z_\mu^{-1}(\mathbf{X})e^{-\beta V(\mathbf{x}) - \mu\beta|\mathbf{X} - \mathbf{q}(\mathbf{x})|^2/2}, \qquad (9.3.10)$$

where

$$Z_\mu(\mathbf{X}) = \int_{\mathbb{R}^{3N}} e^{-\beta V(\mathbf{x}) - \mu\beta|\mathbf{X} - \mathbf{q}(\mathbf{x})|^2/2}\,d\mathbf{x}, \qquad (9.3.11)$$

is a normalization factor. Since

$$-Z_\mu^{-1}(\mathbf{X})\int_{\mathbb{R}^n}\mu(\mathbf{X}-\mathbf{q}(\mathbf{x}))e^{-\beta V(\mathbf{x})-\mu\beta|\mathbf{X}-\mathbf{q}(\mathbf{x})|^2/2}\,d\mathbf{x}=\beta^{-1}Z_\mu^{-1}(\mathbf{X})\nabla_{\mathbf{X}}Z_\mu(\mathbf{X})$$

$$=\beta^{-1}\nabla_{\mathbf{X}}\log Z_\mu(\mathbf{X}),$$

$$(9.3.12)$$

the limiting equations for $\mathbf{X}(\cdot)$ (9.3.6) as $\gamma\to 0$ can be written as

$$\dot{\mathbf{X}}=-\nabla_{\mathbf{X}}F_\mu+\sqrt{2\beta^{-1}}\,\dot{\mathbf{W}},\tag{9.3.13}$$

where

$$F_\mu(\mathbf{X})=-\beta^{-1}\log Z_\mu(\mathbf{X}).\tag{9.3.14}$$

From (9.3.11) and (9.3.14) we have, up to a constant,

$$F_\mu(\mathbf{X})\to F(\mathbf{X})=-\beta^{-1}\log\bar{\rho}(\mathbf{X})\tag{9.3.15}$$

as $\mu\to\infty$. Therefore the limiting equation for $\mathbf{X}(t)$ as $\gamma\to 0$ and then $\mu\to\infty$ is

$$\dot{\mathbf{X}}=-\nabla_{\mathbf{X}}F+\sqrt{2\beta^{-1}}\,\dot{\mathbf{W}}.\tag{9.3.16}$$

This shows that (9.3.6) can be used in this parameter regime to capture the desired free-energy density for $\mathbf{X}$.

A similar argument shows that (9.3.8) is also the reduced density for the extended Hamiltonian system

$$\begin{cases} m\ddot{\mathbf{x}}=-\nabla V(\mathbf{x})+\mu(\mathbf{X}-\mathbf{q}(\mathbf{x}))\nabla\mathbf{q}(\mathbf{x}),\\ M\ddot{\mathbf{X}}=-\mu(\mathbf{X}-\mathbf{q}(\mathbf{x})), \end{cases}\tag{9.3.17}$$

corresponding to the extended Lagrangian

$$\mathcal{L}(\mathbf{x},\mathbf{X})=\tfrac{1}{2}m|\dot{\mathbf{x}}|^2-V(\mathbf{x})+\tfrac{1}{2}M|\dot{\mathbf{X}}|^2-\tfrac{1}{2}\mu|\mathbf{X}-\mathbf{q}(\mathbf{x})|^2.\tag{9.3.18}$$

Here M is an artificial mass parameter associated with the variable $\mathbf{X}$. Therefore the extended Hamiltonian system (9.3.17) can also be used to compute the free energy associated with $\mathbf{X}$. This approach is often called the *extended Lagrangian method*.

As one can see the justification given here applies only to the equilibrium situation. Therefore, at this point it is only safe to use these techniques as sampling techniques for the equilibrium distribution, even though the formulation might also contain useful information for the dynamics.

9.4 Notes

Numerous relevant topics have been omitted in this chapter. Some, such as boosting and projective integration, were discussed in Chapter 6 and so were not repeated here. An interesting perspective on HMM, seamless HMM, boosting and projective integrators was presented in [43]. Also relevant are exponential integrators, the techniques of using integrating factors based on Duhammel's principle, integrators with adaptive time-step selection and so on. These topics have already been covered in many excellent textbooks or review articles, however.

We have focused on coarse-grained molecular dynamics but there are also parallel issues and developments in Monte Carlo algorithms. In particular, coarse-grained Monte Carlo methods have been proposed in [26]. We refer to the review article [27] for details and more recent developments. The basic philosophy is quite similar to that of CGMD. For this reason it is also expected to have the same problems as those discussed above, namely that the memory effects may not be accurately modeled. In principle these difficulties can be overcome using more sophisticated renormalization group techniques in the Mori–Zwanzig formalism [8].

Another interesting idea is the "parareal" algorithm [33]. This was introduced as a strategy for applying domain decomposition methods in time. Obviously the domain decomposition structure is naturally suited for developing multiscale integrators.

References for Chapter 9

[1] A. Abdulle, unpublished.

[2] A. Abdulle, "Fourth order Chebychev methods with recurrence relation," *SIAM J. Sci. Comput.*, vol. 23, pp. 2041–2054, 2002.

[3] A. Abdulle and S. Cirilli, "S-ROCK: Chebychev methods for stiff stochastic differential equations," *SIAM J. Sci. Comput.*, vol. 30, pp. 997–1014, 2008.

[4] A. Abdulle and T. Li, "S-ROCK methods for stiff ITO SDEs," *Comm. Math. Sci.*, vol. 6, pp. 845–868, 2008.

[5] A. Abdulle and A. Medovikov, "Second order Chebychev methods based on orthogonal polynomials," *Numer. Math.*, vol. 90, pp. 1–18, 2001.

[6] A.B. Bortz, M.H. Kalos and J.L. Lebowitz, "A new algorithm for Monte Carlo simulation of Ising spin systems," *J. Comput. Phys.*, vol. 17, pp. 10–18, 1975.

[7] Y. Cao, D. Gillespie and L. Petzold, "The slow scale stochastic simulation algorithm," *J. Chem. Phys.*, vol. 122, pp. 14 116-1–14 116-18, 2005.

[8] A.J. Chorin, "Conditional expectations and renormalization", *Multiscale Modeling and Simulation*, vol. 1, pp. 105–118, 2003.

[9] A.J. Chorin and P. Stinis, "Problem reduction, renormalization and memory", *Comm. Appl. Math. Comp. Sci.*, vol. 1, pp. 1–27, 2005.

[10] A.J. Chorin, O. Hald and R. Kupferman, "Optimal prediction with memory," *Physica D*, vol. 166, nos. 3–4, pp. 239–257, 2002.

[11] W. E and E. Vanden-Eijnden, "Metastability, conformation dynamics, and transition pathways in complex systems," in *Multiscale Modelling and Simulation, Lecture Notes in Computational Science and Engineering*, vol. 39, pp. 35–68, Springer-Verlag, 2004.

[12] W. E, D. Liu and E. Vanden-Eijnden, "Analysis of multiscale methods for stochastic differential equations," *Comm. Pure Appl. Math.*, vol. 58, no. 11, pp. 1544–1585, 2005.

[13] W. E, D. Liu and E. Vanden-Eijnden, "Nested stochastic simulation algorithm for chemical kinetic systems with disparate rates," *J. Chem. Phys.*, vol. 123, pp. 194107-1–194107-8, 2005.

[14] W. E, D. Liu and E. Vanden-Eijnden, "Nested stochastic simulation algorithms for chemical kinetic systems with multiple time scales," *J. Comput. Phys.*, vol. 221, no. 1, pp. 158–180, 2007.

[15] B. Engquist and Y.-H. Tsai, "The heterogeneous multiscale methods for a class of stiff ODEs," *Math. Comp.*, vol. 74, pp. 1707–1742, 2005.

[16] K. Eriksson, C. Johnson and A. Logg, "Explicit time-stepping for stiff ODEs," *SIAM J. Sci. Comput.*, vol. 25, no. 4, pp. 1142–1157, 2003.

[17] I. Fatkullin and E. Vanden-Eijnden, "A computational strategy for multiscale systems with applications to Lorenz 96 model," *J. Comput. Phys.*, vol. 200, no. 2, pp. 605–638, 2004.

[18] C.W. Gear and I.G. Kevrekidis, "Projective methods for stiff differential equations: problems with gaps in their eigenvalue spectrum," *SIAM J. Sci. Comput.*, vol. 24, pp. 1091–1106, 2003.

[19] D.T. Gillespie, "A general method for numerically simulating the stochastic time evolution of coupled chemical reactions," *J. Comput Phys.*, vol. 22, pp. 403–434, 1976.

[20] D.T. Gillespie, "Exact stochastic simulation of coupled chemical reactions," *J. Phys. Chem.*, vol. 81, pp. 2340–2361, 1977.

[21] E. Hairer and G. Wanner, *Solving Ordinary Differential Equations II: Stiff and Differential–Algebraic Problems*, 2nd edition Springer-Verlag, 2004.

[22] O.H. Hald and P. Stinis "Optimal prediction and the rate of decay for solutions of the Euler equations in two and three dimensions," *Proc. Nat. Acad. Sci. USA*, vol. 104, no. 16, pp. 6527–6553, 2007.

[23] E.L. Haseltine and J.B. Rawlings, "Approximate simulation of coupled fast and slow reactions for stochastic kinetics," *J. Chem. Phys.*, vol. 117, pp. 6959–6969, 2002.

[24] P.J. van der Houwen and B.P. Sommeijer, "On the internal stability of explicit, m-stage Runge–Kutta methods for large m-values," *Z. Angew. Math. Mech.*, vol. 60, pp. 479–485, 1980.

[25] M. Iannuzzi, A. Laio and M. Parrinello, "Efficient exploration of reactive potential energy surfaces using Car–Parrinello molecular dynamics", *Phys. Rev. Lett.*, vol. 90, 238 302-1–238 302-4, 2003.

[26] M.A. Katsoulakis, A.J. Majda and D.G. Vlachos, "Coarse-grained stochastic processes and Monte Carlo simulations in lattice systems," *J. Comput. Phys.*, vol. 186, pp. 250–278, 2003.

[27] M.A. Katsoulakis, P. Plechac and L. Rey-Bellet, "Numerical and statistical methods for the coarse-graining of many-particle stochastic systems," *J. Sci. Comput.*, vol. 37, pp. 43–71, 2008.

[28] A. Laio and M. Parrinello, "Escaping free-energy minima," *Proc. Nat. Acad. Sci. USA*, vol. 99, pp. 12 562–12 566, 2002.

[29] V.I. Lebedev, "How to solve stiff systems of differential equations by explicit methods," in *Numerical Methods and Applications*, G.I. Marchuk, ed., pp. 45–80, CRC Press, 1994.

[30] V.I. Lebedev and S.I. Finogenov, "On the utilization of ordered Tchebychev parameters in iterative methods" (in Russian), *Zh. Vychisl. Mat. Fiziki*, vol. 16, no. 4, pp. 895–910, 1976.

[31] T. Li, A. Abdulle and W. E, "Effectiveness of implicit methods for stiff stochastic differential equations," *Commun. Comput. Phys.*, vol. 3, no. 2, pp. 295–307, 2008.

[32] X. Li and W. E, "Variational boundary conditions for molecular dynamics simulations of crystalline solids at finite temperature: treatment of the thermal bath," *Phys. Rev. B*, vol. 76, no. 10, pp. 104 107-1–104 107-22, 2007.

[33] J.-L. Lions, Y. Maday, and G. Turinici. "Résolution d'EDP par un schéma en temps pararéel," *C.R. Acad. Sci. Paris Sér. I Math.*, vol. 332, pp. 661–668, 2001.

[34] L. Maragliano and E. Vanden-Eijnden, "A temperature accelerated method for sampling free energy and determining reaction pathways in rare events simulations," *Chem. Phys. Lett.*, vol. 426, pp. 168–175, 2006.

[35] M. Parrinello and A. Rahman, "Crystal structure and pair potentials: a molecular dynamics study", *Phys. Rev. Lett.*, vol. 45, pp. 1196–1199, 1980.

[36] C.V. Rao and A.P. Arkin, "Stochastic chemical kinetics and the quasi-steady-state assumption: application to the Gillespie algorithm," *J. Chem. Phys.*, vol. 118, pp. 4999–5010, 2003.

[37] R.E. Rudd and J.Q. Broughton, "Atomistic simulation of MEMS resonators through the coupling of length scales," *J. Modeling and Simulation of Microsystems*, vol. 1, pp. 29–38, 1999.

[38] R.E. Rudd and J.Q. Broughton, "Coarse-grained molecular dynamics: nonlinear finite elements and finite temperature", *Phys. Rev. B*, vol. 72, 144 104-1–144 104-32, 2005.

[39] H. Salis and Y. Kaznessis, "Accurate hybrid stochastic simulation of a system of coupled chemical or biochemical reactions," *J. Chem. Phys.*, vol. 122, pp. 054 103-1–054 103-13, 2005.

[40] A. Samant and D.G. Vlachos, "Overcoming stiffness in stochastic simulation stemming from partial equilibrium: a multiscale Monte Carlo algorithm," *J. Chem. Phys.*, vol. 123, pp. 144 114-1–144 114-8, 2005.

[41] R. Srivastava, L. You, J. Summers and J. Yin, "Stochastic vs. deterministic modeling of intracellular viral kinetics," *J. Theor. Biol.*, vol. 218, pp. 309–321, 2002.

[42] M. Tuckerman, B.J. Berne and G.J. Martyna, "Reversible multiple time scale molecular dynamics," *J. Chem. Phys.*, vol. 97, no. 3, pp. 1990–2001, 1992.

[43] E. Vanden-Eijnden, "On HMM-like integrators and projective integration methods for systems with multiple time scales," *Comm. Math. Sci.*, vol. 5, pp. 495–505, 2007.

[44] J.W. Verwer, "Explicit Runge–Kutta methods for parabolic partial differential equations," *Appl. Num. Math.*, vol. 22, pp. 359–379, 1996.

[45] G.A. Voth (ed.), *Coarse-Graining of Condensed Phase and Biomolecular Systems*, CRC Press, 2009.

10

Rare events

10.1 Introduction

We have already discussed various examples of problems with multiple time scales. In this chapter, we will turn to another important class of such problems: rare events. These are the kind of events that happen infrequently, compared with the relaxation time scale of the system; however, when they do happen, they happen rather quickly and have important consequences. Typically, a small amount of noise is present in the system, and it is this that drives these rare events. For such an event to happen, the system has to wait for the different components of the noise to work together to bring the system over some energy barrier or go through a sequence of correlated changes. Chemical reactions, conformation changes of biomolecules, nucleation events and in some cases extreme events that lead to material failure or system failure are all examples of rare events.

It should be emphasized that, in a way, the rare events in which we are interested are not really unusual. For example, conformation changes of biomolecules are rare on the time scale of molecular vibration (which is typically of the order of femtoseconds, 10^{-15} s), but they are not rare on the time scale of our daily lives, which is often measured in minutes, hours or days. After all, all biological processes are driven by such events. See Figure 10.1.

An important concept is that of metastability. Roughly speaking, metastable systems are systems that are stable over a very long time scale compared with the relaxation time scale of the system but ultimately undergo some noise-induced transition. It is helpful to distinguish several different situations:

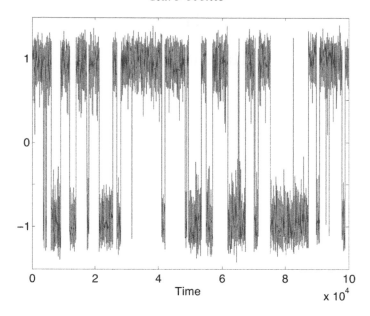

Figure 10.1. Time series from molecular dynamics simulation for the conformation changes of alanie dipeptide. Shown here is a time series for a dihedral angle.

(1) *Gradient and non-gradient systems*. A gradient system has an underlying potential energy and an associated energy landscape. The different stable states are the local minima of the energy. They are separated by barriers or sequences of barriers as well as other intermediate stable states. For this reason, transition events between these locally stable states are also referred to as *barrier-crossing events*. The dynamics of the system is such that it spends most of its time vibrating around different local energy minima, with occasional transitions between the different local minima. For a given transition, we call the local minimum before the transition the *initial state* and the local minimum after the transition the *final state* or the *end state*.

The mechanism of transition in non-gradient systems can be very different from that in gradient systems (see for example [64]). Here we will limit ourselves to gradient systems.

(2) *Entropic barriers*. Entropic barriers refer to cases where the energy landscape is rather flat but the system has to wander through a

very large set of configurations in order to reach the final state. Picture this: you are in a very large room with no lights, and you are supposed to find the small exit which is the only way of getting out of the room.

(3) *Simple or complex energy barriers* This notion is only a relative one. There are two important factors to consider. One is the length scale over which the energy landscape varies. The second is the size of the barriers compared with the thermal noise. If the energy landscape is rather smooth and the individual barriers are much larger than the typical amplitude of the noise (which is characterized by $k_B T$ in physical systems), we refer to it as a *system with a simple* or *smooth energy landscape*, or just a *simple system*. This is the case when the system has a few large barriers. As we will discuss below, in this case the relevant objects for the transition events are saddle points, which serve as the transition states, and minimum energy paths (MEPs), which are the most probable paths. The other extreme is the case when the landscape is rugged, with many saddle points and barriers, and most barriers are of a size comparable with the thermal noise. We refer to this class of systems as *systems with complex or rough energy landscapes*. In this situation the individual saddle points are sometimes not enough to act as an obstacle for the transition: the real obstacle to the transition is the combined effect of many individual barriers. There may not be a well-defined most probable path. Instead, many paths in the transition path ensemble contribute to the transition.

Figure 10.2 illustrates the situation with rough energy landscapes in a temperature regime where the thermal energy is comparable with the small-scale energy fluctuations in the landscape. The two blue regions are the metastable sets. One expects that transitions between the two regions should proceed by a wandering path through the light turquoise region, i.e. most transition paths between the two metastable sets are restricted to a "tube" in this region. One thing that we will do in this chapter is to make such statements more precise.

Below we will first discuss the theoretical background for describing transition between metastable states. We will then discuss numerical algorithms that have been developed to compute the relevant objects,

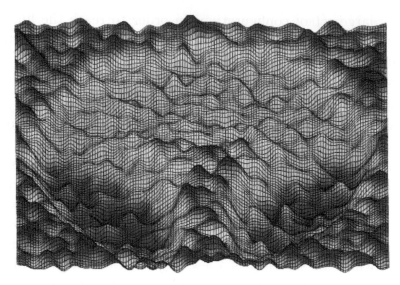

Figure 10.2. An example of a rough energy landscape.

e.g. transition states, transition pathways and transition tubes, for such transitions. These algorithms require one to know something about the final states of the transition. Finally, we discuss some examples of techniques that have been proposed for directly accelerating the dynamics. These techniques do not require one to know anything about the final states.

The setup is as follows. We are given two metastable states of the system, which are respectively the initial and final states of the transition. For a system with a smooth energy landscape, these metastable states are the local minima of the potential energy or sometimes the free energy. In chemistry, one typically refers to the initial state as the reactant state and the final state as the product state. For the same reason, transition events are also referred to as reactions.

Next, we introduce the kind of dynamics that we will consider, following [23]. From a physical viewpoint, it is of interest to study:

(1) *Hamiltonian dynamics*, with no explicit noise:

$$\begin{cases} \dot{\mathbf{x}} = M^{-1}\mathbf{p}, \\ \dot{\mathbf{p}} = -\nabla U, \end{cases} \qquad (10.1.1)$$

where M is the mass matrix; here we have used U (instead of V) to denote the potential energy of the system. In this case the role of noise is played by the chaotic dynamics of the system. For this reason we are not interested in integrable systems.

(2) *Langevin dynamics*:

$$\begin{cases} \dot{\mathbf{x}} = M^{-1}\mathbf{p}, \\ \dot{\mathbf{p}} = -\nabla U - \gamma\mathbf{p} + \sqrt{2\gamma k_B T}M^{1/2}\dot{\mathbf{w}}, \end{cases} \tag{10.1.2}$$

where γ is a friction coefficient, which can also be a tensor although we will only consider the case when it is a scalar and $\dot{\mathbf{w}}$ is the standard Gaussian white noise.

(3) *Overdamped dynamics*:

$$\gamma\dot{\mathbf{x}} = -\nabla V + \sqrt{2\gamma k_B T}\dot{\mathbf{w}}. \tag{10.1.3}$$

All three examples can be cast in the form

$$\dot{\mathbf{z}} = -\mathbf{K}\,\nabla V(\mathbf{z}) + \sqrt{2\varepsilon}\,\boldsymbol{\sigma}\,\dot{\mathbf{w}}(t), \tag{10.1.4}$$

where $\mathbf{K} \in \mathbb{R}^n \times \mathbb{R}^n$ and $\boldsymbol{\sigma} \in \mathbb{R}^n \times \mathbb{R}^m$ are matrices related by

$$\tfrac{1}{2}(\mathbf{K} + \mathbf{K}^{\mathrm{T}}) = \boldsymbol{\sigma}\boldsymbol{\sigma}^{\mathrm{T}} = \mathbf{a}. \tag{10.1.5}$$

For the Hamiltonian dynamics case, we have $\mathbf{z} = (\mathbf{x}, \mathbf{p})^{\mathrm{T}}$, $V(\mathbf{z}) = \tfrac{1}{2}\langle \mathbf{p}, M^{-1}\mathbf{p}\rangle + U(\mathbf{x})$, $\boldsymbol{\sigma} = \mathbf{0}$ and

$$\mathbf{K} = \begin{pmatrix} 0 & -I \\ I & 0 \end{pmatrix}. \tag{10.1.6}$$

For the Langevin dynamics, we have

$$\mathbf{K} = \begin{pmatrix} 0 & -I \\ I & \gamma M \end{pmatrix}, \qquad \boldsymbol{\sigma} = \begin{pmatrix} 0 & 0 \\ 0 & (\gamma M)^{1/2} \end{pmatrix}. \tag{10.1.7}$$

For the overdamped dynamics, we have $\mathbf{z} = \mathbf{x}$, $\mathbf{K} = \gamma^{-1}\mathbf{I}$ and $\boldsymbol{\sigma} = \gamma^{-1/2}\mathbf{I}$. In all three cases we have $\varepsilon = k_B T$. In the following we will focus on (10.1.4). We will assume that the potential energy function V is a Morse function, that is, V is twice differentiable and has nondegenerate critical points:

$$\det \mathbf{H}_j \neq 0, \tag{10.1.8}$$

where $\mathbf{H}_j = \nabla\nabla V(\mathbf{z}_j)$ is the Hessian of V at the critical point $\mathbf{z}_j$. It is relatively straightforward to check that the distribution given by

$$d\mu_s(\mathbf{z}) = Z^{-1}e^{-V(\mathbf{z})/\varepsilon}d\mathbf{z}, \quad \text{where } Z = \int_\Omega e^{-V(\mathbf{z})/\varepsilon}\, d\mathbf{z} \quad (10.1.9)$$

is invariant under the dynamics defined by (10.1.4). We will also assume that the dynamics is ergodic, i.e. the invariant distribution is unique.

10.2 Theoretical background

10.2.1 Metastable states and reduction to Markov chains

In the following, we will discuss *metastable states* or *metastable sets*. These are sets of configurations that are stable during a certain time scale, often some relaxation time. If initiated in that set, the system will remain in it during the specified time scale. This time scale can be the relaxation time of a potential well or of a set of neighboring wells. The important point is that the notion of metastability is one that is relative to certain reference time scales.

In a simple system, each local minimum of the potential is a metastable set if the reference time scale is taken to be the relaxation time of the potential well for that local minimum. The potential well, or the basin of attraction, is the set of points in the configuration space that are attracted to this particular local minimum under the gradient flow

$$\frac{d\mathbf{x}}{dt} = -\nabla V(\mathbf{x}), \quad\quad\quad (10.2.1)$$

where V is the potential of the system. The different basins of attraction are separated by separatrices. The separatrices themselves make up an invariant set for the gradient-flow dynamics (10.2.1). The stable equilibrium points on this invariant set are the saddle points of V.

For complex systems, metastable sets are less well defined since we must deal with a set of wells. But their intuitive meaning is quite clear. In practice, they can be defined with the help of some coarse-grained variables or order parameters. For example, for a molecule we can define metastable sets by restricting the values of a set of torsion angles.

A more precise definition of metastable sets can be found using the spectral theory of transfer operators or the generators associated with the underlying stochastic dynamics. See for example [42, 36]. Roughly speaking, metastability can be related to gaps in the spectrum of the relevant generators of the Markov process and metastable sets can be defined through the eigenfunctions associated with the leading eigenvalues. Indeed, it can be shown that these eigenfunctions are approximately piecewise constant and that the subsets on which the eigenfunctions are approximately constant are the metastable sets.

Going back to the global picture, our objective is not to keep track of the detailed dynamics of the system but rather to capture statistically the sequence of hops or transitions between different local minima or metastable states. This means that, effectively, the dynamics of the system is modeled by a Markov chain: the local minima or the metastable states are the states of the chain and the hopping rates are the transition rates between different metastable states. This idea has been used in two different ways.

(1) First, it has been used as the basis for accelerating molecular dynamics. One example is hyperdynamics, to be discussed below. Here, its purpose has been to capture not the detailed dynamics but the effective dynamics of the Markov chain.

(2) Second, it can be used to design more accurate kinetic Monte Carlo schemes. In standard kinetic Monte Carlo methods, the possible states and the transition rates are predetermined, often empirically. This can be avoided by finding the states and transition rates "on-the-fly" from a more detailed model. See for example [60].

10.2.2 Transition state theory

The classical theory that describes barrier-crossing events in gradient systems is transition state theory (TST), developed by Eyring, Polanyi, Wigner, Kramers, and others [31]. There are two main components in transition state theory: the notion of transition states and an expression for the transition rate. Transition states are the dynamic bottlenecks for the transition. They are important for the following reasons.

(1) With very high probability, transition paths have to pass through a very small neighborbood of the transition states. Exceptional paths have an exponentially small relative probability.

(2) Once the transition state is passed, the system can relax to the new stable state via its own intrinsic dynamics, on the relaxation time scale.

For simple systems, given two neighboring local minima the transition state is simply the saddle point that separates the two minima.

The transition rate, i.e. the average number of transitions per unit time interval, can be computed from the probability flux of particles that pass through the neighborhood of the transition state. As an example, let us consider the canonical ensemble of a Hamiltonian system. For simplicity, we will focus on the example of a one-dimensional system [31]. Let V be a one-dimensional potential with one local minimum at $x = -1$ and another at $x = 1$, separated by a saddle point at $x = 0$. We are interested in computing the transition rate from the local minimum at $x = -1$ to the local minimum at $x = 1$ in a canonical ensemble with temperature T. According to transition state theory, the transition rate is given by the ratio of two quantities: the first is the flux of particles that traverse the transition state region from the side of the negative real axis to the side of the positive real axis. The second is the equilibrium distribution of particles on the side of the negative real axis, which is the basin of attraction of the initial state at $x = -1$. This second quantity is given by

$$Z_A = \frac{1}{Z} \int_{x<0} e^{-\beta H(x,p)} \, dx \, dp,$$

where $H(x,p) = (1/2m)p^2 + V(x)$ is the Hamiltonian of the system, the normalization constant $Z = \int_{\mathbb{R}^2} e^{-\beta H(x,p)} \, dx \, dp$ and $\beta = (k_B T)^{-1}$. The first-mentioned quantity, the particle flux, is given by

$$Z_{AB} = \frac{1}{Zm} \int_{x=0} p\theta(p)e^{-\beta H(x,p)} \, dp$$

where θ is the Heaviside function, $\theta(z) = 1$ if $z > 0$ and $\theta(z) = 0$ if $z < 0$. To leading order we have

$$Z_{AB} \sim \frac{1}{Z\beta} e^{-\beta V(0)}.$$

If we approximate V further by a quadratic function centered at $x = -1$, i.e. by $V(x) \sim V(-1) + (1/2)V''(-1)(x+1)^2$, we obtain

$$Z_A \sim \frac{1}{Z} \frac{2\pi}{\beta} \sqrt{\frac{m}{V''(-1)}} e^{-\beta V(-1)}.$$

The ratio of the two quantities gives the required transition rate:

$$\nu_{\mathrm{R}} = \frac{1}{2\pi}\sqrt{\frac{V''(-1)}{m}}e^{-\beta\delta E}, \qquad (10.2.2)$$

where $\delta E = V(0) - V(-1)$ is the energy barrier.

Note that this rate depends on the mass of the particle and the second derivative of the potential at the initial state, but it does not depend on the second derivative of the potential at the saddle point.

Equation (10.2.2) can obviously be extended to higher dimensions [31].

Transition state theory has been extended in a number of directions. In the most important generalization, one replaces the region near the saddle point by a dividing surface; see for example [63]. In this case, one can still compute the probability flux through the surface and use it as an estimate for the transition rate. Obviously this is only an upper bound for the transition rate, since it does not take into account the fact that particles can re-cross the dividing surface and return to the reactant state. The correct rate is only obtained when the success rate, often called the transmission coefficient, is taken into account [26]. This is conceptually appealing. But, in practice, it is often difficult to obtain the transmission coefficient.

Some of the more mathematically oriented aspects of transition state theory are discussed in [54].

10.2.3 Large-deviation theory

Large-deviation theory is a rigorous mathematical theory for characterizing rare events in general Markov processes [57, 27]. The special case of random perturbations of dynamical systems was considered by Freidlin and Wentzell [27]. This theory is often referred to as the Wentzell–Freidlin theory.

Consider the stochastically perturbed dynamical system

$$\dot{\mathbf{x}} = \mathbf{b}(\mathbf{x}) + \sqrt{\varepsilon}\dot{\mathbf{w}}, \quad \mathbf{x} \in \mathbb{R}^d. \qquad (10.2.3)$$

Fix two points with position vectors $\mathbf{A}$ and $\mathbf{B}$ in $\mathbb{R}^d$ and a time parameter $T > 0$. Let $\boldsymbol{\phi} : [-T, T] \to \mathbb{R}^d$ be a continuous and differentiable path such that $\boldsymbol{\phi}(-T) = \mathbf{A}$, $\boldsymbol{\phi}(T) = \mathbf{B}$. Wentzell–Freidlin theory asserts, roughly speaking, that

Prob{$\mathbf{x}$ stays in a neighborhood of $\boldsymbol{\phi}$ over the interval $[-T, T]$}
$$\sim e^{-S(\boldsymbol{\phi})/\varepsilon}, \tag{10.2.4}$$

where the *action functional* S is defined by

$$S_T(\boldsymbol{\phi}) = \tfrac{1}{2} \int_{-T}^{T} |\dot{\boldsymbol{\phi}}(t) - \mathbf{b}(\boldsymbol{\phi}(t))|^2 \, dt. \tag{10.2.5}$$

A precise statement and its proof can be found in [27]. For our purpose, it is more instructive to understand intuitively (and formally) the origin of this result. From (10.2.3), we have

$$\dot{\mathbf{w}} = (\dot{\mathbf{x}} - \mathbf{b}(\mathbf{x}))/\sqrt{\varepsilon}.$$

thus the probability density for the white noise is approximately $\exp\left(-\tfrac{1}{2}\int |\dot{\mathbf{w}}|^2 \, dt\right)$. This gives (10.2.4) and (10.2.5).

In view of (10.2.4), finding the path with maximum probability becomes a problem of finding the path with minimum action:

$$\inf_{T} \inf_{\phi} S_T(\boldsymbol{\phi}) \tag{10.2.6}$$

subject to the constraint that $\boldsymbol{\phi}(-T) = \mathbf{A}$, $\boldsymbol{\phi}(T) = \mathbf{B}$. Consider the case when $\mathbf{b}(\mathbf{x}) = -\nabla V(\mathbf{x})$. Assume that $\mathbf{A}$ and $\mathbf{B}$ are two neighboring local minima of V separated by the saddle point $\mathbf{C}$. Then we have the following result.

Lemma (Wentzell–Freidlin [27])

(1) *The minimum of* (10.2.6) *is given by*

$$S^* = 2(V(\mathbf{C}) - V(\mathbf{A})).$$

(2) *Consider the paths*

$$\dot{\boldsymbol{\phi}}_1(s) = \nabla V(\boldsymbol{\phi}_1(s)), \quad \boldsymbol{\phi}_1(-\infty) = \mathbf{A}, \quad \boldsymbol{\phi}_1(\infty) = \mathbf{C},$$
$$\dot{\boldsymbol{\phi}}_2(s) = -\nabla V(\boldsymbol{\phi}_2(s)), \quad \boldsymbol{\phi}_2(-\infty) = \mathbf{C}, \quad \boldsymbol{\phi}_2(\infty) = \mathbf{B};$$

then

$$S^* = S_\infty(\boldsymbol{\phi}_1) + S_\infty(\boldsymbol{\phi}_2) = S_\infty(\boldsymbol{\phi}_1).$$

The most probable transition path S^* is thus a combination of $\boldsymbol{\phi}_1$ and $\boldsymbol{\phi}_2$: while $\boldsymbol{\phi}_1$ goes against the original dynamics and therefore requires the action of the noise, $\boldsymbol{\phi}_2$ simply follows the original dynamics and therefore does not require any help from the noise.

It is not difficult to convince oneself that the minimum in T in (10.2.6) is attained when $T = \infty$. To see why the minimization problem in (10.2.6) is solved by the path defined above, note that

$$S_\infty[\boldsymbol{\phi}_1] = 2(V(\mathbf{C}) - V(\mathbf{A})), \quad S_\infty[\boldsymbol{\phi}_2] = 0. \tag{10.2.7}$$

In addition, for any path $\boldsymbol{\phi}$ connecting $\mathbf{A}$ and a point on the separatrix that separates the basins of attraction of $\mathbf{A}$ and $\mathbf{B}$, we have

$$S_\infty[\boldsymbol{\phi}] = \tfrac{1}{2} \int_{\mathbb{R}} \langle \dot{\boldsymbol{\phi}} - \nabla V, (\dot{\boldsymbol{\phi}} - \nabla V) \rangle \, dt + 2 \int_{\mathbb{R}} \dot{\boldsymbol{\phi}} \cdot \nabla V \, dt$$

$$\geq 2 \int_{\mathbb{R}} \dot{\boldsymbol{\phi}} \cdot \nabla V \, dt$$

$$= 2 \int_{\mathbb{R}} \dot{V} \, dt$$

$$\geq 2(V(\mathbf{C}) - V(\mathbf{A})),$$

since $\mathbf{C}$ is the minimum of V on the separatrix.

This result can also be generalized to the case when there are intermediate stable states between $\mathbf{A}$ and $\mathbf{B}$. In that case the most probable transition path is a combination of paths that satisfy

$$\dot{\boldsymbol{\phi}}(s) = \pm \nabla V(\boldsymbol{\phi}(s)) \tag{10.2.8}$$

Paths that satisfy this equation are called *minimum energy paths* (MEPs). One can write (10.2.8) as

$$\nabla V(\boldsymbol{\phi}(s))^\perp = 0, \tag{10.2.9}$$

where $\nabla V(\boldsymbol{\phi}(s))^\perp$ denotes the component of $\nabla V(\boldsymbol{\phi}(s))$ normal to the curve described by $\boldsymbol{\phi}$.

An alternative characterization of MEP is as follows. Given a path $\boldsymbol{\gamma}$, let $\mathbf{z}$ be an arbitrary point on $\boldsymbol{\gamma}$, and $P_{\mathbf{z}}$ be the hyperplane that contains $\mathbf{z}$ and is normal to $\boldsymbol{\gamma}$. Then $\boldsymbol{\gamma}$ is a MEP if, for any point $\mathbf{z}$ on $\boldsymbol{\gamma}$, the point $\mathbf{z}$ is a local minimum of V restricted to $P_{\mathbf{z}}$. As we will see later in this section, this characterization can be directly linked with the transition tubes or principal curves obtained from transition path theory.

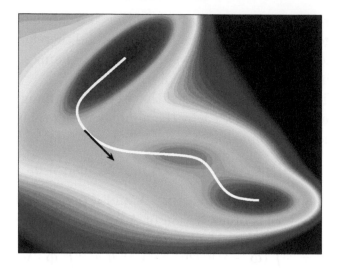

Figure 10.3. Mueller potential and a minimum energy path that connects two local minima.

Example 10.1 A frequently used example is the *Mueller potential*:

$$V(x, y) = \sum_{i=1}^{4} A_i \exp\left(a_i(x - x_i)^2 + b_i(x - x_i)(y - y_i) + c_i(y - y_i)^2\right),$$

$$(10.2.10)$$

where the parameters are as follows:

$$\{A_i\} = (-200, -100, -170, 15), \quad \{a\} = (-1, -1, -6.5, 0.7),$$
$$\{b_i\} = (0, 0, 11, 0.6), \quad\quad\quad\quad \{c_i\} = (-10, -10, -6.5, 0.7),$$
$$\{x_i\} = (1, 0, -0.5, -1), \quad\quad\quad\; \{y_i\} = (0, 0.5, 1.5, 1).$$

There are three local minima in this example. Shown in Figure 10.3 is the minimum energy path that connects two of the local minima. This minimum energy path passes through the third local minimum. In other words, the third local minimum is an intermediate state in the transition between the two end points.

Turning now to the case of a more general dynamics, the main conclusions are as follows. Consider (10.1.4). Define the *optimal exit path* to be a path $\boldsymbol{\phi}^*$ that satisfies

$$\nabla V(\boldsymbol{\phi}^*) - (\nabla V(\boldsymbol{\phi}^*) \cdot \hat{\mathbf{s}})\hat{\mathbf{s}} = 0, \qquad (10.2.11)$$

where

$$\hat{\mathbf{s}} = \mathbf{K}^{-T}\hat{\mathbf{t}}^* / |\mathbf{K}^{-T}\hat{\mathbf{t}}^*| \qquad (10.2.12)$$

and $\hat{\mathbf{t}}^*$ is the unit tangent vector along the curve $\boldsymbol{\phi}^*$. Then

$$S_\infty[\boldsymbol{\phi}^*] = \inf_\phi S_\infty[\boldsymbol{\phi}], \qquad (10.2.13)$$

where the infimum is taken over all paths that satisfy

$$\lim_{t\to-\infty} \boldsymbol{\phi}(t) = \mathbf{z}_0, \qquad \lim_{t\to+\infty} \boldsymbol{\phi}(t) \in \Gamma, \qquad (10.2.14)$$

where Γ is the boundary of the basin of attraction of $\mathbf{z}_0$.

In addition, we have

$$\tau \sim e^{S_\infty[\boldsymbol{\phi}^*]/\varepsilon}. \qquad (10.2.15)$$

This also gives us the transition rate: $\nu_R = \tau^{-1}$.

Similarly to how we defined the optimal exit path, we can also define the *optimal switching path*. Let $\mathbf{z}_1$ and $\mathbf{z}_2$ be two neighboring local minima of V, with basins of attraction B_1 and B_2 respectively. The optimal switching path that goes from $\mathbf{z}_1$ to $\mathbf{z}_2$ satisfies (10.2.11) with

$$\begin{aligned}
\hat{\mathbf{s}} &= \mathbf{K}^{-T}\hat{\mathbf{t}}^*/|\mathbf{K}^{-T}\hat{\mathbf{t}}^*| \quad \text{in } B_1, \\
\hat{\mathbf{s}} &= \mathbf{K}^{-1}\hat{\mathbf{t}}^*/|\mathbf{K}^{-1}\hat{\mathbf{t}}^*| \quad \text{in } B_2.
\end{aligned} \qquad (10.2.16)$$

Note that the optimal switching path coincides with the minimum energy path in the case of overdamped dynamics. But, in general, they are different. In particular the optimal switching path from $\mathbf{z}_1$ to $\mathbf{z}_2$ and that from $\mathbf{z}_2$ to $\mathbf{z}_1$ are in general different. In addition, the optimal switching paths are usually not smooth: they have corners at the saddle points as a result of the change in the definition of $\hat{\mathbf{s}}$ after the separatrix is crossed.

These results can be found in [23].

10.2.4 First-exit times

An alternative viewpoint is to use the Kolmogorov type of equation for the mean first exit time of a domain, say the basin of attraction of a local minimum.

Fix a local minimum $\mathbf{z}_0$ of V. Denote by B its basin of attraction and by S the boundary of B. Let $\tau(\mathbf{z})$ be the mean exit time from B of a trajectory initiated at $\mathbf{z} \in B$:

$$\tau(\mathbf{z}) = \mathbb{E}\min\{t : \mathbf{z}(0) = \mathbf{z},\ \mathbf{z}(t) \in \Gamma,\ \mathbf{z}(s) \in B \text{ if } s < t\}. \qquad (10.2.17)$$

Here τ satisfies the backward Kolmogorov equation [29]

$$-\mathbf{K}\nabla V \cdot \nabla \tau + \varepsilon \mathbf{a} : \nabla\nabla\tau = -1, \qquad (10.2.18)$$

where **a** was defined in (10.1.5), with boundary conditions

$$\tau|_S = 0, \quad \hat{\mathbf{n}} \cdot \nabla\tau|_{\partial B \setminus S} = 0; \qquad (10.2.19)$$

$\hat{\mathbf{n}}$ is the unit outward normal to ∂B. We will be interested in the asymptotic behavior of τ as $\varepsilon \to 0$.

Before discussing the general case, let us consider the overdamped dynamics (10.1.3) in the one-dimensional case with, say, the same potential as that discussed in subsection 10.2.2. We are interested in the mean first-passage time from $x = -1$ to $x = \delta$, where δ is a fixed but small positive number.

For $y \leq \delta$, let $\tau(y)$ be the mean first passage time from y to $x = \delta$; it satisfies the differential equation

$$-V'(y)\tau'(y) + \varepsilon\tau''(y) = -1$$

with boundary conditions $\tau(\delta) = 0$, $\tau(-\infty) = \infty$. The solution to this problem is given simply by

$$\tau(y) = \frac{1}{\varepsilon} \int_y^\delta e^{V(z)/\varepsilon}\, dz \int_{-\infty}^z e^{-V(w)/\varepsilon}\, dw.$$

This can be approximated by

$$\tau(y) \sim \frac{1}{\varepsilon} \int_y^\delta e^{V(z)/\varepsilon}\, dz \int_{-\infty}^0 e^{-V(w)/\varepsilon}\, dw.$$

The main contribution to the first integral in the product comes from the region near $x = 0$ and it can be evaluated asymptotically by Taylor-expanding V near $x = 0$. This gives

$$\int_y^\delta e^{V(z)/\varepsilon}\, dz \sim \sqrt{\frac{2\pi\varepsilon}{-V''(0)}} e^{V(0)/\varepsilon}.$$

The main contribution to the second integral comes from the region near $x = -1$ and it can be evaluated by Taylor-expanding V around $x = -1$. This gives

$$\int_{-\infty}^0 e^{-V(w)/\varepsilon}\, dw \sim \sqrt{\frac{2\pi\varepsilon}{V''(-1)}} e^{-V(-1)/\varepsilon}.$$

Therefore, we have

$$\tau(y) \sim 2\pi \left(|V''(0)||V''(-1)|\right)^{-1/2} e^{(V(0)-V(-1))/\varepsilon}.$$

The transition rate in this case is found to be

$$\nu_R = \frac{1}{2\pi} \left(|V''(0)||V''(-1)|\right)^{1/2} e^{-\delta E/\varepsilon}, \qquad (10.2.20)$$

where $\delta E = V(0) - V(-1)$. Note that, in this overdamped limit, the second derivative at the saddle point comes into play (see (10.2.2)).

We now come to the high-dimensional case, for which we will derive similar results using matched asymptotics on the Kolmogorov equation (10.2.18). Equation (10.2.15) suggests trying an ansatz of the form

$$\tau(\mathbf{z}) = e^{\delta V/\varepsilon} \bar{\tau}(\mathbf{z}), \qquad (10.2.21)$$

where $\bar{\tau}(\mathbf{z})$ depends on ε only algebraically. Inserting (10.2.21) into (10.2.18), we arrive at

$$-e^{-\delta V/\varepsilon} = -\mathbf{K}\nabla V \cdot \nabla\bar{\tau} + \varepsilon \mathbf{a} : \nabla\nabla\bar{\tau}, \qquad (10.2.22)$$

subject to the boundary conditions

$$\bar{\tau}|_S = 0, \quad \hat{\mathbf{n}} \cdot \nabla\bar{\tau}\big|_{\partial B \backslash S} = 0. \qquad (10.2.23)$$

Outer expansion Let

$$\bar{\tau}(\mathbf{z}) = \bar{\tau}_0(\mathbf{z}) + \varepsilon\bar{\tau}_1(\mathbf{z}) + \cdots . \qquad (10.2.24)$$

Substituting into (10.2.22), we obtain, to leading order,

$$-\mathbf{K}\nabla V \cdot \nabla\bar{\tau}_0 = 0. \qquad (10.2.25)$$

This equation says that $\bar{\tau}_0(\mathbf{z})$ is constant along the flow lines of $\mathbf{K}\nabla V$, which, by definition, all lead to $\mathbf{z}_0$ in B. This means that $\bar{\tau}_0(\mathbf{z})$ is constant in B, i.e.

$$\bar{\tau}_0(\mathbf{z}) = C \qquad (10.2.26)$$

for all $\mathbf{z} \in B$, where $C > 0$ is a constant to be determined.

Inner expansion Equation (10.2.26) is inconsistent with the first boundary condition in (10.2.23), since C should not vanish. Therefore, there must be a boundary layer near Γ.

Near Γ we introduce a local coordinate system (ζ, η) where ζ is a set of curvilinear coordinates of Γ and

$$\eta = \frac{\text{dist}(\zeta, S)}{\sqrt{\varepsilon}} \qquad (10.2.27)$$

is the stretched coordinate inside the boundary layer (in the direction normal to Γ). Without loss of generality, we can assume that $(\eta, \zeta) = (0, 0)$ corresponds to the location of the saddle point $\mathbf{z}_s$ on Γ. Assume that, inside the boundary layer,

$$\bar{\tau}(\mathbf{z}) = \hat{\tau}_0(\eta, \zeta) + \varepsilon \hat{\tau}_1(\eta, \zeta) + \cdots . \qquad (10.2.28)$$

Substituting into (10.2.22), we obtain, to leading order,

$$b(\zeta)\eta \frac{\partial \hat{\tau}_0}{\partial \eta} + \hat{a}(\zeta) \frac{\partial^2 \hat{\tau}_0}{\partial \eta^2} - \mathbf{c}(\zeta) \cdot \nabla_\zeta \hat{\tau}_0 = 0. \qquad (10.2.29)$$

Here $\mathbf{c}(\zeta)$ is equal to $\mathbf{K}\nabla V$ evaluated at $(\zeta, 0)$, and

$$b(\zeta) = -\langle \hat{\mathbf{n}}(\zeta), \mathbf{K}\bar{\mathbf{H}}(\zeta)\hat{\mathbf{n}}(\zeta) \rangle,$$
$$\hat{a}(\zeta) = \langle \hat{\mathbf{n}}(\zeta), \mathbf{a}\hat{\mathbf{n}}(\zeta) \rangle > 0, \qquad (10.2.30)$$

where $\hat{\mathbf{n}}(\zeta)$ is the unit normal to Γ at ζ and $\bar{\mathbf{H}}(\zeta)$ is the Hessian of V evaluated at $(\zeta, 0)$. From (10.2.23) we know that (10.2.29) must be solved with the boundary condition

$$\hat{\tau}_0(0, \zeta) = 0. \qquad (10.2.31)$$

In addition, the matching condition with (10.2.27) requires that

$$\lim_{\eta \to \infty} \hat{\tau}_0(\eta, \zeta) = C. \qquad (10.2.32)$$

We will look for a solution of (10.2.29) in the form of

$$\hat{\tau}_0(\eta, \zeta) = g(\eta d(\zeta)) \qquad (10.2.33)$$

for functions g and d, $d(\zeta) > 0$, to be determined later. Letting $u = \eta d(\zeta)$, (10.2.29) implies that

$$\bar{b}(\zeta)u \frac{dg}{du} + \hat{a}(\zeta)d^2(\zeta) \frac{d^2 g}{du^2} = 0, \qquad (10.2.34)$$

where

$$\bar{b}(\zeta) = b(\zeta) - \mathbf{c}(\zeta) \cdot \nabla_\zeta d(\zeta). \tag{10.2.35}$$

Equation (10.2.34) can be solved provided that $\bar{b}(\zeta) = \hat{a}(\zeta)d^2(\zeta)$, which, from (10.2.35), leads to the following equation for $d(\zeta)$:

$$b(\zeta) - \mathbf{c}(\zeta) \cdot \nabla_\zeta d(\zeta) = \hat{a}(\zeta) \, d^2(\zeta). \tag{10.2.36}$$

We only need information about $d(\zeta)$ near $\zeta = 0$. Since $\mathbf{K}\nabla V = \mathbf{0}$ at the saddle point $\mathbf{z}_s$, $c(0) = 0$, from (10.2.36) it follows that

$$d(0) = \sqrt{\frac{b(0)}{\hat{a}(0)}}. \tag{10.2.37}$$

In addition, g must satisfy the following boundary conditions:

$$g(0) = 0, \quad \lim_{u \to \infty} g(u) = C. \tag{10.2.38}$$

Therefore we have

$$g(u) = \frac{2C}{\sqrt{\pi}} \int_0^u e^{-u'^2} \, du'. \tag{10.2.39}$$

In terms of $\hat{\tau}_0$ this becomes

$$\hat{\tau}_0(\eta, \zeta) = \frac{2Cd(\zeta)}{\sqrt{\pi}} \int_0^\eta e^{-(d(\zeta)\eta')^2} \, d\eta'. \tag{10.2.40}$$

Solvability condition It remains to determine the constant C in the expressions (10.2.26) and (10.2.40). We do so using the following solvability condition, obtained by integrating both sides of (10.2.22) against the equilibrium distribution (10.1.9) in B:

$$-e^{-\delta V/\varepsilon} \int_B e^{-V(\mathbf{z})/\varepsilon} d\mathbf{z} = \int_B (-\mathbf{K}\nabla V \cdot \nabla\bar{\tau} + \varepsilon \mathbf{a} \nabla\nabla\bar{\tau}) e^{-V(\mathbf{z})/\varepsilon} \, d\mathbf{z}$$

$$= \varepsilon \int_B \nabla \cdot \left(\mathbf{K}^{\mathrm{T}} \nabla\bar{\tau} \, e^{-V(\mathbf{z})/\varepsilon} \right) d\mathbf{z}$$

$$= -\varepsilon \int_S \hat{\mathbf{n}}(\mathbf{z}) \cdot \mathbf{K}^{\mathrm{T}} \nabla\bar{\tau} \, e^{-V(\mathbf{z})/\varepsilon} \, dS, \tag{10.2.41}$$

where $\hat{\mathbf{n}}(\mathbf{z})$ is the inward unit normal to Γ. The rest of ∂B makes no contribution because of the second boundary condition in (10.2.23). Let us evaluate the various integrals in (10.2.41) in the limit $\varepsilon \to 0$. First we have

$$\int_B e^{-V(\mathbf{z})/\varepsilon}\, d\mathbf{z} = e^{-V(\mathbf{z})/\varepsilon} \left((2\pi\varepsilon)^{n/2} (\det \mathbf{H})^{-1/2} + o(1) \right) \quad (10.2.42)$$

where $\mathbf{H}$ is the Hessian of V evaluated at $\mathbf{z}_0$ and $o(1)$ represents quantities that are vanishingly small as $\varepsilon \to 0$. Next, using (10.2.40) we have

$$\varepsilon \int_S \hat{\mathbf{n}}(\mathbf{z}) \cdot \mathbf{K}^T \nabla \bar{\tau}\, e^{-V(\mathbf{z})/\varepsilon}\, dS$$

$$= \sqrt{\varepsilon} \int_S \left(\hat{a}(\zeta) \frac{\partial \hat{\tau}_0(0,\zeta)}{\partial \eta} + o(1) \right) e^{-\hat{V}(0,\zeta)/\varepsilon}\, d\zeta$$

$$= \sqrt{\varepsilon} \int_S \left(\frac{2C\hat{a}(\zeta)d(\zeta)}{\sqrt{\pi}} + o(1) \right) e^{-\hat{V}(\zeta)/\varepsilon}\, d\zeta, \quad (10.2.43)$$

where $\hat{V}(\zeta)$ denotes the potential evaluated on Γ and expressed in terms of the variable ζ. The final integral in (10.2.43) can be evaluated by Laplace's method and, since $\hat{V}(\zeta)$ attains its minimum at $\zeta = 0$ (corresponding to $\mathbf{z} = \mathbf{z}_s$), this gives

$$\sqrt{\varepsilon} \int_S \left(\frac{2C\hat{a}(\zeta)d(\zeta)}{\sqrt{\pi}} + o(1) \right) e^{-\hat{V}(\zeta)/\varepsilon}\, d\zeta$$

$$= e^{-V(\mathbf{z}_s)/\varepsilon} \left(\frac{\sqrt{2}}{\pi} (2\pi\varepsilon)^{n/2} C\hat{a}(0)d(0)D^{-1/2} + o(1) \right) \quad (10.2.44)$$

where D is the determinant of the Hessian of V evaluated at $\mathbf{z}_s$ after it is restricted to Γ: if $\{\hat{\mathbf{e}}_1, \ldots, \hat{\mathbf{e}}_{n-1}\}$ is an orthonormal basis spanning the hyperplane tangent to Γ at $\mathbf{z}_s$, and $\mathbf{P}$ is the $(n-1) \times n$ tensor whose rows are the $\hat{\mathbf{e}}_j$, then

$$D = \det(\mathbf{P}\mathbf{H}_s\mathbf{P}^T), \quad (10.2.45)$$

where $\mathbf{H}_s$ is the Hessian of V evaluated at $\mathbf{z}_s$. Inserting (10.2.42) and (10.2.44) into (10.2.41) gives an equation for C whose solution is

$$C = \pi \left(\det \mathbf{H} D^{-1} \right)^{-1/2} (\hat{a}(0)d(0))^{-1}$$
$$= \pi \left(\det \mathbf{H} D^{-1} \right)^{-1/2} (|\langle \hat{\mathbf{n}}(0), \mathbf{K}\mathbf{H}_s\hat{\mathbf{n}}(0)\rangle| \langle \hat{\mathbf{n}}(0), a\hat{\mathbf{n}}(0)\rangle)^{-1/2} \quad (10.2.46)$$

where, to get the second equality, we have used (10.2.30) and (10.2.37). This also means that, as $\varepsilon \to 0$,

$$\tau = e^{\delta V/\varepsilon} (C + o(1)). \quad (10.2.47)$$

In the special case of overdamped dynamics, we have $\mathbf{K} = \mathbf{a} = \gamma^{-1}\mathbf{I}$. Therefore (10.2.47) becomes

$$\tau = e^{\delta V/\varepsilon}\left(\pi\gamma\left(\det \mathbf{HD}^{-1}\right)^{-1/2}\left(|\langle\hat{\mathbf{n}}(0), \mathbf{H}_s\hat{\mathbf{n}}(0)\rangle|\langle\hat{\mathbf{n}}(0), \hat{\mathbf{n}}(0)\rangle\right)^{-1/2} + o(1)\right).$$
(10.2.48)

When $d = 1$, we have that $H = V''(\mathbf{A})$, $D = 1$ and $H_s = V''(\mathbf{C})$ and we recover (10.2.20).

10.2.5 Transition path theory

The previous discussions were suited to situations when the system has a smooth energy landscape. We used the ratio of the amplitude of the noise and the barrier height as a small parameter to carry out the asymptotic analysis. Next, we will turn to the situation when the energy landscape has a multiscale structure, i.e. systems with rough energy landscapes (see Figure 10.2). In particular, we have in mind the following situations.

(1) The potential of the system V has a large number of local minima and saddle points, but most of the associated barrier heights δV are comparable with the amplitude of the noise $k_B T$ and therefore, individually, these saddle points do not act as barriers. The real barrier comes from the collective effect of many such saddle points. In this case the notion of transition states has to be replaced, since the bottleneck to the transition is no longer associated with a few isolated saddle points of V.

(2) For the same reason, it no longer makes sense to discuss the most probable path since this is not well defined. Instead one should think about an ensemble of transition paths.

(3) Entropic effects become important. Imagine a situation in which the transition path ensemble is concentrated in two thin tubes connecting the initial and final states, one with a smaller barrier but narrower passage, another with a larger barrier but wider passage. One can imagine that at lower temperatures, the first scenario is more favorable to transitions. At a higher temperature, the second scenario is more favorable. There are also situations where the barrier is of entirely an entropic nature, as mentioned earlier.

From an analytical viewpoint, the main difficulty is the lack of a small parameter that can be used to carry out the asymptotics. Therefore we will proceed in two steps. The first is to formulate an exact theory for the general case, in which we define the relevant objects and derive the relevant formulas for the quantities in which we are interested. This is the purpose of transition path theory (TPT), which was developed to provide an analytical characterization of the transition path ensemble [17]. This theory is quite general and it applies also to non-gradient systems. The second step is to find alternative small parameters with which we can perform asymptotics. As we see below, the small parameter with which we will work is the nondimensionalized width of the effective transition tube.

We will summarize the main results in [17]. Details of the derivations can be found in [17, 16, 55, 18]. Some illustrative examples are presented in [43]. Extensions to discrete Markov processes can be found in [44]. See also the review article [18].

The general setup is as follows. Consider the stochastic process described by the differential equation

$$\dot{\mathbf{x}} = \mathbf{b}(\mathbf{x}) + \sqrt{2}\boldsymbol{\sigma}(\mathbf{x})\dot{\mathbf{w}} \qquad (10.2.49)$$

where $\dot{\mathbf{w}}$ is the standard Gaussian white noise. The generator of this process is given by (see [29])

$$L = \mathbf{b}(\mathbf{x}) \cdot \nabla + \mathbf{a}(\mathbf{x}) : \nabla\nabla = \sum_j b_j(\mathbf{x})\frac{\partial}{\partial x_j} + \sum_{j,k} a_{j,k}(\mathbf{x})\frac{\partial^2}{\partial x_j \partial x_k}$$

$$(10.2.50)$$

where

$$\mathbf{a}(\mathbf{x}) = \boldsymbol{\sigma}(\mathbf{x})\boldsymbol{\sigma}^{\mathrm{T}}(\mathbf{x}).$$

The operator L and its adjoint define the backward and forward Kolmogorov equations associated with the process (10.2.49). We will assume that the stochastic process defined by (10.2.49) has a unique invariant distribution, i.e. the process is ergodic. We will denote this invariant distribution by $\mu(d\mathbf{x}) = m(\mathbf{x})\,d\mathbf{x}$, where m is the density of μ.

Given two metastable sets for the stochastic process, A and B, we are interested in the transition paths from A to B. These are paths that are initiated in A and arrive at B without going back to A in between. One way of defining this ensemble of paths is to consider an infinitely long trajectory $\mathbf{x}(\cdot) : (-\infty, +\infty) \to \mathbb{R}^n$ and to cut out from this trajectory the pieces that satisfy the condition stated above, namely the pieces that leave the boundary of A and arrive at B without going back to A in between. Such a piece is called a *reactive trajectory from A to B*, or an *A–B reactive path* or a *transition path from A to B*. We will call this set of transition paths the *transition path ensemble from A to B*. Given the trajectory $\mathbf{x}(\cdot)$, we will denote by R the subset of times t in which $\mathbf{x}(\cdot)$ belongs to an A–B reactive path.

For this transition path ensemble, we ask the following questions.

(1) Where do these paths spend their time? In the setup described above we are looking for a density m_R such that

$$\lim_{T \to \infty} \frac{1}{2T} \int_{[-T,T]} \delta(\mathbf{x} - \mathbf{x}(t)) \mathbf{1}_R(t) dt = m_R(\mathbf{x}).$$

Here $\mathbf{1}_R$ is the indicator function of the set $R : \mathbf{1}_R(t) = 1$ if $t \in R$ and 0 otherwise. The presence of the factor $\mathbf{1}_R(t)$ indicates that we are only interested in reactive trajectories. Note that if we consider the whole trajectory $\mathbf{x}(\cdot)$ (not just the reactive pieces) and ask the same question then the probability density on the right-hand side should be given by the density for the invariant distribution $m(\mathbf{x})$:

$$\lim_{T \to \infty} \frac{1}{2T} \int_{[-T,T]} \delta(\mathbf{x} - \mathbf{x}(t)) \, dt = m(\mathbf{x}).$$

Note also that $m_R(\mathbf{x})$ is not normalized: in fact, the total integral of m_R gives the probability that a piece of the total trajectory is A–B reactive.

(2) What is the current associated with these reactive trajectories? The current associated with the reactive paths is a vector field $\mathbf{J}_R : \Omega \mapsto \mathbb{R}^n$, $\Omega = \mathbb{R}^n \backslash (A \cup B)$, such that the following holds:

$$\lim_{T \to \infty} \frac{1}{2T} \int_{[-T,T]} \dot{\mathbf{x}}(t) \delta(\mathbf{x} - \mathbf{x}(t)) \mathbf{1}_R(t) dt = \mathbf{J}_R(\mathbf{x}). \qquad (10.2.51)$$

Note that if we consider the whole trajectory $\mathbf{x}(\cdot)$ (not just the reactive pieces), then we have

$$\lim_{T \to \infty} \frac{1}{2T} \int_{[-T,T]} \dot{\mathbf{x}}(t) \delta(\mathbf{x} - \mathbf{x}(t)) \, dt = \mathbf{J}(\mathbf{x})$$

where $\mathbf{J}(\mathbf{x}) = (J_1(\mathbf{x}), \ldots, J_n(\mathbf{x}))^T \in \mathbb{R}^n$ is the "equilibrium current" (or more precisely the steady state current), with

$$J_i(\mathbf{x}) = b_i(\mathbf{x})m(\mathbf{x}) - \sum_{j=1}^{n} \frac{\partial}{\partial x_j}(a_{ij}(\mathbf{x})m(\mathbf{x})). \qquad (10.2.52)$$

This is obtained from the forward Kolmogorov equation, which can be written as

$$\partial_t \rho(\mathbf{x}, t) + \nabla \cdot (\mathbf{b}(\mathbf{x})\rho(\mathbf{x}, t) - \nabla(\mathbf{a}(\mathbf{x})\rho(\mathbf{x}, t))) = 0.$$

The equilibrium current $\mathbf{J}$ is divergence-free, but it does not necessarily vanish for non-gradient systems since the detailed balance condition may not hold.

(3) Transition rate: if N_T is the number of A–B reactive trajectories in a given trajectory $\mathbf{x}(\cdot)$ in $[-T, T]$, what is the limit of $N_T/(2T)$ as $T \to \infty$? We write

$$k_{AB} = \lim_{T \to \infty} \frac{N_T}{2T}.$$

To answer these questions, we need the notion of *committor functions* [15], also known as *capacitance functions* in the mathematics literature.

Let $q_+ : \Omega \mapsto [0, 1]$ be the solution of

$$Lq_+ := \mathbf{b} \cdot \nabla q_+ + \mathbf{a} : \nabla\nabla q_+ = 0, \qquad q_+|_{\partial A} = 0, \quad q_+|_{\partial B} = 1;$$

we call q_+ the *forward committor function* for the transition process from A to B. In a similar fashion, we can define the *backward committor function* $q_- : \Omega \mapsto [0, 1]$ as the solution of

$$L^+ q_- := -\mathbf{b} \cdot \nabla q_- + \frac{2}{m}(\nabla \cdot \mathbf{a}m) \cdot \nabla q_- + \mathbf{a} : \nabla\nabla q_- = 0,$$
$$q_-|_{\partial A} = 1, \quad q_-|_{\partial B} = 0.$$

The functions q_+ and q_- have a very simple interpretation: $q_+(\mathbf{x})$ is the probability that a trajectory initiated at $\mathbf{x}$ will reach B before it hits A, whereas $q_-(\mathbf{x})$ is the probability that a trajectory arriving at $\mathbf{x}$ came from A, i.e. it did not hit B in between. The level sets of these functions are called the *isocommittor surfaces* [15]. The level set $\{q_+ = 1/2\}$ is of particular interest: points on this surface have an equal probability of first reaching A or B. This is the dividing surface that separates the reactant states from the product states.

For gradient systems

$$\dot{\mathbf{x}} = -\nabla V(\mathbf{x}) + \sqrt{2}\,\dot{\mathbf{w}} \qquad (10.2.53)$$

we have $q_+ = 1 - q_-$, where q_+ satisfies the equation

$$\Delta q_+ - \nabla V \cdot \nabla q_+ = 0. \qquad (10.2.54)$$

Consider the simple example of a two-dimensional gradient system with potential

$$V(x_1, x_2) = \tfrac{5}{2}(1 - x_1^2)^2 + 5x_2^2, \qquad (10.2.55)$$

and let us take $A = \{x_1 < -0.8\}$, and $B = \{x_1 > 0.8\}$. It is easy to see that in this case the solution of (10.2.54) is a function of x_1 only: $q_+(x_1, x_2) = 1 - q_-(x_1, x_2) \equiv q(x_1)$, where

$$q(x_1) = \frac{\int_{-0.8}^{x_1} e^{5(1-z^2)^2/2}\,dz}{\int_{-0.8}^{0.8} e^{5(1-z^2)^2/2}\,dz}. \qquad (10.2.56)$$

The questions raised above can all be answered in terms of the quantities just defined. We will merely summarize the results: details of the calculation can be found in [17].

1. Probability density of reactive trajectories We have

$$m_R(\mathbf{x}) = q_+(\mathbf{x})q_-(\mathbf{x})m(\mathbf{x})$$

and

$$\lim_{T\to\infty} \frac{1}{2T} \int_{[-T,T]\cap R} \phi(\mathbf{x}(t))\mathbf{1}_R(t)\,dt = \int_\Omega q_+(\mathbf{x})q_-(\mathbf{x})m(\mathbf{x})\phi(\mathbf{x})\,d\mathbf{x}.$$

If we limit ourselves to the ensemble of reactive trajectories and ask what is the probability density of finding them at $\mathbf{x}$, we have

$$\hat{m}_R(\mathbf{x}) = \frac{1}{Z_R}q_+(\mathbf{x})q_-(\mathbf{x})m(\mathbf{x}),$$

where the normalization factor

$$Z_R = \int_\Omega q_+(\mathbf{x})q_-(\mathbf{x})m(\mathbf{x})\,d\mathbf{x}$$

gives the probability that a trajectory piece is reactive.

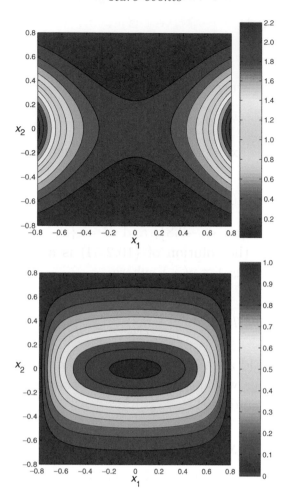

Figure 10.4. Equilibrium density and reactive trajectory density for the example (10.2.55), from [17]. The upper figure shows the equilibrium density for the whole system. The lower figure shows the equilibrium density for the reactive trajectories. It is clear that the equilibrium density for the whole system is concentrated at the wells whereas the equilibrium density for the reactive trajectories is concentrated at the saddle point.

As an illustration consider a simple example, the gradient system discussed above. Shown in Figure 10.4 are the equilibrium probability density m and the probability density for the A–B reactive trajectories m_R. Notice that m peaks at the local minima whereas m_R peaks at the saddle point, owing to the fact that the reactive trajectories are slowed down there (and hence the system spends more time there); the saddle point is also where the drift term vanishes in the stochastic dynamics.

2. The current of A–B reactive trajectories This is given by

$$\mathbf{J}_R = q_+ q_- \mathbf{J} + m q_- \mathbf{a} \nabla q_+ - m q_+ \mathbf{a} \nabla q_-,$$

where $\mathbf{J}$ was given in (10.2.52). Note that $\operatorname{div} \mathbf{J}_R = 0$ in Ω. Note also that

$$\hat{\mathbf{n}}_{\partial A}(\mathbf{x}) \cdot \mathbf{J}_R(\mathbf{x}) = m(\mathbf{x}) \sum_{i,j=1}^{n} \hat{n}_{\partial A,i}(\mathbf{x}) a_{ij}(\mathbf{x}) \frac{\partial q_+(\mathbf{x})}{\partial x_j}$$

$$= m(\mathbf{x}) |\nabla q_+(\mathbf{x})|^{-1} \sum_{i,j=1}^{n} a_{ij}(\mathbf{x}) \frac{\partial q_+(\mathbf{x})}{\partial x_i} \frac{\partial q_+(\mathbf{x})}{\partial x_j} \geq 0,$$

$$(10.2.57)$$

where $\hat{\mathbf{n}}_{\partial A}(\mathbf{x})$ is the unit outward normal of ∂A, which can also be defined as $\nabla q_+ / |\nabla q_+|$. Similarly, one can show that

$$\hat{\mathbf{n}}_{\partial B}(\mathbf{x}) \cdot \mathbf{J}_R(\mathbf{x}) \leq 0, \qquad (10.2.58)$$

where $\hat{\mathbf{n}}_{\partial B}$ is the unit outward normal of ∂B.

3. The transition rate Let S be a *dividing surface*, i.e. a hypersurface in Ω that separates A and B; then

$$\nu_R = \lim_{T \to \infty} \frac{N_T}{2T} = \int_S \hat{\mathbf{n}}_S(\mathbf{x}) \cdot \mathbf{J}_R(\mathbf{x}) \, dS.$$

Since $\mathbf{J}_R$ is divergence-free, i.e. $\operatorname{div} \mathbf{J}_R = 0$, it is easy to see that the integral over S defined above is actually independent of S. In fact, one can show that [17] the transition rate is given by

$$\nu_R = \int_\Omega \langle a \nabla q_+, \nabla q_+ \rangle m(\mathbf{x}) \, d\mathbf{x} = \int_\Omega \sum_{j,k} a_{j,k}(\mathbf{x}) \partial_j q_+(\mathbf{x}) \partial_k q_+(\mathbf{x}) m(\mathbf{x}) \, d\mathbf{x}.$$

For the setting considered in (10.1.3) we have

$$\mathbf{J}_R(\mathbf{x}) = Z_\varepsilon^{-1} e^{-V(\mathbf{x})/\varepsilon} \nabla q_+(\mathbf{x}),$$

where $Z_\varepsilon = \int_\Omega e^{-V(\mathbf{x})/\varepsilon} \, d\mathbf{x}$. The transition rate ν_R is now given by

$$\nu_R = \frac{\varepsilon}{Z_\varepsilon} \int_{\Omega \setminus (A \cup B)} e^{-V(\mathbf{x})/\varepsilon} |\nabla q_+(\mathbf{x})|^2 \, d\mathbf{x}.$$

In fact, it is easy to see that the transition rate admits a variational formulation: define

$$I(q) = \int_\Omega e^{-V(\mathbf{x})/\varepsilon} |\nabla q(\mathbf{x})|^2 \, d\mathbf{x};$$

then

$$\nu_R = \frac{\varepsilon}{Z_\varepsilon} \min_q I(q)$$

subject to the boundary condition that q is 0 and 1 at ∂A and ∂B respectively.

4. Streamlines of the reactive current Given the vector field $\mathbf{J}_R$, we can define the dynamics

$$\frac{d\mathbf{x}(\tau)}{d\tau} = \mathbf{J}_R(\mathbf{x}(\tau)), \tag{10.2.59}$$

where τ is an artificial time parameter (which should not be confused with the physical time duration t of the original trajectory $\mathbf{x}(\cdot)$). Solutions to this equation define the streamlines for the reactive current $\mathbf{J}_R$. They are useful for defining the transition tube.

5. The transition tube So far our statements are exact. Next, we will define, for a complex system, an object analogous to a minimum energy path (or the most probable paths). For this to make any sense the system must be such that the reactive trajectories are localized in the configuration or phase space. This is not the case, for example, when the barrier is mainly entropic in nature (an example is discussed below). Here we will limit ourselves to the situation when the reactive trajectories are indeed localized, in the sense to be made more precise below.

Suppose that we identify a (localized) region on ∂A, say $S_A \subset \partial A$, which contains a portion $1 - \delta$ of the reactive current that goes out of A:

$$\int_{S_A} \hat{\mathbf{n}}_{\partial A}(\mathbf{x}) \cdot \mathbf{J}_{AB}(\mathbf{x}) \, dS = (1 - \delta) \int_{\partial A} \hat{\mathbf{n}}_{\partial A}(\mathbf{x}) \cdot \mathbf{J}_{AB}(\mathbf{x}) \, dS = (1 - \delta)\nu_R. \tag{10.2.60}$$

Let $S(\tau)$ be the image of S_A under the flow map defined by (10.2.59) at (the artificial) time τ. Using the divergence theorem, we have

$$\int_{S_A} \hat{\mathbf{n}}_{\partial A}(\mathbf{x}) \cdot \mathbf{J}_R(\mathbf{x}) \, dS = \int_{S(\tau)} \hat{\mathbf{n}}_{S(\tau)}(\mathbf{x}) \cdot \mathbf{J}_R(\mathbf{x}) \, dS. \tag{10.2.61}$$

We define the union of the images $S(\tau)$ as the *transition tube* that carries the portion $1 - \delta$ of the probability flux of reactive trajectories.

When discussing the transition tube, we have implicitly made the *localization assumption*, i.e. that the diameter of the sets $S(\tau)$ is much smaller than the radius of curvature of the centerline curve of the tube. As was shown in [16], it is possible to establish an approximate characterization of the transition tube using the ratio of the tube width and the radius of curvature of the centerline as the small parameter to develop the asymptotics. Given a curve $\boldsymbol{\gamma}$ that connects A and B, let $\mathbf{z}$ be an arbitrary point along $\boldsymbol{\gamma}$, let $P_\mathbf{z}$ be the hyperplane normal to $\boldsymbol{\gamma}$ and denote by $\mu_\mathbf{z}(dS)$ the restriction of the equilibrium distribution $\mu(d\mathbf{x})$ to $P_\mathbf{z}$. It was shown in [16] that under the localization assumption, the centerline of the transition tube γ^* is approximately a *principal curve* with respect to the family of distributions $\{\mu_\mathbf{z}\}$: for every point $\mathbf{z}$ along the centerline $\boldsymbol{\gamma}^*$, if $\mathbf{C}(\mathbf{z})$ is the center of mass of the distribution $\mu_\mathbf{z}$ on $P_\mathbf{z}$ then $\mathbf{C}(\mathbf{z})$ coincides with $\mathbf{z}$, up to an error of size ε. That is,

$$\mathbf{C}(\mathbf{z}) = \langle \mathbf{x} \rangle_{\mu_\mathbf{z}} = \mathbf{z} + \mathcal{O}(\varepsilon). \tag{10.2.62}$$

This characterization is the basis for the finite-temperature string method, to be discussed in the next section. Once the centerline curve is found, one can determine the width of the transition tube by identifying the standard deviation of the distribution $\mu_\mathbf{z}$ on $P_\mathbf{z}$.

In the zero-temperature limit, $\mathbf{C}(\mathbf{z})$ becomes a local minimizer of V restricted to $P_\mathbf{z}$. Therefore the principal curve becomes the MEP.

The use of hyperplanes is not necessary in this characterization. Other objects, such as isocommittor surfaces or the sets $\{S(\tau)\}$, can also be used.

One may ask: how realistic is the localization assumption? If it is violated, it means that the transition paths are very diffusive. This is the case when entropic barriers are dominant. To characterize the transition path ensemble in this situation, one has to find some coarse-grained variables to represent the entropic factors. At the present time there is no systematic way of doing this. Nevertheless, it should be noted that transition path theory is equally valid in this case even though the localization assumption is violated.

6. Transition state ensemble In the language of transition path theory, the transition state ensemble is simply the distribution $\mu_R(d\mathbf{x}) = m_R(\mathbf{x})\,d\mathbf{x}$. For simple systems, this distribution is sharply peaked at the transition states, the saddle points of the energy; therefore, one may replace the notion of the transition state ensemble by that of the transition state. For complex systems, however, this is not necessarily the case. In fact, when entropic barriers dominate the transition state ensemble can be a rather flat distribution, as we will see later. When the energy barrier dominates, the transition state ensemble should still be peaked. The region that contains most of the mass of μ_R is the bottleneck for the transition and should be considered as the analog of the transition states in transition state theory.

7. Reaction coordinates The notion of reaction coordinates has been used widely and often incorrectly. Reaction coordinates are variables that fully characterize the progress of a reaction. There is an obvious reaction coordinate, namely the committor function q_+. The family of isocommittor surfaces gives a geometric characterization of the progress of the reaction. A given set of functions $\mathbf{r} = (r_1, \ldots, r_m)$, where $r_j = r_j(\mathbf{x})$, can be used as reaction coordinates if and only if it distinguishes the different isocommittor surfaces, i.e. if, for any two points with position coordinates $\mathbf{x}_1$ and $\mathbf{x}_2$ in Ω, $\mathbf{r}(\mathbf{x}_1) = \mathbf{r}(\mathbf{x}_2)$ implies that $q_+(\mathbf{x}_1) = q_+(\mathbf{x}_2)$.

We should distinguish the notion of reaction coordinates from that of slow variables. The following example should illustrate this point (see also [18]). Consider a two-dimensional configuration space with energy function

$$V(x,y) = \frac{1}{\varepsilon^2}(1 - x^2)^2 + \frac{1}{2}y^2,$$

where as usual $\varepsilon \ll 1$. This is a bistable system with local minima $(-1, 0)$ and $(1, 0)$. The slow variable in this problem is y but the reaction coordinate for the transition between the two local minima is x.

An example with an entropic barrier [43]. Consider diffusion in the S-shaped region shown in Figure 10.5. We define the sets A and B to be the little squares at the upper-right and bottom-left corners respectively. In this example, there are no potential barriers, in fact, $V = 0$. However, there is an *entropic barrier* (see Section 10.1): transition paths have to find their way between two obstacles, shown as thin white rectangles. Suppose we start the dynamics in A. The closer the trajectory gets to the middle region enclosed by the obstacles the higher the probability that it will finally reach the lower-left corner before returning to A, because the probability for it to end up in B depends roughly on the distance between the current position and the set B. Figure 10.5 shows the probability density for the reactive trajectories and the committor function $q(x,y)$ for this example. In Figure 10.6 we depict a typical reactive trajectory.

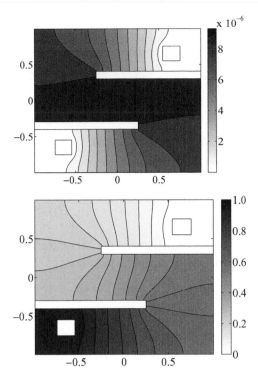

Figure 10.5. Probability density of reactive trajectories (upper panel) and committor function (lower panel) for an example with an entropic barrier. Reprinted with permission from [43]. (Copyright 2006, American Institute of Physics.)

One can see that the trajectory spends most of its time between the obstacles, here shown in black. This is consistent with the behavior of the probability density function of reactive trajectories. Although not illustrated here, the probability current of the reactive trajectories is as expected: basically, the streamlines follow the S-shape (see [43]).

10.3 Numerical algorithms

In this section, we will discuss numerical algorithms that have been developed for computing transition states, minimum energy paths and transition tubes. We will restrict our attention to gradient systems.

10.3.1 Finding transition states

In the simplest case, when the energy landscape is smooth, the most important objects for the transition between the locally stable states are

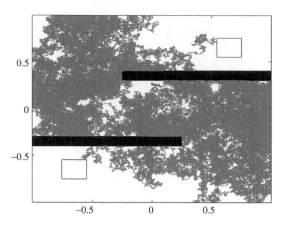

Figure 10.6. A typical reactive trajectory for an example with an entropic barrier. Reprinted with permission from [43]. (Copyright 2006, American Institute of Physics.)

the transition states, which are the saddle points separating the stable states. These saddle points are special critical points for the potential V. They satisfy the equation

$$\nabla V(\mathbf{x}) = 0. \qquad (10.3.1)$$

If we know roughly the location of a particular saddle point, we can use Newton's method to solve (10.3.1) and find the relevant saddle point. This is quite efficient but requires one to make a very good initial guess.

A more general approach is the dimer method [33]. This is an iterative procedure that involves a dimer, which consists of two images of the system connected by a short line segment, say of length δ. The energy of the dimer is the sum of the energies of the two images. The dynamics of the dimer is determined by the translation of the mid-point of the line segment and the rotation of the line segment. Each iteration consists of two steps.

(1) *Rotating the dimer.* The best orientation of the dimer is found by minimizing the energy of the dimer with the mid-point fixed. This can be done using, for example, Newton's method. The energy differences for the different orientations of the dimer are related to the curvature in different directions. The minimum energy orientation is given by the lowest curvature direction.

(2) *Translating the dimer.* First we define the total force on the dimer, $\mathbf{F} = (\mathbf{F}_1 + \mathbf{F}_2)/2$, where $\mathbf{F}_1$ and $\mathbf{F}_2$ are the forces acting on the first and second images respectively. If we use this force to move the dimer, the dimer will end up near the local minimum state for the potential. In order to reach the saddle point, one idea is to reverse the force in the direction of the dimer, i.e. let

$$\mathbf{F}^* = \mathbf{F} - 2\mathbf{F}_{||}, \qquad (10.3.2)$$

where $\mathbf{F}_{||}$ is the component of $\mathbf{F}$ in the direction of the dimer. The dimer is then parallel-translated: its mid-point is moved by $\Delta t\, \mathbf{F}^*$, where Δt is some appropriate time-step size.

The dimer method is fairly effective but it is not fool-proof. There is no guarantee that the relevant saddle point for the transition in which we are interested will be reached, since no information about the final state of the transition is used.

Besides the two methods discussed above, there are many other methods (see for example [35]) for computing saddle points. The nudged elastic band method discussed below, particularly the climbing-image nudged elastic band method (CI-NEB), is also targeted toward computing saddle points.

10.3.2 Finding the minimal energy path

If the stable states in which we are interested are separated by multiple saddle points and intermediate stable states, then it becomes quite difficult to use direct saddle-point-searching methods to find the sequence of relevant saddle points. In this case, we really should look for the minimum energy path (MEP).

There are many different ways of finding the MEP. We will discuss two representative ideas: the elastic band method, and in particular the nudged elastic band method, and the string method. The former is a representative of the so-called "chain-of-states" approach. The latter is a representative of the class of methods based on evolving curves in configuration space. Even though, after discretization, the string method also has the appearance of a chain-of-states method, there are important differences in whether the continuous character is used in an essential way. The elastic band methods, on the one hand, adjust the distance between the states on the chain using an artificial spring force between the states.

The string method, on the other hand, redefines the configurations continuously using interpolation, which amounts to reparametrizing the string. The continuous character provided by the curves becomes even more essential in the finite-temperature string method for finding transition tubes in complex systems, to be discussed later.

Elastic band method The basic idea is to connect two stable states $\mathbf{A}$ and $\mathbf{B}$ by a chain of states (replicas or images) and evolve this chain of states. In an early attempt, Pratt proposed to use Monte Carlo methods to sample chains of states between the initial and final states in order to find the transition state region. Pratt's idea has been developed in two directions. One is Monte Carlo algorithms for sampling true dynamical trajectories between the initial and final states. This is the well-known transition path sampling algorithm developed by Bolhuis, Chandler, Dellago and Geissler. The second is a class of optimization algorithms for finding the MEP using a chain of states. The elastic band method is an example of this type of algorithm.

Denote by $\mathbf{A}$ and $\mathbf{B}$ the initial and final states for the MEP respectively. Given a chain of states $\{\mathbf{x}_0, \mathbf{x}_1, \ldots, \mathbf{x}_N\}$, where $\mathbf{x}_0 = \mathbf{A}$, $\mathbf{x}_N = \mathbf{B}$, let us define an energy for the chain:

$$E = E(\mathbf{x}_0, \mathbf{x}_1, \ldots, \mathbf{x}_N) = \sum_j V(\mathbf{x}_j) + \frac{k\Delta s}{2} \sum_{j=1}^{N} \left| \frac{\mathbf{x}_j - \mathbf{x}_{j-1}}{\Delta s} \right|^2,$$

where Δs is some numerical parameter that is comparable with the distance between neighboring images on the chain. Alternative energy functions have been proposed by Elber and Karplus [25]. See also [13, 51].

In the elastic-band method (also called the plain elastic band method, to be contrasted with the nudged elastic band method discussed below [38]), the idea is to move the chain of states according to the gradient flow of the energy E [30]:

$$\dot{\mathbf{x}}_j = -\frac{\partial E}{\partial \mathbf{x}_j} = -\nabla V(\mathbf{x}_j) + k\Delta s \frac{\mathbf{x}_{j+1} - 2\mathbf{x}_j + \mathbf{x}_{j-1}}{\Delta s^2}, \ j = 1, \ldots, N-1.$$

$$(10.3.3)$$

The first term on the right-hand side is the potential force and the second term is the spring forces. Note that the scaling in the coefficient of the second term is chosen so that if we use an explicit ODE solver to evolve

(10.3.3) then the time-step size must be $\Delta t \sim \Delta s$ to guarantee numerical stability. However, in this scaling the second term drops out in the continuum limit as $\Delta s \to 0$.

The elastic band method is extremely simple and intuitive. However, it has been noticed that it may fail to converge to the MEP [30]. The most common problem is corner-cutting (see [38]).

To overcome this problem, Jónsson *et al.* introduced the *nudged elastic band method* (NEB) [38]. This is a very simple modification of the elastic band method, but it has made the method truly useful. Instead of using the total potential force and spring force to move the chain, one uses only the normal component of the potential force and the tangential component of the spring force:

$$\dot{\mathbf{x}}_j = -\nabla V(\mathbf{x}_j)^{\perp} + (\mathbf{F}_j^s, \hat{\boldsymbol{\tau}}_j)\hat{\boldsymbol{\tau}}_j, \quad j = 1, \dots, N-1, \qquad (10.3.4)$$

where $\mathbf{F}_j^s = k(\mathbf{x}_{j+1} - 2\mathbf{x}_j + \mathbf{x}_{j-1})/\Delta s$, $\hat{\boldsymbol{\tau}}_j$ denotes the tangent vector along the elastic band at $\mathbf{x}_j$ and $\mathbf{F}^{\perp}$ denotes the normal component of the vector $\mathbf{F}$, i.e. $\mathbf{F}^{\perp} = \mathbf{F} - (\mathbf{F}, \hat{\boldsymbol{\tau}}_j)\hat{\boldsymbol{\tau}}_j$.

It is easy to see that if the chain converges to a steady state then it should be a good approximation to the MEP. In fact, from (10.3.4), we see that if the left-hand side vanishes then

$$-\nabla V(\mathbf{x}_j)^{\perp} + (\mathbf{F}_j^s, \hat{\boldsymbol{\tau}}_j)\hat{\boldsymbol{\tau}}_j = 0, \quad j = 1, \dots, N-1.$$

Since the two terms in this equation are normal to each other, each has to vanish. In particular, we have

$$-\nabla V(\mathbf{x}_j)^{\perp} = 0, \quad j = 1, \dots, N-1.$$

The choice of the elastic constant k affects the performance of NEB. If k is too large then the elastic band is too stiff and one has to use very small time-steps to solve the set of ODEs in (10.3.4). If k is too small then there is not enough force to prevent the states on the chain from moving away from the saddle point; hence the accuracy of location of the saddle point will be reduced.

String method One main motivation for the string method is to formulate a strategy using the intrinsic dynamics of curves connecting the stable states $\mathbf{A}$ and $\mathbf{B}$. The dynamics of the curves are determined by the normal velocity at each point along the curve. The tangential component only affects the parametrization of the curve. Therefore, fixing a particular parametrization also fixes the tangential component of the dynamics. The simplest choice is the equal-arclength parametrization. One can also add an energy-dependent weight function along the curve in order to enhance the accuracy near the saddle points [19]. This kind of idea has been used quite extensively since the 1980s, for example, in the work of Brower and co-workers on geometric models of interface evolution [9].

Let $\boldsymbol{\gamma}$ be a curve that connects $\mathbf{A}$ and $\mathbf{B}$. We look for solutions of

$$- (\nabla V)^{\perp} (\boldsymbol{\gamma}) = 0, \tag{10.3.5}$$

where $(\nabla V)^{\perp}$ is the component of ∇V normal to $\boldsymbol{\gamma}$:

$$(\nabla V)^{\perp} (\boldsymbol{\gamma}) = \nabla V(\boldsymbol{\gamma}) - (\nabla V(\boldsymbol{\gamma}), \hat{\boldsymbol{\tau}})\hat{\boldsymbol{\tau}}. \tag{10.3.6}$$

Here $\hat{\boldsymbol{\tau}}$ denotes the unit tangent vector of the curve $\boldsymbol{\gamma}$, and $(\cdot, \cdot)$ denotes the Euclidean inner product. The basic idea of the string method is to find the MEP by evolving curves in configuration spaces that connect $\mathbf{A}$ and $\mathbf{B}$, under the potential force field. The simplest dynamics for the evolution of such curves is given abstractly by

$$v_{\mathrm{n}} = - (\nabla V)^{\perp}, \tag{10.3.7}$$

where v_{n} denotes the normal component of velocity of the curve. This formulation guarantees that it is gauge invariant, i.e. it is invariant under a change in parametrization of the curve. To translate (10.3.7) to a form that can be readily used in numerical computations, we assume that we have picked a particular parametrization of the curve $\boldsymbol{\gamma}$:

$$\boldsymbol{\gamma} = \{\boldsymbol{\varphi}(\alpha) : \alpha \in [0, 1]\}.$$

Then we have $\hat{\boldsymbol{\tau}}(\alpha) = \partial_{\alpha}\boldsymbol{\varphi}/|\partial_{\alpha}\boldsymbol{\varphi}|$. If we choose the equal-arclength parametrization then α is a constant multiple of the arclength from $\mathbf{A}$ to the point $\boldsymbol{\varphi}(\alpha)$. In this case, we also have that $|\partial_{\alpha}\boldsymbol{\varphi}|$ is constant (the value of the constant may depend on time: it is the length of the curve $\boldsymbol{\gamma}$).

Two slightly different forms of the string method have been suggested [19, 24]. In the simpler form one solvs

$$\dot{\boldsymbol{\varphi}} = -\nabla V(\boldsymbol{\varphi}) + \mu\hat{\boldsymbol{\tau}}, \tag{10.3.8}$$

where $\dot{\boldsymbol{\varphi}}$ denotes the time-derivative of $\boldsymbol{\varphi}$ and $\mu\hat{\boldsymbol{\tau}}$ is a Lagrange multiplier term for the purpose of enforcing the particular parametrization of the string, such as the equal-arclength parametrization.

In actual computations, the string is discretized into a number of images $\{\boldsymbol{\varphi}_i(t), \ i = 0, 1, \ldots, N\}$. The images along the string are evolved by iterating upon the following two-step procedure.

(1) *Evolution step.* The images on the string are evolved over some time interval Δt according to the potential force at that point:

$$\dot{\boldsymbol{\varphi}}_i = -\nabla V(\boldsymbol{\varphi}_i). \tag{10.3.9}$$

Equation (10.3.9) can be integrated in time by any stable ODE solver, e.g. the forward Euler or the Runge–Kutta methods. The time-step should be chosen such that neighboring images do not cross each other:

$$|\nabla V(\boldsymbol{\varphi}_{j+1}) - \nabla V(\boldsymbol{\varphi}_j)|\Delta t < \Delta s.$$

This gives the condition

$$|\nabla^2 V|\Delta t < 1.$$

(2) *Reparametrization step.* The points (images) are redistributed along the string using a simple interpolation procedure (see [24]). Take the example of equal-arclength parametrization. Then this reparametrization step again involves two stages. In the first one computes the total arclength of the string by summing the distances between neighboring images. This allows one to compute the new parametrization, i.e. the values of $\{\alpha_j, \ j = 0, 1, \ldots, N\}$ which make up an equal-distance parametrization. In the second stage one computes the images at these new parameter values using standard piecewise-polynomial interpolation.

It is not necessary to invoke the reparametrization procedure at every time-step: one can reparametrize after a number of evolution steps.

One problem with this approach is that, even after the string has converged to a steady state, the images on the string may still move back and forth along the string. To avoid this, we may modify (10.3.8) and consider

$$\dot{\boldsymbol{\varphi}} = -\nabla V(\boldsymbol{\varphi})^\perp + \lambda\hat{\boldsymbol{\tau}}. \tag{10.3.10}$$

By projecting out the tangential component of the potential force, one eliminates the tendency for the images to move back and forth along the string.

There is a subtle point associated with the discretization of (10.3.10). This was first noted in [38, 19] in connection with NEB. To see this more clearly, let us write (10.3.10) as

$$\dot{\boldsymbol{\varphi}} = -\nabla V(\boldsymbol{\varphi}) + (\nabla V(\boldsymbol{\varphi}), \partial_\alpha \boldsymbol{\varphi}) \frac{\partial_\alpha \boldsymbol{\varphi}}{|\partial_\alpha \boldsymbol{\varphi}|^2} + \lambda \hat{\boldsymbol{\tau}}. \qquad (10.3.11)$$

This has the form of a hyperbolic partial differential equation – the second term on the right-hand side is a convection term. Therefore one should be careful when discretizing that term in order to avoid numerical instability. The simplest idea is to use one-sided differencing. The coefficient of the convection term is given by $\partial_\alpha V(\boldsymbol{\varphi}(\alpha))$, up to a positive factor. This means that when V is increasing one should use backward differencing and when V is decreasing one should use forward differencing [34, 47]:

$$\partial_\alpha \boldsymbol{\varphi}_j = \begin{cases} \dfrac{\boldsymbol{\varphi}_{j+1} - \boldsymbol{\varphi}_j}{\Delta\alpha} & \text{if } V(\boldsymbol{\varphi}_{j+1}) > V(\boldsymbol{\varphi}_j) > V(\boldsymbol{\varphi}_{j-1}), \\[2mm] \dfrac{\boldsymbol{\varphi}_j - \boldsymbol{\varphi}_{j-1}}{\Delta\alpha} & \text{if } V(\boldsymbol{\varphi}_{j+1}) < V(\boldsymbol{\varphi}_j) < V(\boldsymbol{\varphi}_{j-1}). \end{cases}$$

To put it differently, one should not use divided differencing across the saddle point. For this reason, it is harder to design a uniformly high-order-accurate string method in this form [47].

It is natural to ask whether the elastic band method can also be formulated in terms of the evolution of continuous curves. As we remarked earlier, if we use the original scaling of the spring constant then the term that represents the elastic force disappears in the continuum limit (which suggests that the elastic term in (10.3.3) may not be strong enough to prevent clustering of the states near the local minima as $\Delta s \to 0$). In order to retain the spring term, one has to replace the spring constant $k\Delta s$ by k. In that case, we can take the continuum limit of the (nudged) elastic band method and obtain

$$\dot{\boldsymbol{\varphi}} = -\nabla V(\boldsymbol{\varphi})^\perp + k(\boldsymbol{\varphi}_{\alpha\alpha}, \hat{\boldsymbol{\tau}})\hat{\boldsymbol{\tau}}. \qquad (10.3.12)$$

However, to evolve this dynamics one has to use much smaller time-steps $\Delta t \sim (\Delta s)^2$, and it takes many more iterations for the dynamics to converge to the MEP.

Another way of designing the string method is to use the alternative characterization for the MEP discussed at the end of Section 10.2.3, namely, every point along the MEP is the local minimum of the potential energy V restricted to the hyperplane normal to the MEP. Using this, we can formulate a string method as follows. At each iteration, one finds the local minima of V on the hyperplanes normal to the current configuration of the string. Naively, one might simply use this set of local minima as the new configuration of the string – resulting in an unstable numerical algorithm. Numerical instability can be avoided if a relaxation procedure is used, however. This idea can be combined with various optimization techniques to improve its efficiency [50]. It is closely related to the finite-temperature string method discussed below.

10.3.3 Finding the transition path ensemble or the transition tubes

Transition path sampling Transition path sampling (TPS) is the first systematic numerical approach for studying the transition between metastable states in systems with complex energy landscapes. It is a Monte Carlo algorithm for sampling the transition path ensemble [14, 4]. In addition to being a numerical tool, it has also helped to clarify some of the most important conceptual issues for understanding rare events in systems with complex energy landscapes. Indeed the significance of the committor functions and isocommittor surfaces, which have now become basic concepts in transition path theory, was first realized in TPS.

Transition path sampling is a two-step procedure: sampling the transition path and analyzing the path ensemble. We examine the former now.

Sampling the transition path The idea is as follows. Fix a time parameter $T > 0$. We start with an initial reactive trajectory over a time interval of length T. We then generate a sequence of reactive trajectories using a standard procedure in the Metropolis algorithm.

(1) *Proposal step.* Generate a new trajectory over the same time interval using some Monte Carlo moves.
(2) *Acceptance or rejection step.* Decide whether the new trajectory should be accepted.

Next we explain some of the details. The first question is how to choose the time parameter T. To understand this, we note that there are three time scales of interest: the relaxation time T_r, the commitment time T_c and the transition time T_h. They are related by $T_r < T_c < T_h$. The meanings of T_r and T_h are obvious. The commitment time is the time that a reactive trajectory needs in order to finish the reaction after it finds the transition state region, i.e. it is the time needed for the system to be committed to the product state. For simple systems, this is simply the time that a reactive trajectory needs in order to cross the saddle point once it comes near to it. For complex systems, particularly for diffusive transitions, the commitment time can be much longer than molecular vibration times but it is usually much shorter than the average time between transitions. The value of T should be much larger than the commitment time T_c but much smaller than the transition time T_h.

Now let us distinguish two different transition state ensembles: the canonical ensemble and the microcanonical ensemble. Denote by A and B the initial and final states respectively. Typically A and B are some regions in the configuration space. If the original dynamics is described by a deterministic Hamiltonian system then the corresponding transition path ensemble should be microcanonical. In this case a trajectory is rejected if it does not have the correct value of the Hamiltonian or if it fails to be reactive, i.e. it does not connect A and B. All acceptable reactive trajectories have equal weights. For a canonical ensemble one prescribes some Gibbs weight for the reactive trajectories. For example, if the dynamics is Langevin then the (unnormalized) probability weight associated with a path $\mathbf{x}(\cdot)$ is given by

$$\exp\left(-\frac{1}{4}\beta\gamma^{-1}\int_0^T |M\ddot{\mathbf{x}} + \gamma\dot{\mathbf{x}} + \nabla V(\mathbf{x})^2\,dt\right). \qquad (10.3.13)$$

For the microcanonical ensemble, it might be non-trivial to find the initial reactive trajectory. One strategy is to fix two points in A and B respectively, say $\mathbf{x}_1$ and $\mathbf{x}_2$, and solve a two-point boundary value problem that satisfies

$$\mathbf{x}(0) = \mathbf{x}_1, \quad \mathbf{x}(T) = \mathbf{x}_2.$$

New trajectories can be obtained by solving the deterministic differential equations (Hamilton's equation) with initial data that are obtained from perturbations of the old trajectory. For example, one can randomize the velocity and position at $t = T/2$ and use the randomized data as initial conditions to launch trajectories for both $t < T/2$ and $t > T/2$. Such a trajectory is guaranteed to be a true trajectory over the whole time interval $[0, T]$ since both the position and the velocity are continuous at $t = T/2$. However, it is not guaranteed to be reactive, i.e. its position at $t = 0$ is not necessarily in A and its position at $t = T$ is not necessarily in B. If it is reactive then the constructed trajectory will be accepted. If it is not reactive, it will be abandoned and the process will be repeated.

For the canonical ensemble, we may start with any path that connects A and B. To generate new trajectories, one can use standard Monte Carlo strategies for sampling polymer paths. For example, one can put the trajectories onto a lattice and move the different segments on the lattice. Other more efficient ways of generating Monte Carlo moves are discussed in [14].

Once a trial trajectory is constructed, it is either accepted or rejected according to a standard Metropolis procedure. If we denote the old trajectory by $\mathbf{x}_o$ and the new trajectory by $\mathbf{x}_n$ then the acceptance probability is given by

$$P_a = \min\left(1, \frac{W(\mathbf{x}_n)P_g(\mathbf{x}_n \to \mathbf{x}_o)}{W(\mathbf{x}_o)P_g(\mathbf{x}_o \to \mathbf{x}_n)}\right),$$

where $W(\mathbf{x}_o)$ is the statistical weight for $\mathbf{x}_o$ and $P_g(\mathbf{x}_o \to \mathbf{x}_n)$ is the probability of generating $\mathbf{x}_n$ from $\mathbf{x}_o$ in the Monte Carlo move, and so on.

Analyzing the path ensemble After the ensemble of the reactive trajectories has been generated, one still has to perform statistical analysis on this ensemble in order to extract objects of interest such as the transition state ensemble. For TPS this is defined as the collection of points, obtained from the intersection of the trajectories harvested with the iso-committor surface $\{q_+ = 1/2\}$, having the same statistical weight as that of the trajectory itself. These intersection points are found using the following procedure.

Take a trajectory from the ensemble of reactive trajectories that we have harvested. At each point of the trajectory, compute the value of q_+ by launching dynamic trajectories from that point and counting the number of trajectories that reach A and B. The value of q_+ at that point is given by the percentage of trajectories that first reach B. This is a monotonic function along the trajectory. So, there is a unique point along the trajectory where q_+ has the value $1/2$. That point belongs to the transition state ensemble with the statistical weight of the trajectory itself. Another object of interest is the transition rate. The calculation of this quantity is quite involved and we simply refer the reader to [15].

Transition path sampling has the advantage that it makes no assumption on the mechanism of transition. However, analyzing the data provided by TPS in order to extract information about reaction mechanisms or rates can be a highly nontrivial matter, even after one has gone through the steps of generating the paths.

The finite-temperature string method The objective of the finite temperature string method is to find the transition tube for the case when the transition path ensemble is localized. This is done by finding the centerline curve, the string, of the transition tube and identifying the equilibrium distribution restricted to the hyperplanes normal to the centerline curve. So, in fact one is looking for a curve and a family of probability distributions on the hyperplanes normal to the curve. As we noted earlier, the centerline curve is a principal curve. Therefore it is not surprising that the finite-temperature string method is similar in spirit to the expectation–maximization (EM) algorithm used in statistics for finding principal curves; see [32].

In the current setting the maximization step is replaced by a relaxation step. Given the current configuation of the string, the new configuration is found through the following steps.

(1) *Expectation step.* Sample on the hyperplanes normal to the current configuation of the string. The constraint that the sampling is limited to the hyperplane can be realized either as a hard constraint (for example, using Lagrange multipliers) or as a soft constraint by adding penalty terms to the energy.

(2) *Relaxation step.* Compute the empirical center of mass on each hyperplane, and move the string to a new configuration according to

$$\boldsymbol{\varphi}^{n+1} = \boldsymbol{\varphi}^n + \Delta t\, \mathbf{C}(\boldsymbol{\varphi}^n), \qquad (10.3.14)$$

where $\boldsymbol{\varphi}^n$ denotes the current configuration of the string, $\mathbf{C}(\boldsymbol{\varphi}^n)$ denotes the center of mass on the hyperplanes and Δt is the step size. Reparametrize the string if necessary.

In practice the strings are discretized, and therefore (10.3.14) is performed for each configuration on the string. In addition, for really rough potentials it is useful to add local smoothing to the new configuration of the string before the next cycle starts (see Figure 10.7).

For more details and numerical examples, see [21, 49].

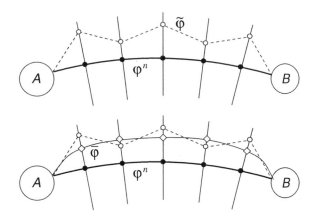

Figure 10.7. Illustration of the finite-temperature string method. Top panel: the new configuration of the string is obtained by moving on the hyperplane according to (10.3.14). Bottom panel: local smoothing is added in order to improve the regularity of the new string.

Figure 10.8 shows the kinds of numerical result obtained using the-finite temperature string method. Shown in red is the centerline of the transition tube computed using the string method. The green curves are obtained by looking at the standard deviation of the equilibrium distribution restricted to the hyperplanes normal to the centerline. The region bounded by the green curves should be regarded as the transition tube. The blue curve is a sample reactive trajectory. One can see that it lies mostly within the transition tube.

In an interesting recent development, one replaces the hyperplanes in the finite temperature string method by Voronoi cells [56]. These cells have the advantage that they are more regular than the hyperplanes and therefore it is easier to perform constraint sampling with them. An illustration of the finite temperature string method based on Voronoi cells is shown in Figure 10.9.

The string method has been extended in a number of ways.

(1) *The growing string method.* The ends of the string are allowed to grow to find the metastable states [45].
(2) *The quadratic string method.* Convergence can be accelerated using quasi-Newton iteration algorithms [10]; see also [48].
(3) *In combination with metadynamics* in order to search more globally in configuration space [8].
(4) *Using collective variables* [41].

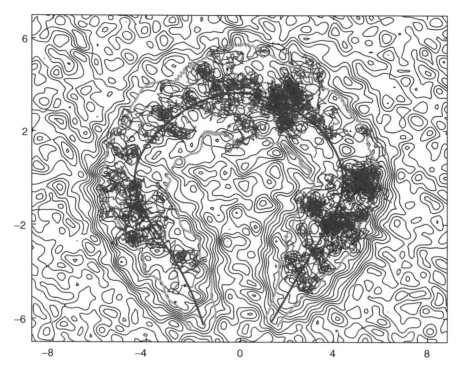

Figure 10.8. Results of the finite temperature string method applied to the rough potential shown in Figure 10.2. The green curves bound the transition tube. The red curve is the centerline of the tube. The blue curve displays an example of a true reactive sample path.

10.4 Accelerated dynamics and sampling methods

The techniques just presented require one to know beforehand the initial and final states of the transition. It is natural to ask whether we can accelerate the original dynamics in such a way that it still captures statistically the correct kinetics of the system. We will discuss very briefly one example of this kind, hyperdynamics [58]. We then consider some examples of improved sampling strategies for computing the free energy, metadynamics [39] and temperature-accelerated molecular dynamics [40].

10.4.1 TST-based acceleration techniques

Hyperdynamics is one example of the approaches, developed mainly by A.F. Voter, for accelerating molecular dynamics using transition state theory (TST). The objective of these approaches is not to reproduce accurately the individual trajectories of the molecular dynamics but,

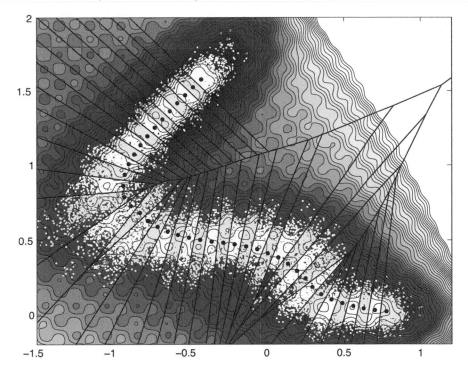

Figure 10.9. Results of a finite temperature string method based on Voronoi cells, for a perturbed Mueller potential. The black dots show the string, the centerline of the transition tube. The cloud of white dots shows the samples in each Voronoi cell computed in the sampling (or expectation) step. (Courtesy of Eric Vanden-Eijnden.)

rather, the Markov chain obtained as a result of a transition between two different stable states (local minima of the potential energy V). This requires approximating accurately, in a statistical sense, the sequence of stable states obtained in the MD simulation, as well as the transition rates (or transition time) between the stable states. Other examples of the approaches developed by Voter under the same umbrella include temperature-accelerated dynamics (TAD) and parallel-replica dynamics, except that the parallel-replica method does not rely on TST [60].

The basic idea of hyperdynamics is to add a bias potential ΔV to the original potential V. The bias potential should satisfy the following requirements.

(1) It should raise the potential wells without modifying the behavior around the transition states. By raising the potential wells, it speeds up the process of escaping the wells.

(2) Transition state theory should still apply. In other words, the bias potential does not introduce additional correlated events near the transition states.

Constructing such a bias potential is in general a rather nontrivial task. Some examples are discussed in [60].

The addition of a bias potential does change the time it takes for the system to move between different stable states. To recover the statistics for the original system, we rescale time, using a formula obtained from TST [58, 59]. Denote by $\mathbf{x}(t_n)$ the state of the modified system at the nth time-step, and by Δt the MD time-step for the modified system. The time it takes for the original system to reach $\mathbf{x}(t_n)$ is given by

$$t_n = \sum_1^n \Delta t \, \exp\left(\Delta V(\mathbf{x}(t_j))/(k_B T)\right). \qquad (10.4.1)$$

Clearly, (10.4.1) is only valid in a statistical sense.

10.4.2 Metadynamics

While hyperdynamics works with the original variables and potential energy landscape of the system, other coarse-grained molecular dynamics methods have been developed for exploring the dynamics of some coarse-grained variables in the free energy landscape, and acceleration methods have also been developed to speed up free energy exploration. Here we will discuss some examples of this type.

In Section 9.3 we showed that, by simulating (9.3.6) at small γ and large μ and monitoring the evolution of $\mathbf{x}(t)$, one can indeed sample the free energy landscape in the variables $\mathbf{q}(\mathbf{x})$. One difficulty remains, though: the dynamics described by (9.3.6) still encounter barriers just as does the original dynamics, and this reduces the efficiency of the entire approach. To deal with this problem in metadynamics, Iannuzzi, Laio and Parrinello suggested further modifying the dynamics to include in (9.3.6) (or, rather, (9.3.17) in the original paper) an additional non-Markovian term which discourages the trajectory from going back to regions that it has already visited. For instance, one may replace (9.3.6) by

$$\begin{cases} \gamma \dot{\mathbf{x}} = -\nabla V(\mathbf{x}) + \mu(\mathbf{X} - \mathbf{q}(\mathbf{x}))\nabla \mathbf{q}(\mathbf{x}) + \sqrt{2\beta^{-1}\gamma}\,\dot{\mathbf{w}}, \\ \dot{\mathbf{X}} = -\mu(\mathbf{X} - q(\mathbf{x})) + \sqrt{2\beta^{-1}}\,\dot{\mathbf{W}} \\ \qquad + A\displaystyle\int_0^t (\mathbf{X}(t) - \mathbf{X}(t'))e^{-|\mathbf{X}(t)-\mathbf{X}(t')|^2/\Delta q^2}\,dt', \end{cases} \qquad (10.4.2)$$

where A and Δq are adjustable parameters. Proceeding as before, it is easy to see that the limiting equation for $\mathbf{X}$ as $\gamma \to 0$ and $\mu \to \infty$ is

$$\dot{\mathbf{X}} = -\nabla_{\mathbf{X}} F + A \int_0^t (\mathbf{X}(t) - \mathbf{X}(t')) e^{-|\mathbf{X}(t) - \mathbf{X}(t')|^2/\Delta q^2} dt' + \sqrt{2\beta^{-1}}\,\dot{\mathbf{W}}.$$
(10.4.3)

The memory term added in (10.4.2) fills up the potential well that the trajectory has already visited. In particular, if we let

$$U(\mathbf{X}, t) = \tfrac{1}{2} A \Delta q^2 \int_0^t e^{-|\mathbf{X} - \mathbf{X}(t')|^2/\Delta q^2} dt' \qquad (10.4.4)$$

then, as $t \to \infty$, we have that $U(\mathbf{X}, t) - U(\mathbf{X}', t)$ converges to an estimate of $F(\mathbf{X}') - F(\mathbf{X})$ when $\gamma^{-1} \gg \mu \gg 1$. The parameters A and Δq control the accuracy for the resolution of the free energy: as they are decreased, the resolution improves but the convergence rate in time deteriorates. Given some accuracy requirement, estimating the optimal choice of the parameters γ, μ, A, and Δq in a metadynamics calculation is a nontrivial question which we will leave aside.

10.4.3 Temperature-accelerated molecular dynamics

Another interesting development is temperature-accelerated molecular dynamics (TAMD); see for example [1, 40]. Here, for simplicity we will focus on the overdamped-dynamics case. The extension to molecular dynamics is quite straightforward. The main idea is again a simple modification of (9.3.6) in Section 9.3:

$$\begin{cases} \gamma \dot{\mathbf{x}} = -\nabla V(\mathbf{x}) + \mu(\mathbf{X} - \mathbf{q}(\mathbf{x}))\nabla \mathbf{q}(\mathbf{x}) + \sqrt{2\gamma\beta^{-1}}\,\dot{\mathbf{w}}, \\ \dot{\mathbf{X}} = -\mu(\mathbf{X} - \mathbf{q}(\mathbf{x})) + \sqrt{2\tilde{\beta}^{-1}}\,\dot{\mathbf{W}}. \end{cases} \qquad (10.4.5)$$

Here $\tilde{\beta} = 1/(k_B \tilde{T})$ and $\tilde{T} > T$. As in Section 9.3, γ is small so we are in a setting where we can apply the averaging theorems for stochastic ODEs. In addition μ is large; therefore the effective dynamics for the coarse-grained variables is a gradient flow for the free energy of the coarse-grained variables.

The advantage of using an elevated temperature for the coarse-grained variables is that it accelerates the exploration of the free energy landscape for those variables. At the same time, it does not affect the sampling of the constrained dynamics for the original system, i.e. it does not affect the accuracy of the gradient of the free energy. See [1] for some very interesting applications of TAMD.

Temperature-accelerated molecular dynamics should not be confused with temperature-accelerated dynamics (TAD), developed by Sorensen and Voter [52]. The former is a technique for accelerating the free energy calculation. The latter is a technique for accelerating the dynamics (kinetics) of the original system, based on TST.

10.5 Notes

Transition pathways of coarse-grained systems The ideas discussed in this chapter can be combined with those discussed in Chapter 6. For example, one can combine HMM and the string method, giving rise to the string method in collective variables (see for example [41]). The macroscale solver is the string method in a set of coarse-grained variables. The data needed in the string method can be estimated using a constrained microscale solver, which is usually some constrained molecular dynamics. In particular the seamless method discussed in Chapter 6 can be applied in this context, resulting in an algorithm in which the string moves simultaneously with the microscopic configurations at each point along the string.

However, it should be emphasized that the notions of slow variables and reaction coordinates are not necessarily the same, as we discussed in subsection 10.2.5.

Non-gradient systems We have mostly focused our discussion on gradient systems. Noise-induced transitions in non-gradient systems can behave quite differently. For one thing, besides critical points, non-gradient systems can have other invariant sets such as limit cycles or even strange attractors. More importantly, the boundaries between the basins of attraction of different invariant sets can be very complex. The implication of these complications in the mechanism of noise-induced transition in non-gradient systems has not been fully investigated yet, and we refer to [64] for a summary of the initial work done in this direction.

Quantum tunneling Quantum tunneling is in many aspects very similar to the noise-induced transitions that we have discussed in this chapter. For example, one can formulate transition state theory in the quantum context. However, the mathematical issues associated with quantum tunneling appear to be much more complex. We refer to [62] for a discussion of some of these issues.

Application to kinetic Monte Carlo methods The ideas discussed in this chapter can be used to design more accurate kinetic Monte Carlo schemes. In standard kinetic Monte Carlo methods, the possible states and the transition rates are predetermined, often empirically. This can be avoided by finding the states and transition rates "on-the-fly" from a more detailed model. See for example [60].

References for Chapter 10

[1] C. Abrams and E. Vanden-Eijnden, "Large-scale conformational sampling of proteins using temperature-accelerated molecular dynamic," *Proc. Nat. Acad. Sci. USA*, **107**, 4961–4966, 2010.

[2] C.H. Bennett, "Exact defect calculations in model substances," in *Algorithms for Chemical Computation*, vol. 63, no. 46, A.S. Nowick and J.J. Burton, eds., ACS Symposium Series, 1977.

[3] B.J. Berne, G. Ciccotti and D.F. Coker, eds., *Classical and Quantum Dynamics in Condensed Phase Simulations*, World Scientific, 1998.

[4] P.G. Bolhuis, D. Chandler, C. Dellago and P.L. Geissler, "Transition path sampling: throwing ropes over mountain passes in the dark," *Annu. Rev. Phys. Chem.*, vol. 53, pp. 291–318, 2002.

[5] P.G. Bolhuis, C. Dellago and D. Chandler, "Reaction coordinates of biomolecular isomerization," *Proc. Nat. Acad. Sci. USA*, vol. 97, pp. 5877–5882, 2000.

[6] A. Bovier, M. Eckhoff, V. Gayrard and M. Klein, "Metastability in reversible diffusion processes Part I: Sharp estimates for capacities and exit times," WIAS preprints, vol. 767, 2003.

[7] A. Bovier, V. Gayrard and M. Klein, "Metastability in reversible diffusion processes. Part 2: Precise estimates for small eigenvalues," WIAS-preprints, vol. 768, 2003.

[8] D. Branduardi, F.L. Gervasio and M. Parrinello, "From A to B in free energy space," *J. Chem. Phys.*, vol. 126, p. 054 103, 2007.

[9] R.C. Brower, D.A. Kessler, J. Koplik and II. Levine, "Geometric models of interface dynamics," *Phys. Rev. A*, vol. 29, pp. 1335–1342, 1984.

[10] S.K. Burger and W. Yang, "Quadratic string method for determining the minimum-energy path based on multiobjective optimization," *J. Chem. Phys.*, vol. 124, pp. 054 109-1–054 109-13, 2006.

[11] E.A. Carter, G. Ciccotti, J.T. Hynes and R. Kapral "Constrained reaction coordinate dynamics for the simulation of rare events," *Chem. Phys. Lett.*, vol. 156, pp. 472–477, 1989.

[12] D. Chandler, "Statistical mechanics of isomerization dynamics in liquids and the transition state approximation," *J. Chem. Phys.*, vol. 68, pp. 2959–2970, 1978.

[13] R. Czerminski and R. Elber, "Self avoiding walk between two fixed points as a tool to calculate reaction paths in large molecular systems," *Int. J. Quantum Chem.*, vol. 24, pp. 167–186, 1990.

[14] C. Dellago, P.G. Bolhuis and D. Chandler, "Efficient transition path sampling: Application to Lennard-Jones cluster segments," *J. Chem. Phys.*, vol. 108, no. 22, pp. 9236–9245, 1998.

[15] C. Dellago, P.G. Bolhuis, P.L. Geissler, "Transition path sampling," *Adv. Cheml. Phys.*, vol. 123, pp. 1–84, 2002.

[16] W. E and E. Vanden-Eijnden, "Metastability, conformation dynamics, and transition pathways in complex systems,", in *Multiscale Modelling and Simulation, Lecture Notes in Computational Science and Engineering*, vol. 39, pp. 35–68, Springer-Verlag, 2004.

[17] W. E and E. Vanden-Eijnden, "Towards a theory of transition paths," *J. Stat. Phys.*, vol. 123, no. 3, pp. 503–523, 2006.

[18] W. E and E. Vanden-Eijnden, "Transition path theory and path-finding algorithms for the study of rare events," *Ann. Rev. Phys. Chem.*, vol. 61, pp. 391–420, 2010.

[19] W. E, W. Ren and E. Vanden-Eijnden, "String method for the study of rare events," *Phys. Rev. B*, vol. 66, no. 5, pp. 052 301-1–052 301-4, 2002.

[20] W. E, W. Ren and E. Vanden-Eijnden, "Energy landscape and thermally activated switching of submicron-sized ferromagnetic elements," *J. Appl. Phys.*, vol. 93, no. 4, pp. 2275–2282, 2003.

[21] W. E, W. Ren and E. Vanden-Eijnden, "Finite temperature string method for the study of rare events," *J. Phys. Chem. B*, vol. 109, no. 14, pp. 6688–6693, 2005.

[22] W. E, W. Ren and E. Vanden-Eijnden, "Transition pathways in complex systems: reaction coordinates, iso-committor surfaces and transition tubes," *Chem. Phys. Lett.*, vol. 143, no. 1–3, pp. 242–247, 2005.

[23] W. E, W. Ren and E. Vanden-Eijnden, "String method for rare events. I: Smooth energy landscapes", unpublished.

[24] W. E, W. Ren and E. Vanden-Eijnden, "Simplified and improved string method for computing the minimum energy paths in barrier-crossing events," *J. Chem. Phys.*, vol. 126, no. 16, pp. 164 103-1–164 103-8, 2007.

[25] R. Elber and M. Karplus, "A method for determining reaction paths in large molecules: application to myoglobin," *Chem. Phys. Lett.*, vol. 139, pp. 375–380, 1987.

[26] H. Eyring, "The activated complex in chemical reactions," *J. Chem. Phys.*, vol. 3, pp. 107–115, 1935.

[27] M. I. Freidlin and A. D. Wentzell, *Random Perturbations of Dynamical Systems*, 2nd edition, Springer-Verlag, 1998.

[28] D. Frenkel and B. Smit, *Understanding Molecular Simulation: From Algorithm to Applications*, 2nd edition, Elsevier, 2001.

[29] C.W. Gardiner, *Handbook of Stochastic Methods*, 2nd edition, Springer-Verlag, 1997.

[30] R.E. Gillilan and K.R. Wilson, "Shadowing, rare events, and rubber bands. A variational Verlet algorithm for molecular dynamics," *J. Chem. Phys.*, vol. 97, pp. 1757–1772, 1992.

[31] P. Hänggi, P. Talkner and M. Borkovec, "Reaction-rate theory: fifty years after Kramers," *Rev. Mod. Phys.*, vol. 62, pp. 251–341, 1990.

[32] T. Hastie, R. Tibshirani and J. Friedman, *The Elements of Statistical Learning*, Springer, 2001.

[33] G. Henkelman and H. Jónsson, "A dimer method for finding saddle points on high dimensional potential surfaces using only first derivatives," *J. Chem. Phys.*, vol. 111, pp. 7010–7022, 1999.

[34] G. Henkelman and H. Jónsson, "Improved tangent estimate in the NEB method for finding minimum energy paths and saddle points," *J. Chem. Phys.*, vol. 113, pp. 9978–9985, 2000.

[35] G. Henkelman, G. Jóhannesson, and H. Jónsson, "Methods for finding saddle points and minimum energy paths," in *Progress in Theoretical Chemistry and Physics*, S.D. Schwartz, ed., pp. 269–300, Kluwer Academic Publishers, 2000.

[36] W. Huisinga, S. Meyn and C. Schütte, "Phase transitions and metastability in Markovian and molecular systems," *Ann. Appl. Probab.*, vol. 14, pp. 419–458, 2004.

[37] M. Iannuzzi, A. Laio and M. Parrinello, "Efficient exploration of reactive potential energy surfaces using Car–Parrinello molecular dynamics," *Phys. Rev. Lett.*, vol. 90, pp. 238 302-1–238 302-4, 2003.

[38] H. Jónsson, G. Mills and K.W. Jacobsen, "Nudged elastic band method for finding minimum energy paths of transitions," in *Classical and Quantum Dynamics in Condensed Phase Simulations*, B.J. Berne, G. Ciccotti, and D.F. Coker, eds., pp. 385–404, World Scientific, 1998.

[39] A. Laio and M. Parrinello, "Escaping free-energy minima," *Proc. Nat. Acad. Sci. USA*, vol. 99, pp. 12 562–12 566, 2002.

[40] L. Maragliano and E. Vanden-Eijnden, "A temperature accelerated method for sampling free energy and determining reaction pathways in rare events simulations," *Chem. Phys. Lett.*, vol. 426, pp. 168–175, 2006.

[41] L. Maragliano, A. Fischer, E. Vanden-Eijnden and G. Ciccotti, "String method in collective variables: minimum free energy paths and isocommittor surfaces," *J. Chem. Phys.*, vol. 125, pp. 024 106-1–024 106-15, 2006.

[42] B.J. Matkowsky and Z. Schuss, "Eigenvalues of Fokker–Planck operator and the approach to equilibrium for diffusion in potential fields," *SIAM J. Appl. Math.*, vol. 40, no. 2, pp. 242–254, 1981.

[43] P. Metzner, C. Schütte and E. Vanden-Eijnden, "Illustration of transition path theory on a collection of simple examples," *J. Chem. Phys.*, vol. 125, pp. 084 110-1–084 110-17, 2006.

[44] P. Metzner, C. Schütte and E. Vanden-Eijnden, "Transition path theory for Markov jump processes," *Multiscale Model. Sim.*, vol. 7, no. 3, pp. 1192–1219, 2009.

[45] B. Peters, A. Heyden, A.T. Bell and A. Chakraborty, "A growing string method for determining transition states: comparison to the nudged elastic band and string methods," *J. Chem. Phys.*, vol. 120, pp. 7877–7886, 2004.

[46] L.R. Pratt, "A statistical method for identifying transition states in high dimensional problems," *J. Chem. Phys.*, vol. 9, pp. 5045–5048, 1986.

[47] W. Ren, "Higher order string method for finding minimum energy paths," *Comm. Math. Sci.*, vol. 1, pp. 377–384, 2003.

[48] W. Ren, Numerical methods for the study of energy landscapes and rare events, Ph.D. thesis, New York University, 2002.

[49] W. Ren, E. Vanden-Eijnden, P. Maragakis and W. E, "Transition pathways in complex systems: application of the finite temperature string method to the alanine dipeptide," *J. Chem. Phys.*, vol. 123, no. 13, pp. 134 109-1–134 109-12, 2005.

[50] A. Samanta and W. E, "Optimization-based string method," preprint.

[51] E.M. Sevick, A.T. Bell and D.N. Theodorou, "A chain of states method for investigating infrequent event processes occurring in multistate, multidimensional systems," *J. Chem. Phys.*, vol. 98, pp. 3196–3212, 1993.

[52] M.R. Sorensen and A. Voter, "Temperature-accelerated dynamics for simulation of infrequent events," *J. Chem. Phys.*, vol. 112, pp. 9599–9606, 2000.

[53] J.E. Straub, "Reaction rates and transition pathways," in *Computational Biochemistry and Biophysics*, O.M. Becker, A.D. MacKerell, Jr., B. Roux and M. Watanabe, eds., pp. 199–221, Marcel Dekker, 2001.

[54] F. Tal and E. Vanden-Eijnden, "Transition state theory and dynamical corrections in ergodic systems," *Nonlinearity*, **19**, 501–509, 2006.

[55] E. Vanden-Eijnden, "Transition path theory," in *Computer Simulations in Condensed Matter: From Materials to Chemical Biology*, M. Ferrario, G. Ciccotti and K. Binder, eds., vol. 1, pp. 439–478, Springer, 2006.

[56] E. Vanden-Eijnden and M. Venturoli, "Revisiting the finite temperature string method for the calculation of reaction tubes and free energies," *J. Chem. Phys.*, vol. 130, p. 194 103, 2009.

[57] S.R.S. Varadhan, *Large Deviations and Applications*, SIAM, 1984.

[58] A.F. Voter, "Hyperdynamics: accelerated molecular dynamics of infrequent events," *Phys. Rev. Lett.*, vol. 78, pp. 3908–3911, 1997.

[59] A.F. Voter, "A method for accelerating the molecular dynamics simulation of infrequent events," *J. Chem. Phys.*, vol. 106, pp. 4665–4677, 1997.

[60] A.F. Voter, F. Montalenti and T.C. Germann, "Extending the time scale in atomistic simulation of materials," *Ann. Rev. Mater. Res.*, vol. 32, pp. 321–346, 2002.

[61] D.J. Wales, *Energy Landscapes*, Cambridge University Press, 2004.

[62] U. Weiss, *Quantum Dissipative Systems*, World Scientific, 1999.

[63] E. Wigner, "The transition state method," *Trans. Faraday Soc.*, vol. 34, pp. 629, 1938.

[64] X. Zhou, Study of noise-induced transition pathways in non-gradient systems using the adaptive minimum action method, Ph.D. thesis, Princeton University, 2009.

11
Other perspectives

11.1 Open problems

Despite the explosive growth of activities in multiscale modeling in recent years, this is still a largely open territory. Many important issues are not clearly understood. Let us mention a few problems that are of general interest.

(1) *A better understanding of renormalization group analysis.* The renormalization group method is one of the most popular analytical techniques in physics for handling multiscale problems. However, at a more rigorous level it is still somewhat of a mystery. Since the renormalization group analysis carried out in the physics literature typically involves drastic truncations, it is very desirable to have a better understanding of the nature of the error caused by these truncations.

(2) *A systematic analysis of memory effects using the Mori–Zwanzig formalism.* As discussed in Chapter 2, the Mori–Zwanzig formalism is a systematic way of eliminating variables in a system. However, it often leads to very complicated models with memory terms. Therefore it is natural to ask: how do we make systematic approximations for the memory effects? We will discuss one example below for the case when the interaction involving the eliminated variables is linear. However, it would be very useful to develop more general strategies. In particular, it would be of interest to use this kind of approximation technique to assess the accuracy of memory-dependent hydrodynamic models of polymeric fluids such as Maxwell models [3].

(3) *A numerical analysis for the effect of noise.* It is possible to carry out the kind of analysis of the quasicontinuum method presented in Chapter 7 because the effect of noise can be ignored. This is not the case for most multiscale problems. Indeed, indicated in the notes in Chapter 7, a major question in formulating coupled atomistic–continuum models is whether we should add noise terms to the continuum models. Therefore, it is very important to develop a better framework for assessing the numerical error in the presence of noise.

(4) *A systematic analysis of the error in QM–MM.* From the point of view of numerical analysis, a major difficulty is associated with the fact that we do not have an obvious small parameter (such as the grid size or a ratio of time scales) with which we can speak about the order of accuracy or assess the accuracy asymptotically.

(5) *Adaptive algorithms.* Adaptivity is crucial for the success of multi-scale algorithms. In fact, multiscale, multi-physics modeling introduces another dimension for adaptivity: besides adaptively selecting the grid sizes and algorithms, one also has to adaptively select the models, the coarse-grained variables and the partitioning of the set of slow and fast components of the dynamics.

Most of these problems are geared toward a better or more rigorous understanding. This is natural since, as we said at the beginning, one main motivation for multiscale modeling is to put more rigor into modeling.

There are many other problems, of course. Instead of going through more problems we will provide some personal perspectives on problems without scale separation.

We have seen throughout this book that scale separation is a key property that one can use in order to derive reduced models or develop sublinear scaling algorithms. Unfortunately many multiscale problems encountered in practice do not have this property. The most well-known example is provided by fully developed turbulent flows [20].

The first thing one should ask, in this kind of problem, is whether there are other special features of which one can take advantage. We have discussed the case when the problem has (statistically) self-similar behavior. In this case, one can either use the renormalization group method, discussed in Section 2.6, or HMM as we illustrated in the notes of Chapter 6.

This allows us to obtain reduced models or develop sublinear scaling algorithms.

Fully developed turbulent flows are expected to be approximately self-similar in a statistical sense but, because of intermittency, one does not expect self-similarity to hold exactly. Characterizing the more sophisticated scaling behavior in turbulent flows is still among the most important and grand challenges in classical physics [20].

This brings us to the inevitable question: what if the problem does not have any special features of which we can take advantage? One answer to this question is simply that there is not much one can do and one just has to solve the original microscale problem by brute force. There are no alternatives even if one is only interested in the macroscale behavior of the solution.

However, it is often more productive to formulate a different question: what is the best appproximate solution one can get, given the maximum cost one can afford? The answer to this question goes under the general umbrella of *variational model reduction*. This is consistent with the philosophy of *optimal prediction* advocated by Chorin *et al.* [4, 5] and can indeed be regarded as being a poor man's version of optimal prediction.

We will discuss some general guidelines for this kind of problem. A theoretical guideline is the Mori–Zwanzig formalism. As was explained in Section 2.7, this is a universal tool for eliminating degrees of freedom in any kind of problem. However, the complexity of a problem is not substantially reduced unless we make approximations. Therefore, in practice the task is to find the best approximation within a specified class of models.

11.1.1 Variational model reduction

There are two main ingredients in variational model reduction [12]:

(1) the class of models in which the reduced model lies, including the choice of the reduced (or coarse-grained) variables;
(2) an objective function with which we can discuss optimality.

The class of models can be as follows:

(1) *Some specific form of model*, as in the case of HMM (see Section 6.8). For example, if the microscale model is the Boltzmann equation, we may consider the class of models that are nonlinear conservation laws for the zeroth-, first-, second- and third-order

moments, and where the fluxes depend only on the local values of these moments.

(2) *Models of limited complexity.* For example, if the microscale model is given by a large $N \times N$ matrix A, we may consider the class of rank-M matrices, where $M \ll N$.

The objective function is usually some measurement of the error for a class of solutions to the original microscale model. Obviously the quality of the reduced model crucially depends on these two factors.

After these ingredients are specified, we still have the practical problem of how to obtain the reduced model. We will illustrate these practical issues using the example of *operator compression* [10]: given a large $N \times N$ symmetric positive definite matrix A and a number M which is much smaller than N, we would like to find the rank-M matrix that best approximates A, in the sense that

$$I(B) = \|A^{-1} - B^*\|_2$$

is minimized. Here B is a $N \times N$ matrix whose rank is no more than M and B^* is its pseudo-inverse.

It is well known that if A has the diagonalized form

$$A = O^{\mathrm{T}} \Lambda O,$$

where O is an $N \times N$ orthogonal matrix, $\Lambda = \mathrm{diag}(\lambda_1, \dots, \lambda_N)$ and $\lambda_1 \leq \dots \leq \lambda_N$ then the optimal approximation is given by

$$A_M = O^{\mathrm{T}} \Lambda_M O,$$

where $\Lambda_M = \mathrm{diag}(\lambda_1, \dots, \lambda_M, 0, \dots, 0)$. We can also regard A_M as the projection of A onto an eigensubspace corresponding to its M smallest eigenvalues. We will denote this subspace by V_M.

To find A_M, the most obvious approach is to diagonalize A. This is a general solution. However, its computational complexity scales as N^3 or at least $M^2 N$ [14, 8].

In many cases, A comes from the discretization of some differential operator which is local, or it is the generator of some Markov chain whose dynamics is mostly local. In this case, it might be possible to develop much more efficient algorithms, as was shown in [10]. The key idea is to choose a localized basis for V_M in order to represent A_M. This is possible in many situations of practical interest.

Let us first look at a simple example, the one-dimensional Laplace operator with periodic boundary conditions of period 2π. Let $M = 2n+1$, where n is an integer. The eigensubspace of interest is

$$V_M = \text{span}\left\{ e^{-inx}, e^{-i(n-1)x}, \ldots, e^{inx} \right\}.$$

This expression also provides a basis for V_M. Obviously, these basis functions are nonlocal.

One can also choose other basis sets. For example, if we let

$$u_j(x) = \sum_{k=-n}^{n} e^{ik(x-x_j)}, \quad j = -n, -n+1, \ldots, n, \qquad (11.1.1)$$

where $x_j = 2j\pi/(2n+1)$, then $\{u_j\}$ forms a basis of V_M. This is called the Dirichlet basis since the basis functions correspond to the Dirichlet kernel in Fourier series. A feature of this basis set is that the basis functions decay away from their centers. Therefore we should be able to obtain a set of approximate basis functions by truncating the functions $\{u_j\}$ away from the centers.

However, the Dirichlet kernel does not decay very quickly – it has a slowly decaying oscillatory tail. To understand this, we can view the Dirichlet kernel as the projection of a δ-function, which is the most localized function, to the subspace V_M. Let $\delta(x) = \sum_{k=-\infty}^{\infty} e^{ikx}$; its projection onto V_M is

$$\delta_n(x) = D_n(x) = \sum_{k=-n}^{n} e^{ikx} = \frac{\sin(n + \frac{1}{2})x}{\sin \frac{1}{2}x}, \quad -\pi \le x \le \pi,$$

which is the Dirichlet kernel. The slow decay of the oscillatory tail of D_n is a result of the Gibbs phenomenon caused by the lack of smoothness of $\delta(x)$. An obvious idea for obtaining basis functions with better decay properties is to use filtering [15]. We define

$$\delta_n^\sigma(x) = \sum_{k=-n}^{n} \sigma\left(\frac{k}{n}\right) e^{ikx}, \qquad (11.1.2)$$

where $\sigma(\cdot)$ is a filter. It is well known that the decay properties of δ_n^σ depend on the smoothness of the filter σ. The smoother the filter, the faster the decay of δ_n^σ. See [15] or [10] for a precise formulation of these statements.

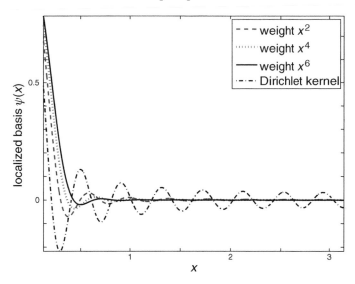

Figure 11.1. The localized bases produced by the weight functions x^2 (red broken curve), x^4 (blue dotted curve) and x^6 (black solid curve) and the Dirichlet kernel with $n = 16$ (black broken-and-dotted curve). Here the best localization is obtained with the weight function x^6.

This discussion only applies to the particular example that we have just discussed. In the general case, one obvious way of obtaining basis functions with fast-decay properties is to solve the optimization problem

$$\min_{\psi \in V_M, \ \psi \neq 0} \mathcal{B}[\psi] = \frac{\int w(x)|\psi(x)|^2 dx}{\int |\psi(x)|^2 dx}, \qquad (11.1.3)$$

where $w(\cdot)$ is an even nonnegative weight function. By centering w at various locations, one obtains a set of basis functions that decay away from the centers. The rate of decay depends on the choice of w. Such ideas were explored in [19] for electronic structure analysis when $w(x) = x^2$ is used. Other examples of weight functions include (see Figure 11.1):

(1) $w(x) = x^{2p}$, where p is an integer [10];
(2) $w(x) = 1 - \mathbf{1}_{[-1,1]}(x)$, where $\mathbf{1}_{[-1,1]}$ is the indicator function of the set $[-1, 1]$ [13].

It is natural to ask how w determines the decay property of the basis functions selected, in this way. This question was addressed in [10] for the example discussed above. It was shown there that the decay rate of the basis function is determined by how fast $w(x)$ approaches 0 as $x \to 0$.

Lemma *Assume that w is smooth in $(-\pi, \pi)$. If $w^{(2l)}(0)$ is the first nonzero term in $\{w^{(2m)}(0) \mid m \in \mathbb{N}\}$ then*

$$|\psi(x)| \le \frac{Cn}{|nx|^{l+1}}, \quad x \in (-\pi, \pi), \ x \ne 0,$$

where C is a constant that is independent of n and x.

This lemma suggests that if we take $w(x) = x^{2p}$ then the larger the value of p, the faster the basis functions decay. However, choosing large values of p is bad for numerical stability and accuracy, since the condition number of the problem (11.1.3) increases very rapidly as a function of p [10]. Therefore the best weight function has to be a compromise between the decay rate and numerical stability. Numerical results suggest that the weight functions x^6 and x^8 are good choices.

Qualitatively similar results are expected to hold for more general cases, such as the case when V_M is a high-dimensional eigensubspace of some elliptic differential operator. But this has not been investigated.

The standard numerical technique for finding eigensubspaces is the subspace iteration method [14, 21]. This is a generalization of the power method for computing leading eigenvalues and eigenvectors of a matrix. Subspace iteration consists of two main steps: multiplication of a set of approximate basis vectors by a matrix A, followed by orthogonalization [14, 21]. The cost of this algorithm is at least $\mathcal{O}(NM^2)$. In [10] a localized version of the subspace iteration algorithm was proposed, the localized subspace iteration (LSI), in which the orthogonalization step is replaced by a localization step. The resulting algorithm has an $\mathcal{O}(N)$ complexity.

Consider for example the operator

$$\mathbf{A} = -\nabla \cdot (\mathbf{a}^\varepsilon(\mathbf{x})\nabla). \tag{11.1.4}$$

When $\mathbf{a}^\varepsilon$ has the form $\mathbf{a}^\varepsilon(\mathbf{x}) = \mathbf{a}(\mathbf{x}/\varepsilon)$, one can apply homogenization theory to reduce the complexity of this operator. In the most general case, one can ask: given an integer M, what is the $M \times M$ matrix that best approximates this operator? Localized subspace iteration can be used to find this matrix and the associated eigensubspaces of $\mathbf{A}$; see [10] for details. In the case when $\mathbf{a}^\varepsilon$ has scale separation, the generalized finite element method, the residual-free bubble element method and MsFEM can all be regarded as other ways of finding approximately this optimal

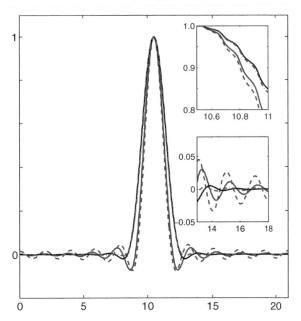

Figure 11.2. The homogenization problem: the localized bases produced by the weight functions x^2 (blue solid curve), x^6 (black solid curve), $1 - \mathbf{1}_{[-1,1]}$ (red broken curve) and $1 - \mathbf{1}_{[-2,2]}$ (magenta dotted-and-broken curve). The lower inset shows the tail region while the upper inset shows the oscillatory behavior of the basis functions in more detail. (Courtesy of Jianfeng Lu.)

subspace (see Chapter 8). See Figures 11.2 and 11.3 for some illustrative results.

Another example, the optimal reduction of Markov chains, was discussed in [11].

In many cases, this type of poor man's optimal prediction procedure can produce very poor results. For example coarse-grained molecular dynamics (CGMD), discussed in Section 9.3, can be regarded as a variational model-reduction technique applied to molecular dynamics. As indicated there, the results of CGMD can be qualitatively incorrect since it neglects memory effects entirely.

11.1.2 Modeling memory effects

From a theoretical viewpoint, one can always eliminate degrees of freedom and obtain a "reduced system" using the Mori–Zwanzig procedure, as was explained in Section 2.7. Here we put "reduced system" in quotation

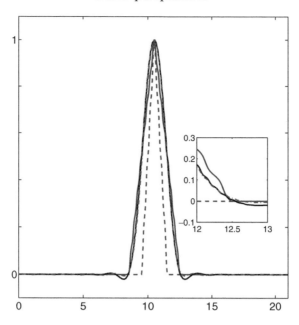

Figure 11.3. The homogenization problem: the localized bases produced by the weight functions x^6 (black solid curve) and $1 - \mathbf{1}_{[-2,2]}$ (magenta broken-and-dotted curve) together with the basis function of the generalized finite element method (or MsFEM) with support size 2 (red broken curve) or 4 (blue solid curve). The inset shows the details of these basis functions at two coarse grids away from the center. (Courtesy of Jianfeng Lu.)

marks since the complexity of a model obtained from the Mori–Zwanzig procedure is not necessarily reduced: it contains the same amount of information as the original model except that the initial data for the eliminated degrees of freedom is replaced by a random process. The complexity coming from the extra degrees of freedom in the original model is replaced by the complexity from the memory kernels in the Mori–Zwanzig formalism. Therefore unless we make approximations for the memory kernels, these "reduced models" may just be as complicated as the models with which we began. This does not mean that the Mori–Zwanzig formalism is useless. In fact, it provides a natural starting point for making approximations to the memory kernels.

From the viewpoint of the Mori–Zwanzig formalism, when the time scales for the eliminated degrees of freedom are much shorter than those for the variables retained, the memory terms are very small and can be neglected. We have already seen many examples of this kind.

○ ○ ○ ○ ● ● ● ●

u_{-1} u_0 u_1 u_2

Figure 11.4. The eliminated and retained degrees of freedom. The degrees of freedom associated with the solid circles are kept. The others are eliminated.

In the other extreme, when the memory kernels are long-ranged and slowly varying, one can make a different kind of approximation, as was done in [6, 16] to obtain the *t*-model.

In between these two extreme cases we find the most difficult situation. Consider the simplest example of this kind: a one-dimensional chain of atoms with nearest-neighbor interactions [1],

$$m\ddot{x}_j = \phi'(x_{j+1} - x_j) - \phi'(x_j - x_{j-1}), \quad -\infty < j < \infty, \quad (11.1.5)$$

where ϕ is, say the Lennard-Jones potential,

$$\phi(r) = 4\varepsilon \left(\left(\frac{\sigma}{r} \right)^{12} - \left(\frac{\sigma}{r} \right)^6 \right). \quad (11.1.6)$$

The lattice parameter for this system is $a_0 = \sqrt[6]{2}\sigma$, and the displacement is defined as $u_j = x_j - ja_0$; the phonon dispersion relation is given by

$$\omega(k)^2 = \frac{\kappa^2}{m}(2 - 2\cos ka_0), \quad k \in \left[-\frac{\pi}{a_0}, \frac{\pi}{a_0} \right], \quad (11.1.7)$$

where $\kappa^2 = \phi''(a_0)$.

We split the chain of atoms into two groups, one consisting of the atoms in whose dynamics we are interested and the other consisting of atoms that play the role of a heat bath. We define the heat bath as the set of atoms with $j \leqslant 0$ (see Figure 11.4). After adopting the harmonic approximation for the heat bath atoms, we obtain the following Hamiltonian for the whole system:

$$H = \sum_{j \leqslant 0} \frac{\kappa^2}{2}(u_{j+1} - u_j)^2 + \sum_{j > 0} \phi(x_{j+1} - x_j) + \sum_j \frac{1}{2}mv_j^2;$$

here v_j is the velocity of the jth atom. If we apply the Mori–Zwanzig formalism to eliminate the heat bath atoms, we obtain the *generalized*

Langevin equation (GLE) for the retained atoms (see [1]):

$$m\ddot{u}_1 = \phi'(x_2 - x_1) - \int_0^t \theta(t-s)\dot{u}_1(s)\,ds + F(t),$$

$$m\ddot{u}_j = \phi'(x_{j+1} - x_j) - \phi'(x_j - x_{j-1}), \quad j > 1, \qquad (11.1.8)$$

where

$$\theta(t) = \frac{\sqrt{m}\kappa}{t} J_1\left(\frac{2\kappa t}{\sqrt{m}}\right), \qquad (11.1.9)$$

$$F(t) = \sum_{j \leqslant 0} \big(c_j(t)\big(u_{j+1}(0) - u_j(0)\big) + s_j(t)v_j(0)\big). \qquad (11.1.10)$$

Here J_1 is a Bessel function of the first kind. The coefficients $c_j(t)$ and $s_j(t)$ are governed by the following relations:

$$\dot{s}_j = c_{j-1} - c_j,$$
$$m\dot{c}_j = \kappa^2(s_j - s_{j+1}),$$
$$s_0(t) = 0, \quad s_j(0) = 0, \quad c_{-1}(0) = -\kappa^2, \quad c_j(0) = 0, \; j \leq -2.$$

The term $F(\cdot)$ is often treated as a noise term. More examples of this kind of analysis can be found in [17].

The GLE is still quite complicated, owing to the presence of long-range kernels. For high-dimensional models, these long-range kernels also have spatial components that couple all the boundary atoms together. Modeling such memory-dependent dynamics is still a highly demanding task. Therefore it is desirable to make approximations [9]. A systematic approach to this was proposed in [18], on the basis of the following principles:

(1) *Efficiency.* The kernels should be as local as possible.
(2) *Accuracy.* Given the size of the time interval and the set of neighboring atoms on which the kernels depend, we are interested in finding the most accurate approximation to the original dynamics.
(3) *Stability.* The approximate GLEs should be stable. This will constrain the approximate kernels to be positive definite.
(4) *The noise term should be approximated appropriately.* For example, when applicable the fluctuation–dissipation theorem should be preserved.

One interesting observation made in [18] is that one can "use space to buy time," i.e. by increasing the set of neighboring atoms on which the memory kernels depend, one can increase the rate of decay of the

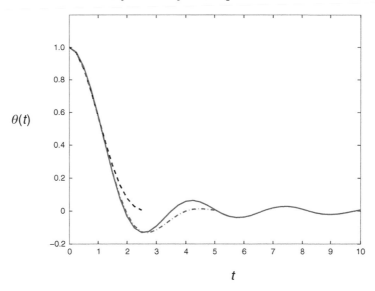

Figure 11.5. A typical memory kernel (solid line) and two different local approximations (the broken and broken-and-dotted lines).

latter. For this reason, it is of interest to increase the spatial dependence of the memory kernels since this is often more convenient than dealing with long-range memory terms. More details can be found in [9] and [18] (see Figure 11.5).

The work in [9, 18] may be an interesting start, but it is limited to the situation when the interaction involving the eliminated degrees of freedom is approximately linear. The problem is basically open for more general cases.

References for Chapter 11

[1] S.A. Adelman and J.D. Doll, "Generalized Langevin equation approach for atom/solid-surface scattering: collinear atom/harmonic chain model," *J. Chem. Phys.*, vol. 61, pp. 4242–4245, 1974.

[2] S.A. Adelman and J.D. Doll. "Generalized Langevin equation approach for atom/solid-surface scattering: general formulation for classical scattering off harmonic solids," *J. Chem. Phys.*, vol. 64, pp. 2375–2388, 1976.

[3] R.B. Bird, C.F. Curtiss, R.C. Armstrong and O. Hassager, *Dynamics of Polymeric Liquids, Vol. 2: Kinetic Theory*, John Wiley, 1987.

[4] A.J. Chorin, O. Hald and R. Kupferman, "Optimal prediction with memory," *Physica D*, vol. 166, nos. 3–4, pp. 239–257, 2002.

[5] A.J. Chorin, A.P. Kast, and R. Kupferman, "Optimal prediction of underresolved dynamics," *Proc. Nat. Acad. Sci. USA*, vol. 95, no. 8, pp. 4094–4098, 1998.

[6] A.J. Chorin and P. Stinis, "Problem reduction, renormalization, and memory," *Comm. Appl. Math. Comp. Sci.*, vol. 1, pp. 1–27, 2005.

[7] S. Curtarolo and G. Ceder, "Dynamics of an inhomogeneously coarse grained multiscale system," *Phys Rev Lett.*, vol. 88, pp. 255 504-1–255 504-4, 2002.

[8] P. Drineas, R. Kannan and W. Mahoney, "Fast Monte Carlo algorithms for matrices II: computing low-rank approximations to a matrix," *SIAM J. Sci. Comp.*, vol. 36, pp. 158–183, 2006.

[9] W. E and Z. Huang, "Matching conditions in atomistic-continuum modeling of materials," *Phys. Rev. Lett.*, vol. 87, no. 13, pp. 135 501-1–135 501-4, 2001.

[10] W. E, T. Li and J. Lu, "Localized basis of eigen-subspaces," *Proc. Nat. Acad. Sci. USA*, vol. 107, pp. 1273–1278, 2010.

[11] W. E, T. Li and E. Vanden-Eijnden, "Optimal partition and effective dynamics of complex networks," *Proc. Nat. Acad. Sci. USA*, vol. 105, pp. 7907–7912, 2008.

[12] W. E, T. Li and E. Vanden-Eijnden, "Variational model reduction," in preparation.

[13] W. Gao and W. E, "Orbital minimization with localization," *Discrete Contin. Dyn. Syst. Ser. A*, vol. 23, pp. 249–264, 2009.

[14] G. Golub and C. Van Loan, *Matrix Computations*, The Johns Hopkins University Press, 1996.

[15] D. Gottlieb and C.W. Shu, "On the Gibbs phenomenon and its resolution," *SIAM Rev.*, vol. 39, pp. 644–668, 1997.

[16] O.H. Hald, and P. Stinis, "Optimal prediction and the rate of decay for solutions of the Euler equations in two and three dimensions," *Proc. Nat. Acad. Sci. USA*, vol. 104, no. 16, pp. 6527–6553, 2007.

[17] E.G. Karpov, G.J. Wagner and W.K. Liu, "A Green's function approach to deriving wave-transmitting boundary conditions in molecular dynamics," *Int. J. Numer. Meth. Engrg*, vol. 62, pp. 1250–1262, 2005.

[18] X. Li and W. E, "Variational boundary conditions for molecular dynamics simulations of crystalline solids at finite temperature: treatment of the thermal bath," *Phys. Rev. B*, vol. 76, no. 10, pp. 104 107-1–104 107-22, 2007.

[19] N. Marzari and D. Vanderbilt, "Maximally localized generalized Wannier functions for composite energy bands," *Phys. Rev. B*, vol. 56, pp. 12 847–12 865, 1997.

[20] A.S. Monin and A.M. Yaglom, *Statistical Fluid Mechanics, Vol. I: Mechanics of Turbulence*, Dover Publications, 2007.

[21] Y. Saad, *Numerical Methods for Large Eigenvalue Problems*, Manchester University Press, 1992.

Subject index

461